Sergei N. Magonov, Myung-Hwan Whangbo

Surface Analysis with STM and AFM

Also of Interest

- E. Lifshin (ed.) **Characterization of Materials**
  Volumes 2A and 2B from the series *Materials Science and Technology,* edited by R. W. Cahn, P. Haasen, E. J. Kramer
  Volume 2A, VCH 1992. Volume 2B, VCH 1994.

- N. J. DiNardo **Nanoscale Characterization of Surfaces and Interfaces**, VCH 1994.

- S. Amelinckx, D. Van Dyck, J. F. Van Landuyt, G. Van Tendeloo (eds.)
  Handbook of Microscopy. Volume 1. Methods. Volume 2. Applications in Materials Science, VCH 1996.

- D. A. Bonnell **Scanning Tunneling Microscopy and Spectroscopy. Theory, Techniques and Applications**, VCH 1995.

- K. Wetzig, D. Schulze **In Situ Scanning Electron Microscopy in Materials Research**, Akademie Verlag 1995.

© VCH Verlagsgesellschaft mbH, D-69451 Weinheim (Federal Republic of Germany) 1996

Distribution:

VCH, P.O. Box 101161, D-69451 Weinheim (Federal Republic of Germany)

Switzerland: VCH P.O. Box, CH-4020 Basel (Switzerland)

United Kingdom and Ireland: VCH (UK) Ltd., 8 Wellington Court, Cambridge CB1 1HZ (England)

USA and Canada: VCH, 220 East 23rd Street, New York, NY 10010-4606 (USA)

Japan: VCH, Eikow Building, 10-9 Hongo 1-chome, Bunkyo-ku, Tokyo 113 (Japan)

ISBN 3-527-29313-2

Sergei N. Magonov, Myung-Hwan Whangbo

Surface Analysis with STM and AFM

Experimental and Theoretical Aspects
of Image Analysis

Weinheim · New York · Basel · Cambridge · Tokyo

Dr. S. N. Magonov
Digital Instruments
520 E. Montecito St.
Santa Barbara, CA 93103
USA

Dr. M.-H. Whangbo
Department of Chemistry
North Carolina State Univ.
Raleigh, NC 27695-8204
USA

This book was carefully produced. Nevertheless, authors and publishers do not warrant the information contained therein to be free of errors. Readers are advised to keep in mind that statements, data, illustrations, procedural details or other items may inadvertently be inaccurate.

Published jointly by
VCH Verlagsgesellschaft mbH, Weinheim (Federal Republic of Germany)
VCH Publishers, Inc., New York, NY (USA)

Editorial Directors: Dr. Peter Gregory, Dr. Ute Anton
Production Manager: Dipl.-Wirt.-Ing. (FH) Bernd Riedel

Every effort has been made to trace the owners of copyrighted material; however, in some cases this has proved impossible. We take this opportunity to offer our apologies to any copyright holders whose rights we may have unwittingly infringed.

Library of Congress Card No. applied for.

A catalogue record for this book is available from the British Library.

Die Deutsche Bibliothek Cataloguing-in-Publication Data:

Magonov, Sergei N.:
Surface analysis with STM and AFM : experimental and
theoretical aspects of image analysis / Sergei N. Magonov ;
Myung-Hwan Whangbo. – Weinheim ; New York ; Basel ;
Cambridge ; Tokyo : VCH, 1996
   ISBN 3-527-29313-2
NE: Whangbo, Myung-Hwan:

© VCH Verlagsgesellschaft mbH, D-69451 Weinheim (Federal Republic of Germany), 1996

Printed on acid-free and chlorine-free paper.

All rights reserved (including those of translation into other languages). No part of this book may be reproduced in any form – by photoprinting, microfilm, or any other means – nor transmitted or translated into a machine language without written permission from the publishers. Registered names, trademarks, etc. used in this book, even when not specifically marked as such, are not to be considered unprotected by law.

Composition: Filmsatz Unger & Sommer GmbH, D-69469 Weinheim
Printing: Strauss Offsetdruck GmbH, D-69509 Mörlenbach
Bookbinding: Wilh. Osswald & Co., D-67433 Neustadt

Printed in the Federal Republic of Germany

Preface

Scanning tunneling microscopy (STM) and atomic force microscopy (AFM) are powerful tools for the examination of surfaces. The research, development, and application of the STM and AFM methods are currently making rapid progress. As a result, a large number of papers are being published every year on diverse subjects, from theory to experiment as well as on applications to a variety of materials. This makes it rather difficult for an individual to keep up with such fast development. The physical concepts employed in the instrumentation of STM and AFM are simple, but the interpretation of the STM and AFM results can be complicated because of the convolution of several interactions in the measurement process. This complication exists in the large-scale imaging of surface morphology as well as in the molecular- and atomic-scale images. Thus, many STM and AFM studies can be misinterpreted. To help to alleviate this problem, we felt it necessary to bring together into a book the essential components of STM and AFM studies, namely the practical aspects of STM and AFM, the image simulation by surface electron density plot calculations, and the qualitative evaluation of tip force induced surface corrugations.

The primary goal of this book is to describe how the surfaces of various materials are characterized by employing STM and AFM, and what physical/chemical features can be deduced from their images. The text consists of three parts. The first part is concerned with the backgrounds and fundamentals of STM and AFM, the physical phenomena leading to these methods, and the practical aspects of imaging. The second part describes the theoretical aspects of image analysis, the density plot calculations, and the accommodation of tip–sample force interactions. The third part deals with the experimental STM and AFM images and their interpretation for a variety of materials, which include layered inorganic materials, organic conducting salts, organic adsorbates at solid/liquid interfaces, self-assembled amphiphiles, and polymers. The experimental examples described here have mostly been taken from our own original publications.

This book is designed to be a reference work for researchers already involved in STM and AFM as well as for newcomers to the field. It can also be used as a text for a one-semester special-topic course on STM/AFM applications at the graduate level.

The authors would like to thank their colleagues Dr. Georg Bar, Hardy Bengel, Konrad Crämer, Alexander Wawkuschewski, Dr. Igor Tuzov, Dr. Jingqing Ren, Dr. Weigen Liang, Jeffrey Paradis, Professor Dongwoon Jung, and Dong-Kyun Seo. The writing of this book would not have been possible without their experimental and theoretical studies. Professor H.-J. Cantow is specially thanked for his continuous support for and interest in our work. The authors also thank Professor R. Brec, Dr. M. Evain, Professor G. Koßmehl, Professor M. Möller, Professor R. Mülhaupt, Professor J. Peterman, Dr. B. Pfannemüller, Professor M. Schwörer, Professor G. Thiele and Professor E. Yagubskii for making their samples available to us. Dr. V. Elings, Professor P. Hansma and Professor B. Parkinson are acknowledged for invaluable discussions concerning various aspects of STM and AFM. Hardy Bengel and Konrad Crämer are thanked for their help in preparing the figures and diagrams used in this book. M.-H. W. thanks the Alexander von Humboldt Foundation for a Humboldt Research Award for Senior US Scientists, which made possible his extended visit to the Materials Research Center, Albert-Ludwigs University, where the writing was completed. Finally, the authors thank their wives, Elena and Jin-Ok, and their children, Katja, Masha, Jennifer, and Albert, for their patience and moral support.

Our joint research activities have been supported by the US Department of Energy, Office of Basic Sciences, Division of Materials Sciences, under Grant DE-FG05-86ER45259, and by the European Community under the Human Capital and Mobility Project (ERBCHRXCT940675).

Freiburg, Germany *Sergei N. Magonov*
Raleigh, North Carolina, USA *Myung-Hwan Whangbo*
November 1995

Contents

	Preface	V
1	**Introduction**	1
1.1	**Development of Scanning Probe Microscopy**	1
1.2	**Key Problems of STM and AFM Applications**	2
1.2.1	Image Interpretation	2
1.2.2	Tip–Sample Interactions	4
1.2.3	Surface Relaxation and Local Hardness	5
1.2.4	Surface Forces and AFM	6
1.3	**Objectives**	7
	References	7
2	**Physical Phenomena Relevant to STM and AFM**	9
2.1	**Electron Transport Processes**	9
2.1.1	Conventional Electron Tunneling Regime	10
2.1.2	Electronic and Mechanical Contact Regimes	10
2.1.3	STM in Different Environments	11
2.2	**Survey of Force Interactions**	11
2.2.1	Force-vs.-Distance Curves	12
2.2.2	Short-Range Forces and Sample Deformation	13
2.2.3	Long-Range and Other Forces	16
2.2.3.1	Long-Range Forces	16
2.2.3.2	Adhesion and Capillary Forces	18
	References	18
3	**Scanning Probe Microscopes**	21
3.1	**Operating Principles and Main Components**	22
3.1.1	Scanner	23
3.1.2	Tip–Sample Approach and Electronic Feedback	23
3.1.3	Scanning Modes and Parameters	24
3.1.4	Images and Filtering	25
3.1.5	Isolation of Vibrational Noise	27
3.2	**Scanning Tunneling Microscope**	27
3.2.1	STM Tips and Current Detection	27
3.2.2	Bias Voltage	28
3.2.3	Scanning Tunneling Spectroscopy	30

3.3	**Atomic Force Microscope**	31
3.3.1	Contact Mode and Force Detection	33
3.3.2	AFM Probes	35
3.3.3	Dynamic AFM Measurements	37
3.3.3.1	AFM Operation in the Attractive Force Regime	38
3.3.3.2	Tapping Mode	39
3.3.3.3	Force-Modulation Techniques	39
3.3.3.4	Magnetic Force Microscopy	39
3.4	**STM and AFM as Metrology Tools**	40
3.4.1	Resolution in STM and AFM	40
3.4.2	Metrological Applications	43
	References	44
4	**Practical Aspects of STM and AFM Measurements**	47
4.1	**Samples**	47
4.2	**Optimization of Experiments**	48
4.2.1	Optimization of STM Experiments	48
4.2.2	Optimization of Contact-Mode AFM Experiments	50
4.2.3	Optimization of Tapping-Mode AFM Experiments	53
4.3	**STM and AFM Measurements**	55
4.3.1	Large-Scale Imaging	55
4.3.2	Atomic-Scale Imaging	57
4.3.3	Image Artifacts	58
	References	62
5	**Simulations of STM and AFM Images**	65
5.1	**Electronic Structures of Solids**	65
5.2	**Theoretical Aspects of STM**	68
5.2.1	Tunneling Between Metals	68
5.2.2	Tunneling Between Metal and Semiconductor	69
5.2.3	Tersoff–Hamman Theory and its Extension	72
5.2.4	Other Theories	73
5.3	**Theoretical Aspects of AFM**	74
5.4	**Image Simulation by Density Plot Calculations**	74
5.4.1	STM Image Simulation	74
5.4.2	AFM Image Simulation	76
5.4.3	STM and AFM Images of Graphite	77
	References	80
6	**STM and AFM Images of Layered Inorganic Compounds**	83
6.1	**Layers from MX_6 Trigonal Prisms and Octahedra**	83

6.2	**Images of Layered Compounds**	86
6.2.1	2H-MoS$_2$	86
6.2.2	MoOCl$_2$	88
6.2.3	WTe$_2$	89
6.2.4	NbTe$_2$	92
6.2.5	β-Nb$_3$I$_8$	94
6.2.6	1T-TaSe$_2$	98
6.3	**Charge Density Waves of MC$_8$ (M = K, Rb, Cs)**	105
6.3.1	Observations	105
6.3.2	Origin of Nonuniform Charge Distribution	107
6.4	**Concluding Remarks**	109
	References	110
7	**STM Images Associated with Point Defects of Layered Inorganic Compounds**	113
7.1	**Imperfections in Compounds with Metal Clusters**	113
7.2	**Point Defects in Semiconductor 2H-MoS$_2$**	116
7.3	**Cases Tractable by Electronic Band Structure Calculations**	118
7.3.1	Ligand-Atom Vacancy	119
7.3.2	Metal-Atom Vacancy	120
7.3.3	Donor Substitution at the Metal Site	120
7.4	**Cases Intractable by Electronic Band Structure Calculations**	123
7.4.1	Donor Substitution at the Ligand Site	123
7.4.1.1	The Case of Negative Bias	124
7.4.1.2	The Case of Positive Bias	125
7.4.2	Acceptor Substitution at the Ligand Site	125
7.4.2.1	The Case of Positive Bias	126
7.4.2.2	The Case of Negative Bias	127
7.4.3	Acceptor Substitution at the Metal Site	127
7.5	**Survey of Image Imperfections Observed for d^2 2H-MX$_2$ Systems**	128
7.5.1	Atomic-Scale Images	129
7.5.2	Nanometer-Scale Images	131
7.6	**Concluding Remarks**	133
	References	134
8	**Tip–Sample Interactions**	135
8.1	**Electronic Interactions in STM**	135
8.1.1	Tip Electronic States	135
8.1.2	Tip-Induced Local States	136
8.2	**Force Interactions in STM**	137
8.2.1	Force Interactions in Ambient Conditions	138
8.2.2	Force Interactions in Ultra High Vacuum (UHV)	140

8.3	Tip–Sample Interactions in AFM	145
8.3.1	Force Interactions on the Atomic Scale	145
8.3.2	Surface Deformation	146
8.4	**Concluding Remarks**	148
	References	148
9	**Surface Relaxation in STM and AFM Images**	151
9.1	**Tip Force Induced Deformation in HOPG**	151
9.1.1	Three-for-Hexagon Pattern of HOPG	151
9.1.2	Hexagonal Moiré Patterns in STM Images	154
9.2	**Wagon-Wheel Patterns of $MoSe_2$ Epilayers on MoS_2**	157
9.3	**STM and AFM Images of α-$RuCl_3$ and α-$MoCl_3$**	159
9.3.1	Images of α-$RuCl_3$ at Low Applied Force	160
9.3.2	Images of α-$RuCl_3$ at High Applied Force	163
9.3.3	Tip Force Induced Surface Deformation in α-$RuCl_3$	163
9.3.4	AFM Images of α-$MoCl_3$	167
9.4	**Layered Transition-Metal Tellurides MA_xTe_2**	169
9.4.1	Atomic-Scale Deformation in the Commensurate Tellurides	169
9.4.2	Structure of Incommensurate Telluride $TaGe_{0.355}Te_2$	176
9.5	**Tip Force Induced Changes in AFM Images of $NbTe_2$**	177
9.6	**Nanoscale Ring Structure of MoS_2 and WSe_2**	180
9.7	**Concluding Remarks**	184
	References	185
10	**Organic Conducting Salts**	189
10.1	**Crystal and Electronic Structures**	189
10.2	**Early STM Studies of Organic Conductors**	193
10.3	**STM and AFM Imaging of Organic Conductors**	194
10.3.1	Surface Processes During Imaging	194
10.3.2	Molecular-Scale Images	196
10.4	**Analysis of the Images of TCNQ Salts**	199
10.4.1	TTF-TCNQ	199
10.4.2	$Qn(TCNQ)_2$	201
10.4.3	$4EP(TCNQ)_2$	202
10.4.4	$TEA(TCNQ)_2$	204
10.4.5	TCNQ Salts with Substituted Phenylpyridines	205
10.5	**Analysis of the Images of BEDT-TTF Salts**	206
10.5.1	Cation-Layer Images of α-Phases	206
10.5.2	HOMO Density of β-$(BEDT-TTF)_2I_3$	209
10.5.3	Cation-Layer Images of κ-Phases	212
10.5.4	Anion-Layer Images of κ-Phases	212

10.6	Concluding Remarks	216
	References	217

11	**Organic Adsorbates at Liquid/Solid Interfaces**	219
11.1	**STM of Organic Adsorbates**	219
11.1.1	Organic Compounds and Substrates	219
11.1.2	STM Imaging at Liquid/Solid Interfaces	221
11.2	**STM of Normal and Cyclic Alkane Layers**	223
11.2.1	Images of Normal Alkanes on HOPG	223
11.2.2	Molecular Order of Cycloalkane Adsorbates on HOPG	228
11.3	**Influence of Substrate on Adsorbate Structure**	233
11.3.1	Molecular-Scale Images of Normal Alkanes on β-Nb_3I_8	233
11.3.2	4-Alkyl-4'-cyanobiphenyls on HOPG	235
11.3.3	4-Alkyl-4'-cyanobiphenyls on β-Nb_3I_8	237
11.4	**Concluding Remarks**	241
	References	241

12	**Self-Assembled Structures**	243
12.1	**Scanning Probe Microscopy Studies of Thin Organic Films**	243
12.1.1	Morphology and Molecular Order	243
12.1.2	Nanomechanical Properties	244
12.2	**Self-Organization of Amphiphiles**	245
12.2.1	Basic Principles	245
12.2.2	Sample Preparation and AFM Imaging	246
12.3	**AFM Study of N-(n-Alkyl)-D-gluconamides**	250
12.3.1	Crystal Structures	250
12.3.2	Layers with Crystal-Like Order	253
12.3.2.1	Thin Overlayers	253
12.3.2.2	Double Layers	255
12.3.3	Supramolecular Assemblies	257
12.3.3.1	Micellar Structures	257
12.3.3.2	Fiber-Like Assemblies	259
12.3.3.3	Rod-Like Assemblies	261
12.3.4	Structural Models	264
12.4	**AFM Study of N-(n-alkyl)-N'-D-maltosylsemicarbazones**	266
12.4.1	Self-Assembled Structures of 10MS	266
12.4.2	Self-Assembled Structures of 16MS	269
12.4.3	Structural Models	273
12.5	**Concluding Remarks**	274
	References	275

13	**Polymers**	277
13.1	**General Considerations**	277
13.1.1	Polymer Structure	277
13.1.2	Analysis of Polymer Surfaces	279
13.1.3	Applying STM and AFM	279
13.2	**STM of Polymer Samples**	281
13.2.1	Conducting Polymers	281
13.2.2	Metal-Coated Polymer Surfaces	282
13.2.3	Polymer Layers on Conducting Substrates	283
13.3	**AFM of Polymer Crystal Surfaces**	284
13.3.1	Polydiacetylene Single Crystal	284
13.3.2	Polyethylene Single Crystal	286
13.3.3	Polymer Spherulites	291
13.4	**AFM of Oriented Polymers**	294
13.4.1	Imaging of Molecular Chain Order	294
13.4.2	Nanostructure of Polyethylene Tapes and Fibers	296
13.4.3	Other Oriented Polymer Samples	303
13.5	**AFM of Di-Block Copolymers**	303
13.5.1	Poly(styrene-b-isoprene) Films	305
13.5.2	Poly(styrene-b-methyl methacrylate) and Poly(styrene-b-2-vinyl-pyridine) Films	306
13.6	**Concluding Remarks**	308
	References	310
14	**Future Outlook**	313
	Acknowledgements	317
	Index	319

1 Introduction

1.1 Development of Scanning Probe Microscopy

The invention of a scanning tunneling microscope nicely exemplifies the creation of a new research tool by innovative implementation of scientific and technological knowledge, thereby further advancing fundamental science and technology. The quantum-mechanical phenomenon of electron tunneling had been known for a long time, but the use of this phenomenon for the imaging of a conducting surface on atomic scale was realized only in 1982 when the first scanning tunneling microscope was built by Binnig et al [1]. At present, scanning tunneling microscopy (STM) is a powerful tool for analyzing metallic and semiconducting surfaces. The most important feature of STM is the real-space visualization of surfaces on atomic scale. What is converted into an image in STM is either the spatial variation of the tunneling current or the spatial variation of the tip height. The tunneling current decreases exponentially with increasing tip–sample distance. Thus, at any given location of the tip over the sample surface, the electron transfer involves only one atom, or only a few atoms, at the tip apex and on the surface closest to them. This gives rise to the local character of STM measurements, which makes it possible to visualize surface structures with sub-angstrom resolution and to detect various atomic-scale defects that are inaccessible by diffraction and spectroscopic techniques [2–4]. In addition, STM is used to examine adsorbate structures and dynamic phenomena on surfaces (e.g., diffusion and chemical reactions).

Since a tunneling current is employed in STM, the application of this method is mostly limited to metals and semiconductors. To enable the detection of atomic-scale features of insulating surfaces, an atomic force microscope was invented [5]. In atomic force microscopy (AFM) it is commonly the repulsive force between the tip (located at the end of a cantilever) and sample that is measured, on the basis of the cantilever deflection. In this contact-mode AFM, the spatial variation of the tip–sample repulsive force or that of the tip height is converted into an image. Because the repulsive force is universal, AFM is applicable to conducting as well as insulating materials. In general, AFM enables one to detect surface morphology, nanoscale structures, and molecular- and atomic-scale lattices.

Contact-mode AFM was originally introduced for high-resolution surface profiling. With the progress in AFM applications, it became clear that for many materials this objective can be achieved only by minimizing tip–sample force interactions, because the latter may modify the topography of a sample surface. In addition, it was also realized that these interactions can be utilized to probe the mechanical properties of surfaces such as indentation, adhesion and friction. For example, the tip may cause elastic or inelastic surface deformations [6], which can be recognized from the images obtained with high forces. In imaging with low force, the influence of the weak surface forces (e.g., van der Waals, hydrophilic, hydrophobic, and electrostatic interactions) on the cantilever movement becomes significant [7]. It is a challenging task to deconvolute the contributions of these forces to the image contrast. Invaluable information about the tip–sample force interactions can be obtained by analyzing the force versus tip–sample distance curves (hereinafter referred to as force-vs.-distance curves).

The success of STM and AFM led to a new family of scanning probe techniques in which different types of tip–sample interactions are utilized. For further details of these methods, the reader is referred to several reviews and books [2–4]. So far, STM and AFM are the most advanced scanning probe methods and the only ones providing atomic-resolution images. AFM has found much broader application than STM and is currently the dominant scanning probe technique.

1.2 Key Problems of STM and AFM Applications

The development of instrumentation and the availability of commercial microscopes at moderate price have accelerated the use of STM and AFM and demonstrated their unique potential in surface characterization. In applying STM and AFM to surfaces of chemical interest, common problems to be faced are how to distinguish genuine features from experimental artifacts in observed images, how to improve the image resolution, how to collect comprehensive experimental information, and how to interpret observed images (especially those with atomic or molecular resolution). A close interplay between experiment and theory is essential in answering these questions and in making STM and AFM truly indispensable analytical tools for surface science and nanotechnology.

1.2.1 Image Interpretation

Atomic-scale STM and AFM images are routinely recorded for many crystalline surfaces, but their interpretation is by no means straightforward. It is tempting to assign the atomic-size spots of STM and AFM images to the surface atomic or molecular structures. Such a correspondence has been found in several cases, but this interpretation can be misleading, especially for STM, because the electron tunneling involves only the energy levels of the sample lying in the vicinity of the Fermi level e_f. When the tip–sample interactions are neglected, the STM image is described by the partial electron density plot $\rho(r_0, e_f)$ of the sample surface [8]. In contact-mode AFM measurements, all the electrons of the surface atoms are involved in the repulsive interactions with the tip, so that the AFM image is described by the total electron density plot $\rho(r_0)$ of the surface. Consequently, it is reasonable to assign the AFM images to the surface topography, but this is not necessarily the case in STM.

When the geometry of a sample is known, it is straightforward to calculate the $\rho(r_0, e_f)$ plots from its electronic structure. For a layered material, the surface reconstruction is negligible so that the geometry of the surface layer is well approximated by the layer geometry of the bulk crystal structure. For a large number of organic and inorganic layered materials, the atomic- and molecular-scale features of their STM images [9] have been successfully interpreted on the basis of the $\rho(r_0, e_f)$ plots calculated with the extended Hückel tight binding (EHTB) electronic band structure method [10].

In general, the contribution of an atom to the $\rho(r_0, e_f)$ plot increases as its distance to the tip decreases and as its electronic contribution to the energy levels around the Fermi level increases. Since the more-protruding atoms do not necessarily make more contributions to the energy levels near the Fermi level, it is difficult to interpret STM images unless appropriate partial density plots are calculated, even when the geometry of the sample surface is known. The essential findings of the STM studies carried out in conjunction with $\rho(r_0, e_f)$ plot calculations [9] can be summarized as follows:

(a) When the height corrugation of the surface is of the order of 0.5 Å, the lower-lying atoms of the surface can dominate the $\rho(r_0, e_f)$ plot and hence the STM image.
(b) If the subsurface atoms lie more than 1 Å (0.1 nm) below the surface atoms, the STM patterns are dominated by the topmost surface atoms, even when the energy levels around the Fermi level are dominated by the subsurface atoms.
(c) Insulating molecules adsorbed on a metallic substrate are detected by STM, because their orbitals mix slightly into the Fermi level of the metallic substrate and because they are close to the tip.

It is challenging to characterize surface reconstruction on the basis of high-resolution STM and AFM images. For this purpose, it is necessary to calculate the $\rho(r_0, e_f)$ and $\rho(r_0)$ plots for a number of model structures until a good match is found between theory and experiment.

To help interpret the STM images of semiconducting materials on the basis of partial density plot calculations, it is desirable to carry out imaging with bias voltages of different magnitudes and polarities (i.e., positive or negative). Such measurements, as well as the recording of current-vs.-voltage curves, are the subject of scanning tunneling spectroscopy (STS), which was critically reviewed by Trompt [11].

1.2.2 Tip-Sample Interactions

To be precise, the image interpretation described above is valid when the tip is point-like and the tip-sample interactions are negligible. In practice, these conditions are hardly realized. The tip-shape anisotropy can induce various artifacts in large- and atomic-scale images. Even the use of tips with perfect shape might lead to a nontrivial image perturbation because of the inevitable tip-sample interactions in STM and AFM (especially in ambient-condition experiments). The scanning tip can exert strong vertical and lateral forces on the sample, thereby causing surface deformation and removal of weakly bound and defective layers.

Consideration of the tip-sample force interactions is critical in the imaging of soft organic materials. AFM possesses a unique potential for the characterization of these materials, as has been demonstrated in studies of the self-assembled structures of saccharide-based amphiphiles [12]. On the basis of the AFM images of their adsorbates on mica, it was possible to determine the molecular packing in the bilayers, detect a variety of micellar nanostructures (disks, grains, cylinders, etc.) and characterize the topography of the supramolecular assemblies. In general, the imaging of organic compounds requires minimization of the applied force to avoid surface damage and reduce the tip-sample contact area. In contact-mode AFM the applied force is significantly diminished when the tip and sample are both immersed in liquids (e.g., water, ethanol) [7a]. To lessen the surface modification induced by the lateral force of the tip, one may employ dynamic modes (e.g., tapping-mode AFM) in which the cantilever vibrates for the tip to make an intermittent contact with the sample [13]. In the AFM of polymers, minimization of the tip-sample force interactions makes it possible to image mechanically weak surface nanostructures that are inaccessible by electron microscopy (e.g., 2–3 nm-wide nanofibrils of stretched polyethylene tapes under an applied force of ca. 2 nN [14]).

The resolution of contact-mode AFM strongly depends on the sample. On poorly ordered surfaces, the resolution is limited by the tip-sample contact area. For example, the detection of the polyethylene nanofibrils mentioned above suggests a tip-sample contact diameter smaller than 2 nm, which is consistent with the theoretical estimate [15a]. In contact-mode AFM studies of crystalline surfaces, atomic- and molecular-scale lattices are routinely observed in images obtained with different applied forces. This indicates that the image resolution for crystalline surfaces does not critically depend on the size of the contact area. In contrast to the case of STM, atomic-size defects are hardly detected in AFM images.

Strong tip-sample force interactions are common for ambient-condition STM measurements. In air, a sample surface is coated with a liquid contamination layer (mostly made up of water). Therefore, while scanning the surface, the STM tip is in contact with the contamination layer and exerts a load (i.e., force) on the surface via this layer. This load, which is in the order of several hundreds of nanonewtons, is much higher than the typical operating force in contact-mode AFM [16]. Consequently, the STM images recorded at ambient condition are more likely to represent "distorted surfaces" under the action of the tip force rather than the "ideal surfaces" one expects in the absence of the tip–sample interactions. To understand the influence of such force interactions on STM images, it is necessary to perform measurements as a function of the tip–sample separation. Typically, the force interactions become stronger with decreasing tip–sample distance. In terms of the operational parameters of STM experiments, the tip–sample distance is reduced by decreasing $R_{gap} = |V_{bias}|/I_{set}$, where V_{bias} is the bias voltage and I_{set} is the set-point tunneling current. Therefore, by conducting STM measurements at different levels of tip–sample interactions, one can also learn about the surface mechanical properties.

1.2.3 Surface Relaxation and Local Hardness

In general, the surface of a material consists of atoms with different local environments. The local area of a surface atom (i.e., the atom and its closest neighbors) can have a different hardness, depending on its local structural and chemical bonding environment. It is expected that under the tip force a harder surface region (or atom) is depressed less, and hence appears more elevated in the image, than a softer region (or atom). One can investigate the spatial variation of the local hardness in a given sample surface by analyzing depression pattern induced by the tip force, on the basis of STM or AFM measurements under several different levels of force interactions. Macroscopic deformations of materials are discussed in terms of hardness and stiffness. The hardness is defined as the load (i.e., applied force) divided by the contact area, and the stiffness as the slope of the load-vs.-depression curve [15b]. As will be discussed below, the surface local hardness (or stiffness) is a nanoscale analog of the corresponding macroscopic properties.

AFM is indispensable for the detection of surface relaxation on the nanometer and subnanometer scales. On the images of stretched polyethylene tapes, the force-dependent contrast variations reveal the presence of hard and soft nanoscale regions within a nanofibril core [14a]. The tip force might also cause conformational changes in biological macromolecules [17]. In the AFM studies of layered transition-metal tellurides MA_xTe_2 (M = Nb, Ta; A = Si, Ge), strong applied forces of up to several hundred nanonewtons induce reversible contrast changes in their atomic-scale images. In a number of layered materials (e.g., graphite, transition-metal halides) the interlayer arrangement can cause a specific hardness pattern in the topmost layer, which appears as a moiré pattern in the AFM and STM images [18]. All these observations are explained in terms of the local hardness concept. The tip–sample interactions leading to a macroscopic deformation of the tip and sample (in the region of their contact surface) have been described by the continuum theory, which neglects the presence of discrete atoms and molecules in the interacting macroscopic bodies.

The microscopic surface deformation described by the local hardness concept is not covered by the continuum theory. It is a challenging problem to bring the prediction of local hardness to a quantitative level.

What the $\rho(r_0, e_f)$ and $\rho(r_0)$ plots simulate are ideal STM and AFM images expected for a defect-free surface in the absence of tip–sample interactions. Therefore, for certain compounds, it may be inadequate to interpret the STM images in terms of electron density plots calculated for non-relaxed surface structures. In principle, a comprehensive analysis requires knowledge of the surface deformation and the associated change in the $\rho(r_0, e_f)$ plot. To estimate a possible surface reconstruction, a systematic study of AFM images with different applied forces can be useful. AFM contrast variations might indicate a local hardness variation of the surface, from which one can make a reasonable guess of the deformed surface structure needed for the electron density plot calculations. At tip forces of several hundreds of nanonewtons, the surface deformations will be similar to those expected in STM. Such an interplay between theory and experiment would be invaluable for self-consistent STM and AFM image interpretations.

1.2.4 Surface Forces and AFM

A rational examination of the tip–sample interactions in low-force imaging requires consideration of the surface forces. Tip–sample adhesion, which leads to hysteresis in the force-vs.-distance curve, is a direct consequence of the surface forces. In ambient-condition experiments this hysteresis is enhanced by the capillary forces associated with the contamination layer. In subliquid measurements, in which the capillary forces are absent, the magnitude of the hysteresis can be assigned to the tip–sample adhesion. To determine the spatial variation of the tip–sample adhesion, one may measure the force-vs.-distance curves at a large number of sample surface locations [7c, d]. Alternatively, one may determine the variation of the lateral force (rather than the vertical force employed in the force-vs.-distance curve measurements), assuming that the lateral force increases with the tip–sample adhesion [7e]. It is appealing to correlate the adhesion and lateral force variations with the chemical nature of particular surface regions. However, the relationship between the chemical nature of a surface, its morphology, and the adhesion is not well understood. Recently, hydrophilic and hydrophobic forces have attracted considerable attention. Systematic studies aimed at understanding the effects of these forces on AFM imaging are currently under active scrutiny [19]. These efforts can be facilitated by operating AFM with chemically modified probes and in different environments so as to alter deliberately the strengths of the specific surface forces. Such an approach can help, for example, to identify nanoscale surface regions with different hydrophilic and hydrophobic properties.

1.3 Objectives

This book is primarily concerned with how the surfaces of various materials are characterized by employing STM and AFM, and what physical/chemical features can be deduced from their images. To achieve these objectives, two important steps in the application of STM and AFM should be considered. One is the recording of experimental images as a function of tunneling parameters and applied forces to extract information about the surface morphology, nanostructure, and atomic-scale features. The other is the interpretation of the observed images in terms of the topographic, electronic, and mechanical properties of the surfaces.

Several monographs concerning scanning probe studies have already appeared [2-4], but their overlap with this book is minimal. Most of the examples given in this book were taken from the experimental and theoretical studies carried out by the research groups at Freiburg and Raleigh. It is hoped that the experimental and theoretical aspects of STM and AFM studies presented in this book are practical and interesting for researchers already involved in STM and AFM, as well as for newcomers to the field.

References

[1] G. Binnig, H. Rohrer, Ch. Gerber, E. Weibel, *Phys. Rev. Lett.* **1982**, *49*, 57.
[2] R. Wiesendanger, H.-J. Güntherodt (Eds.), *Scanning Tunneling Microscopy I, II* and *III*, Springer, Heidelberg, **1992** and **1993**.
[3] C. J. Chen, *Introduction to Scanning Tunneling Microscopy*, Princeton University Press, Princeton, **1993**.
[4] D. A. Bonnell (Ed.), *Scanning Tunneling Microscopy and Spectroscopy*, VCH, New York, **1993**.
[5] G. Binnig, C. Quate, Ch. Gerber, *Phys. Rev. Lett.* **1986**, *56*, 930.
[6] (a) M. Radmacher, R. W. Tillmann, M. Fritz, H. E. Gaub, *Science* **1992**, *257*, 1900. (b) M. Salmeron, G. Neubauer, A. Folch, M. Tomitori, D. F. Ogletree, P. Sautet, *Langmuir* **1993**, *9*, 3600. (c) B. Bhushan, V. N. Koinkar, *Appl. Phys. Lett.* **1994**, *64*, 1653. (d) J. P. Aime, C. Elkaakour, C. Odin, T. Bouhacina, D. Michel, J. Curely, J. Dautant, *J. Appl. Phys.* **1994**, *76*, 754.
[7] (a) A. L. Weisenhorn, P. Maivald, H.-J. Butt, P. K. Hansma, *Phys. Rev. B* **1992**, *45*, 11 226. (b) N. A. Burnham, D. D. Dominguez, R. L. Mowery, R. J. Colton, *Phys. Rev. Lett.* **1990**, *64*, 1931. (c) K. O. van der Werf, C. A. J. Putman, B. G. de Grooth, J. Greve, *Appl. Phys. Lett.* **1994**, *65*, 1195. (d) D. R. Baselt, J. D. Baldeschwieler, *J. Appl. Phys.* **1994**, *76*, 33. (e) C. D. Frisbie, L. F. Rozsnyai, A. Noy, M. S. Wrighton, C. M. Lieber, *Science* **1994**, *265*, 2071.
[8] J. Tersoff, D.R. Hamman, *Phys. Rev. B* **1985**, *31*, 805.
[9] For a review, see: S. N. Magonov, M.-H. Whangbo, *Adv. Mater.* **1994**, *6*, 355.
[10] M.-H. Whangbo, R. Hoffmann, *J. Am. Chem. Soc.* **1978**, *100*, 6093.

[11] R. M. Trompt, *J. Phys. Condens. Matt.* **1989**, *1*, 10211.
[12] I. Tuzov, K. Crämer, B. Pfannemüller, S. N. Magonov and M.-H. Whangbo, *New J. Chem.*, in press.
[13] (a) Q. Zhong, D. Innis, K. Kjoller, V. B. Elings, *Surf. Sci. Lett.* **1993**, *290*, L688. (b) M. Dreier, D. Anselmetti, H.-J. Güntherodt, *J. Appl. Phys.* **1994**, *76*, 5095.
[14] (a) A. Wawkuschewski, H.-J. Cantow, S. N. Magonov, *Adv. Mater.* **1994**, *6*, 476. (b) A. Wawkuschewski, K. Crämer, H.-J. Cantow, S. N. Magonov, *Ultramicroscopy* **1995**, *58*, 185.
[15] (a) T. P. Weihs, Z. Nawaz, S. P. Jarvis, J. B. Pethica, *Appl. Phys. Lett.* **1991**, *59*, 3536. (b) J. B. Pethica, W. C. Oliver, *Phys. Scr.* 1987, *T19*, 61.
[16] (a) C. M. Mate, R. Erlandsson, G. M. McClelland, S. Chiang, *Surf. Sci.* **1989**, *208*, 473. (b) M. Salmeron, D. F. Ogletree, C. Ocal, H.-C. Wang, G. Neubauer, W. Kolbe, G. Meyers, *J. Vac. Sci. Technol. B* **1991**, *9*, 1347.
[17] D. J. Müller, G. Bült, A. Engel, *J. Mol. Biol.* **1995**, *249*, 239.
[18] (a) M.-H. Whangbo, W. Liang, J. Ren, S. N. Magonov, A. Wawkuschewski, *J. Phys. Chem.* **1994**, *98*, 7602. (b) H. Bengel, H.-J. Cantow, S. N. Magonov, L. Monconduit, M. Evain, M.-H. Whangbo, *Surf. Sci. Lett.* **1994**, *321*, L170. (c) H. Bengel, H,-J. Cantow, S. N. Magonov, H. Hillebrecht, G. Thiele, W. Liang, M.-H. Whangbo, *Surf. Sci.*, in press. (d) H. Bengel, H.-J. Cantow, S. N. Magonov, L. Monconduit, M. Evain, W. Liang, M.-H. Whangbo, *Adv. Mater.* **1994**, *6*, 649.
[19] (a) V. Y. Yaminsky, B. W. Ninham, *Langmuir* **1993**, *9*, 3618. (b) W. A. Ducker, D. R. Clarke, *Colloid. Surf.* **1994**, *94*, 275.

2 Physical Phenomena Relevant to STM and AFM

The range of applications of the scanning probe technique is determined by the nature of the probing interaction. The electron tunneling and force interactions between macroscopic bodies, which are relevant for STM and AFM, have been the subject of numerous studies in the past [1, 2]. Theoretical and experimental aspects of the physical phenomena involved in STM and AFM are discussed in this chapter.

2.1 Electron Transport Processes

One of the important factors controlling the nature of the electron transfer between the tip and the sample (i.e., at the electrodes) is the tip–sample separation. In the conventional view of STM, it is assumed that the tip–sample separation is large (Fig. 2.1 (a)) [3], although this is not necessarily true in practical applications. Therefore, it is important to consider electron transfer processes at small tip–sample separations. The theoretical studies of Ciraci [4] show that as the tip–sample separation decreases, the electron transfer process changes from the conventional tunneling regime to the electronic contact regime and to the mechanical contact regime. This analysis provides a starting point for the interplay between theory and experiment aimed at better understanding the STM and AFM operations and the associated tip–sample interactions.

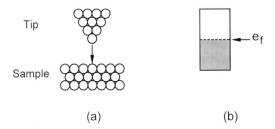

Figure 2.1 (a) Ideal tip–sample arrangement in STM. (b) Partially filled band of a metallic sample, where the Fermi level is indicated by e_f.

2.1.1 Conventional Electron Tunneling Regime

The electron transfer process in metallic tip–insulator–metal systems is classified into three mechanisms on the basis of the current-vs.-voltage (I–V) relationship [5, 6]. In tunnel emission, when the applied voltage (V) is much less than the effective barrier height (ϕ) of the insulator ($V \ll \phi/e$), the current is proportional to the applied voltage ($I \propto V$). In field emission, with $V \gg \phi/e$, the I–V dependence is given by

$$I \propto V^2 \exp(-\text{const}/V)$$

In Schottky emission, where the potential barrier is low and $V \gg k_B T/e$, the current is proportional to $\exp(\text{const}/V^{1/2})$. In the topografiner (the predecessor of the scanning tunneling microscope), which employs the field emission phenomenon, the tip–sample separation is around 100 nm. In the electron tunneling process which occurs between electrodes separated by 0.5–1 nm, the I–V dependence is linear [5].

In the tunneling regime, the tip and sample can be regarded as independent and the electron transfer between them is described by the perturbation approach, using the wave functions of the free electrodes. Representing the tip by an atom with a single s-orbital and assuming a small bias voltage between the tip and sample, Tersoff and Hamman [7] showed on the basis of the perturbation approach that the spatial variation of the tunneling current I_tun is described by that of the partial electron density distribution $\rho(r, e_f)$ of the sample as $I_\text{tun} \propto \rho(r, e_f)$. The partial electron density is associated with the energy levels lying in the vicinity of the Fermi level e_f (Fig. 2.1 (b)). The evaluation of this density distribution at the tip–sample distance of r_0 (from the sample surface) gives rise to the partial density plot $\rho(r_0, e_f)$, which simulates the observed STM image.

2.1.2 Electronic and Mechanical Contact Regimes

When the tip–sample separation is decreased below a certain distance, the overlap of their wave functions increases and the potential barrier between the electrodes is gradually lowered. This causes rearrangement of the electron density distribution and induces short-range attractive forces (adhesion) and displacements of the atoms in the tip and in the sample [4]. The local electronic and structural modifications are significant but they are reversible. Furthermore, the transport of current takes place via tunneling although the electronic states are substantially modified.

As the tip–sample distance decreases, direct tip–sample contact begins with a quantum dot contact, for which the diameter of the contact area is smaller than the mean free path of an electron. In this regime, electron transport occurs in the absence of any barrier (i.e., ballistic conduction). On further shortening of the tip–sample distance, the gap resistance R_{gap} (i.e., the resistance between the tip and the sample) reaches the limiting value of the Sharvin resistance (i.e., $4\pi^2 e/h^2 = 13$ KΩ). Below this R_{gap} value, ohmic conductance is observed (i.e., the $I-V$ curve is linear). The current-vs.-distance dependence in the ballistic regime is rather sharp [8], so that high-resolution imaging can be achieved in this regime [9]. However, in the ballistic and ohmic contact regimes, the tip and sample may undergo an irreversible deformation due to strong force interactions (see Section 2.2).

2.1.3 STM in Different Environments

The scanning tunneling microscope was initially designed for operation in ultrahigh vacuum (UHV). It soon became clear that this instrument can be used in ambient conditions, under liquid, and in an electrochemical environment, to obtain surface images with atomic resolution. Quite strikingly, it was found that organic molecules, which are insulating, can be imaged by STM when they are adsorbed on conducting substrates [10]. So far the electron transfer mechanisms responsible for atomic-scale imaging are not well understood. One cannot exclude the possibility that the actual electron transfer mechanism can be different from conventional tunneling. The $I-V$ dependence found for ambient-condition STM measurements is consistent with Schottky emission [6], which suggests that electron transfer is facilitated by much lower barrier heights in air than in vacuum. This is believed to result from the contamination layer present between the tip and the surface. Even in UHV, surfaces can be contaminated by oxides, hydroxides, and water, thereby affecting the relationship between the current and tip–sample separation [11, 12]. It has been suggested that the low barrier heights and atomic-resolution imaging originate from intermediate-state tunneling (i.e., via the localized states of the adsorbates in the tip–sample gap) [12]. Electron transfer through the water overlayer may become a dominant process in low-current (picoamp-range) STM imaging of DNA macromolecules on mica, which was performed in humid air [13]. These observations clearly demonstrate a need for systematic experimental and theoretical studies of STM in different environments.

2.2 Survey of Force Interactions

A thorough consideration of tip–sample interactions is essential in STM and AFM, because the tip is placed very close to the sample surface. In contact-mode AFM, the atomic-scale image contrast originates from the variation of the tip–sample repulsive force and therefore contains local information due to the short-range character of this force. It is not clear whether long-range attractive forces can be used to image surfaces with atomic resolution. Nevertheless, the attractive forces are important in AFM and STM because they contribute to the overall tip–sample interactions and hence to the size of the tip–sample contact area. The forces that the tip exerts on the sample can induce a surface deformation and add complexity to the resulting image [14].

2.2.1 Force-vs.-Distance Curves

Force-vs.-distance curves describing the interactions between two macroscopic bodies are traditionally measured with surface force apparatus [15]. The corresponding curves for the tip/sample system can be obtained with an atomic force microscope [16]. When the tip comes close to the sample, the cantilever is deflected from its equilibrium position in response to the force experienced by the tip (Fig. 2.2). It bends towards the sample when the force is attractive, and away from it when the force is repulsive. As the sample approaches the tip in the nontouching regime, the van der Waals (VDW) attraction bends the cantilever towards the sample. When the sample is moved further, at a certain point (the jump-in contact point) the attraction force gradient exceeds the spring constant of the cantilever, and the tip jumps onto the sample surface, thereby making a contact with the sample. As the sample moves further towards the tip, the cantilever is deflected as the sample is moved (the touching regime). In addition to the bending of the cantilever, the tip and sample may undergo elastic (reversible) or plastic (irreversible) deformations.

When the sample is retracted from the tip in the touching regime, the cantilever moves again with the sample. It may even deflect towards the sample before the tip breaks the contact with the sample due to the adhesive and capillary forces. The latter arise from the contamination liquid layer covering the sample surface in air. The tip loses contact with the sample surface at the jump-out point, where the transition from touching to nontouching occurs, and the force-vs.-distance curve returns to the

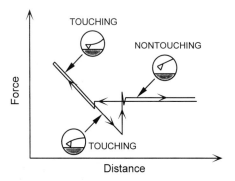

Figure 2.2 Typical force-vs.-distance curve observed in AFM experiments in air.

nontouching line. The difference between the minimum point of the "retrieval" force-vs.-distance curve and the nontouching line is defined as the pull-out force, which becomes identical to the adhesive force when the capillary force vanishes.

In the attractive force region where the force-vs.-distance curve exhibits hysteresis, the behavior of the probe is strongly influenced by the long-range forces. Details of the curve can vary significantly, depending on the nature of the tip and sample (e.g., metallic or insulating) as well as on the medium in which they are immersed. For example, when the nonconducting tip and sample are immersed in water, the pull-out forces are reduced to the level of 1 nN or less [16]. In the repulsive force region, the mechanical response of the sample depends on the hardness of the sample, that of the tip, and the force constant of the cantilever.

2.2.2 Short-Range Forces and Sample Deformation

Short-range force interactions between atoms are often described on the basis of atom–atom pair potentials such as Lennard-Jones or Morse type potentials [2a]. They are strongly repulsive at a short distance and slightly attractive at a long distance, so the short-range repulsive interaction is more sensitive to a small change in the distance than is the long-range attractive interaction. This leads to a sharp force-vs.-distance relationship and provides the basis for a high-resolution surface imaging in contact-mode AFM.

An important question concerning contact AFM is: which property of the sample controls the tip–sample repulsive force? At the computational level, Gordon and Kim provided an efficient and reliable method to calculate the forces between closed-shell atoms and molecules in the regions of the attractive well and the repulsive wall [17]. In this method, the interaction energies are calculated on the basis of the electron densities of the interacting systems. To a first approximation, therefore, the spatial variation of the tip–sample repulsive force F_{rep} should be described by that of the total electron density $\rho(r)$ of the sample, as $F_{rep} \propto \rho(r)$. The evaluation of this density distribution at the tip–sample distance of r_0 (from the sample surface) gives rise to the total density plot $\rho(r_0)$, which simulates the observed AFM image.

In many contact-mode AFM experiments and in ambient-condition STM measurements, the tip–sample repulsive interactions are strong enough to induce a surface deformation during the scanning. The simple case of an elastic contact between a sphere pressed into a flat surface (Fig. 2.3) was analyzed by Hertz [18], who correlated the contact characteristics (e.g., the load, contact area, and indentation depth) with the mechanical properties of two interacting bodies (e.g., Young's moduli and Poisson's ratios). Hertz theory, commonly used in the analysis of macroscopic hardness tests, was also applied by Pethica and co-workers [19] to describe the tip–sample interactions in STM and AFM. This study allows one to estimate the size of the tip–sample contact area (e.g., for a tungsten tip of 20 nm radius acting on an or-

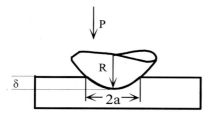

Figure 2.3 Deformation of the sample surface produced by the tip with apex radius r, where a is the radius of the contact area and δ is the surface depression.

ganic sample under an applied force of 1 nN, the diameter of the contact area is about 2.3 nm). It should be recalled that Hertz theory is valid only for an elastic deformation in a nonadhesive contact. In reality, contacting bodies deform according to their elastic as well as inelastic properties, and the surface forces bring about adhesion [20], thereby leading to a contact area greater than that given by Hertz theory. Several theories concerning the deformations and adhesion of contacting bodies predict [21, 22] a nonzero contact area even at zero applied force. In general, it is difficult to describe solids in contact even for bodies of ideal geometrical shape, because the stress distribution in the solids depends on the surface interaction, which in turn depends on the exact shape of the deformed surfaces [20]. All these continuum theories deal with the experimental results obtained by the surface force apparatus and the hardness tester (nanoindenter) [23].

Macroscopic contacts between bodies are involved in the operation of both STM and AFM instruments, and the diameter of the tip–sample contact area is on the order of several nanometers. In addition, both scanning tunneling and atomic force microscopes provide atomic-scale images of the surfaces deformed by the tip force during the scanning. Therefore, invaluable information about the nanomechanical surface properties can be obtained by these methods. How the sample surface is deformed under the tip force (when the tip is harder than the sample) can be tentatively envisioned as follows. The region of the sample in contact with the tip undergoes a macroscopic deformation as predicted by the continuum theories in which the sample is regarded as a body of homogeneous density distribution. In most cases of chemical interest, the surface of a sample consists of atoms with several different environments, so that the local hardness of the surface varies from place to place. Therefore, the strong tip force induces a surface deformation according to the local hardness variation [14, 24, 25]. Such a deformation, which can be on the nanometer or even atomic scale, is superposed on the macroscopic one (Fig. 2.4). The experimental evidence for this phenomenon is presented in Chapter 9.

During scanning in contact-mode AFM, the tip moves laterally so that the cantilever experiences vertical (normal) as well as lateral (friction) forces. Therefore, measurements of the lateral force variation in AFM experiments allow one to examine the

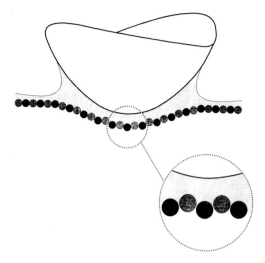

Figure 2.4 Superposition of the macroscopic and microscopic deformations of the sample surface under the tip force.

tip-surface friction from the micron scale down to the atomic scale. Currently, it is the subject of intensive studies to test the feasibility of using AFM for the analysis of friction. The recent study of Salmeron and co-workers [26] with a silicon nitride (Si_3N_4) tip on mica showed that the frictional forces are proportional to the normal forces (i.e., loads) in the moderate range (10–80 nN), in agreement with the friction behavior of macroscopic systems [27]. (At low loads a nonlinear behavior of the friction versus load is observed, due to the surface forces and the presence of the weakly adsorbed layers on the surface.) There have been numerous observations of atomic-scale friction behavior (e.g., the dependence of the friction force on the load as well as the scanning speed and direction) [28]. To analyze such experimental data, one should consider the atomic-scale surface corrugations and the possible tip-force induced surface relaxation. The importance of the surface corrugations to the friction behavior has been demonstrated theoretically and experimentally [29–31]. The tip-force induced surface relaxation was estimated to be small (below 0.1 nm) for the combination of a diamond-tip with a diamond surface [28], but the same cannot be expected for other systems. In analyzing the frictional data from AFM experiments, one should consider the tip-force induced corrugations – the macroscopic ones described by the continuum theories as well as the microscopic ones expected from the variation of the surface local hardness.

2.2.3 Long-Range and Other Forces

2.2.3.1 Long-Range Forces

Bodies at separations well beyond the chemical bonding distances experience VDW interactions, which can occur even when the electronic states of the separated bodies are decoupled. The VDW forces are important in STM and AFM because, depending on the shape of the tip, the atoms at the tip apex experience strong repulsion and deformation due to the VDW attraction between the sample and the back of the tip apex [32–34]. The magnitude of the VDW forces for the model tip/sample systems relevant to STM and AFM is estimated to be in the 1–20 nN range [32].

According to Lifshitz theory [35], the VDW forces are considered as dispersion forces associated with the electromagnetic fluctuations. This theory shows that the forces between bodies interacting through a medium depend on the dielectric properties of the bodies and the medium, and they are generally attractive but can also be repulsive. (The latter happens for the interaction of different bodies immersed in a medium, when the refractive index and dielectric constant of the medium have values lying between the corresponding values of the bodies.) Measurements of the forces between macroscopic bodies separated in the 50–250 nm range show that Lifshitz theory is in good agreement with experiment [2b].

It should be noted that not all long-range interactions follow the predictions of Lifshitz theory. For example, the attractive VDW forces between hydrophobic solids immersed in aqueous media exceed the expected theoretical values by one or two orders of magnitude [36]. This hydrophobic attraction was explained by Yaminsky and Ninham [37], who showed that fluctuations in molecular density give rise to long-range effects similar to the electromagnetic effects associated with charge density fluctuations. Their analysis of the molecular fluctuation shows that the approach of two surfaces in a poorly wetting liquid decreases the density of the liquid in the gap, due to enhancement of the thermal fluctuations in the lateral direction (Fig. 2.5 (a)). This lateral decompression manifests itself as the attractive force acting between two surfaces. For the case of two hydrophilic solids immersed in a medium (Fig. 2.5 (b)), the density of the liquid in the gap increases if the components of the medium possess higher affinity for the solid than for each other. The compression effect manifests itself as the repulsive forces acting between the two solid surfaces. This hydro-

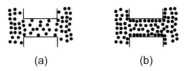

(a)　　　　　　(b)

Figure 2.5 (a) Density decrease in the gap between two hydrophobic bodies immersed in an aqueous medium. (b) Density increase in the gap between two hydrophilic bodies immersed in an aqueous medium. (Adapted from Ref. 37)

philic repulsion (also referred to as "hydration force") is of shorter range than the hydrophobic attraction because molecular compression into a film, which occurs in hydrophilic repulsion, is limited to the thickness of the adsorbed layers on the surface. Consideration of the hydrophobic attraction and hydrophilic repulsion is necessary in the AFM analysis of polymers, organic amphiphiles, and biological samples.

When the tip and sample are immersed in an electrolyte solution, they are subject to a force, not expected from Lifshitz theory, because ions of opposite charges assemble near the tip and sample surfaces to form double layers [2a]. As the double layers start to overlap, a strong repulsive force develops. For two planar surfaces immersed in electrolyte and separated by a distance D larger than the Debye length L (the characteristic decay length of repulsion associated with the medium), the electrostatic repulsive force between the double layers varies exponentially as $\exp(-D/L)$. For smaller separations, the double-layer interaction is perturbed because of the VDW forces and the charge regulation (i.e., re-adsorption of counterions reducing the surface charge density) [38]. Consequently, at $D < 3$ nm, the net interaction becomes attractive and the two surfaces suddenly touch. The total interaction between two charged surfaces in solution, which includes charge regulation, VDW, and electrostatic double-layer forces, is described by Derjaguin-Landau-Verwey-Overbeek (DLVO) theory [39]. In practice, the observed force behavior can be complicated. For example, the force-vs.-distance curve observed for two mica surfaces in dilute sodium chloride (NaCl) solution (e.g., 10^{-3} M or less) follows the DLVO theory, but the curve obtained in a more concentrated solution (10^{-2} M) remains repulsive all the way to contact [40]. The latter effect is caused by the hydrophilic repulsion, which arises from the hydrated ions bound to the surfaces.

It is interesting to consider the theoretical estimates of the VDW interactions for model tip/sample systems. On the basis of Lifshitz theory, Hartmann [33] calculated the VDW interactions between a silica (SiO_2) probe and a metal surface immersed in various liquids for the tip-sample separation of $1-10^4$ nm. Immersion of the tip/sample system in highly polar liquids is found to reduce the VDW forces, at small tip-sample distances, by up to more than two orders of magnitude with respect to the value in vacuum. It is also found that the VDW forces can be made either attractive or repulsive by an appropriate choice of the immersion medium. For the operation of Coulomb charge [41] or magnetic force microscopy (MFM) [42] in which weak long-range forces are employed, it may be important to use an immersion medium that nearly eliminates the VDW forces.

In biological applications of AFM, it is important to recognize that in water many surfaces are charged. The surface charging can come about from the dissociation of the surface groups (e.g., the dissociation of carboxylic groups into protons and carboxylate anions) or from the adsorption of ions onto the surface. The surface charges attract counterions, thereby forming a double layer in the vicinity of the surface, and the overall ion concentration increases near the sample. The theoretical analysis by Butt [43] shows that the electrostatic force between the tip and the sample is repulsive, increases with surface charge density, and decreases roughly exponentially with distance. To bring the tip in contact with the sample surface, the external applied force should overcome this electrostatic repulsion. Otherwise, what is imaged would reflect the surface charge distribution rather than the topography of the sample. The electrostatic repulsion can be reduced by imaging in a medium with a high salt concentration.

2.2.3.2 Adhesion and Capillary Forces

Adhesion between two macroscopic bodies is a consequence of the long- and short-range force interactions. AFM allows one to measure the variation in the local adhesion of the surface by studying the hysteresis of the force-vs.-distance curves [44]. In the experimental evaluation of adhesion, it is necessary to consider the possible influence of the capillary forces on the hysteresis. In ambient-condition experiments, the tip apex is wetted by the liquid contamination layer present on the surface, thereby forming a capillary between the tip and the sample (see Fig. 2.4). This gives rise to the capillary force, which is estimated to be in the 10–100 nN range under ambient conditions [16]. The capillary force increases the pull-out force and hence leads to an additional load on the sample surface during AFM imaging, thereby preventing low-force operation (see Chapter 4). It is preferable to conduct adhesion measurements as well as AFM imaging of soft materials under conditions in which the capillary forces are negligible (e. g., under vacuum, under dried gas, or by immersing the tip/sample in a liquid medium).

References

[1] B. Duke, *Tunneling in Solids*, F. Seitz, D. Turnbull (Eds.), Academic Press, New York, **1969**.
[2] (a) J. N. Israelachvili, *Intermolecular and Surface Forces*, 2nd ed., Academic Press, New York, **1992**. (b) Derjaguin B. V., Churaev N. V., Muller V. M., *Surface Forces* (In Russian), Nauka, Moscow, 1987.
[3] G. Binnig, H. Rohrer, Ch. Gerber, E. Weibel, *Phys. Rev. Lett.* **1982**, *49*, 57.
[4] C. Ciraci, in *Scanning Tunneling Microscopy III*, R. Wiesendanger, H.-J. Güntherodt (Eds.), Springer, Heidelberg, **1993**, p. 139.
[5] P. R. Emtage, W. Tantraporn, *Phys. Rev. Lett.* **1962**, *8*, 267.
[6] J. Jahanmir, P. E. West, T. N. Rhodin, *Appl. Phys. Lett.* **1988**, *52*, 2086.
[7] J. Tersoff, D. R. Hamman, *Phys. Rev. B* **1985**, *31*, 805.
[8] (a) Y. V. Sharvin, *Zh. Eksp. Teor. Phys.* **1965**, *48*, 984. (b) M. Salmeron, D. F. Ogletree, C. Ocal, H.-C. Wang, G. Neubauer, W. Kolbe, G. Meyers, *J. Vac. Sci. Technol. B* **1991**, *9*, 1347.
[9] D. P. E. Smith, G. Binnig, C. F. Quate, *Appl. Phys. Lett.* **1986**, *49*, 1166.

[10] (a) H. Ohtani, R. J. Wilson, S. Chiang, C. M. Mate, *Phys. Rev. Lett.* **1988**, *60*, 2398. (b) J. S. Foster, J. Frommer, *Nature* **1988**, *333*, 542.
[11] (a) M. Binggeli, D. Carnal, R. Nyffenegger, H. Sigenthaler, R. Christoph, H. Rohrer, *J. Vac. Sci. Technol. B* **1991**, *9*, 1985. (b) M. Grundner, J. Halbritter, *J. Appl. Phys.* **1980**, *51*, 5396. (c) W. Lisowski, A. H. J. van den Berg, C. A. M. Kip, L. I. Hanekamp, *Fresenius J. Anal. Chem.* **1991**, *341*, 196.
[12] (a) R. Berthe, J. Halbritter, *Phys. Rev. B* **1991**, *43*, 6880. (b) G. Repphun, J. Halbritter, *J. Vac. Sci. Technol. A*, **1995**, *13*, 1693.
[13] R. Guckenberger, M. Heim, G. Cevc, H. P. Knapp, W. Wiegräbe, A. Hillebrand, *Science* **1994**, *266*, 1538.
[14] M.-H. Whangbo, W. Liang, J. Ren, S. N. Magonov, A. Wawkuschewski, *J. Phys. Chem.* **1994**, *98*, 7602.
[15] J. N. Israelachvili, D. Tabor *Proc. R. Soc. London Ser. A* **1972**, *331*, 19.
[16] A. L. Weisenhorn, P. Maivald, H.-J. Butt, P. K. Hansma, *Phys. Rev. B* **1992**, *45*, 11226.
[17] R. G. Gordon, Y. S. Kim, *J. Chem. Phys.* **1972**, *56*, 3122.
[18] H. Hertz, *J. Reine Angew. Math.* **1882**, *92*, 156.
[19] (a) J. B. Pethica, W. C. Oliver, *Phys. Scr.* **1987**, *T19*, 61. (b) T. P. Weihs, Z. Nawaz, S. P. Jarvis, J. B. Pethica, *Appl. Phys. Lett.* **1991**, *59*, 3536.
[20] R. G. Horn, J. N. Israelachvili, F. Pribac, *J. Colloid Interface Sci.* **1987**, *115*, 480.
[21] (a) K. L. Johnson, K. Kendall, A. D. Roberts, *Proc. R. Soc. London Ser. A* **1971**, *324*, 301. (b) B. V. Derjaguin, V. M. Muller, Y. P. Toporov, *J. Colloid Interface Sci.* **1975**, *53*, 314.
[22] (a) V. M. Muller, V. S. Yushchenko, D. V. Derjaguin, *J. Colloid Interface Sci.* **1980**, *77*, 91. (b) B. D. Hughes, L. R. White, *J. Mech. Appl. Math.* **1979**, *32*, 445.
[23] J. B. Pethica, R. Hutchings, W. C. Oliver, *Phil. Mag. A* **1983**, *48*, 593.
[24] H. Bengel, H.-J. Cantow, S. N. Magonov, L. Monconduit, M. Evain, M.-H. Whangbo, *Surf. Sci.* **1994**, *321*, L170.
[25] H. Bengel, H.-J. Cantow, S. N. Magonov, H. Hillebrecht, G. Thiele, W. Liang, M.-H. Whangbo, *Surf. Sci.* **1995**, in press.
[26] J. Hu, X.-D. Xiao, D. F. Ogletree, M. Salmeron, *Surf. Sci.* **1995**, *327*, 358.
[27] J. L. Singer, H. M. Pollock (Eds.), *Fundamentals of Friction: Macroscopic and Microscopic Processes*, Kluwer, Dordrecht, **1992**.
[28] G. J. Germann, S. R. Cohen, G. Neubauer, G. M. McCleland, D. Coulman, *J. Appl. Phys.* **1993**, *73*, 163.
[29] C. M. Mate, G. McClelland, R. Erlandsson, S. Chiang, *Phys. Rev. Lett.* **1987**, *59*, 1942.
[30] (a) J. Krim, D. H. Solina, R. Chiarello, *Phys. Rev. Lett.* **1991**, *66*, 181. (b) M. Ciepak, E. D. Smith, M. O. Robbins, *Science* **1994**, *265*, 1209.
[31] H.-J. Günterodt, D. Anselmetti, E. Meyer (Eds.), *Forces in Scanning Probe Techniques*, Kluwer, Dordrecht, **1995**.
[32] F. O. Goodman, N. Garcia, *Phys. Rev. B* **1991**, *43*, 4728.
[33] U. Hartmann, *Phys. Rev. B* **1991**, *43*, 2404.
[34] C. Girard, D. van Labeke, J. M. Vigoureux, *Phys. Rev. B* **1989**, *40*, 12133.
[35] E. M. Lifshitz, *Soviet Phys. JETF (Engl. Ed.)* **1956**, *2*, 73.
[36] H. K. Christenson, P. M. Claesson, *Science* **1988**, *239*, 390.
[37] V. Y. Yaminsky, B. W. Ninham, *Langmuir* **1993**, *9*, 3618.

[38] R. M. Pashley, *J. Colloid Interface Sci.* **1981**, *83*, 531.
[39] (a) B. V. Derjaguin, L. D. Landau, *J. Exp. Theor. Phys.* **1945**, *15*, 633. (b) E. J. W. Verwey, J. Th. G. Overbeek, *Theory of Stability of Lyophobic Colloids*, Elsevier, Amsterdam, **1948**.
[40] R. M. Pashley, *J. Colloid Interface Sci.* **1981**, *83*, 531.
[41] B. D. Terris, J. E. Stern, D. Rugar, H. J. Mamin, *Phys. Rev. Lett.* **1989**, *63*, 2669.
[42] (a) D. V. Abraham, C. C. Williams, H. K. Wickramasinghe, *Appl. Phys. Lett.* **1988**, *53*, 1446. (b) U. Hartman, *J. Vac. Sci. Technol. A* **1990**, *8*, 411. (c) A. Wadas, P. Rice, J. Moreland, *Appl. Phys. A* **1994**, *59*, 63.
[43] H.-J. Butt, *Biophys. J.* 1991, *60*, 777.
[44] (a) G. S. Blackman, C. M. Mate, M. R. Philpott, *Phys. Rev. Lett.* **1990**, *65*, 2270. (b) H. A. Mizes, K.-G. Loh, R. J. D. Miller, S. K. Ahuja, E. F. Grabowski, *Appl. Phys. Lett.* **1991**, *59*, 2901. (c) D. R. Baselt, J. D. Baldeschwieler, *J. Appl. Phys.* **1994**, *76*, 33.

3 Scanning Probe Microscopes

The topografiner [1], invented for the examination of surface roughness by employing the field emission phenomenon, is the predecessor of the scanning tunneling microscope. The atomic force microscope incorporates the basic principles of the surface force apparatus [2] and the mechanical profilometer [3]. Scanning tunneling and atomic force microscopes have the same components (i.e., piezodrive and tip) and operation principles as the topografiner. However, both microscopes have a far superior resolution than the topografiner and profilometer. The lateral resolution achieved by the topografiner is around 20 nm, whereas atomic-scale features are routinely detected in STM and AFM. In the profilometer, the tip exerts a relatively high force (hundreds of micronewtons) on the sample. This limits its application to hard materials, and the large tip–sample contact area lowers the spatial resolution of the imaging. These problems are avoided in contact-mode AFM, which operates at much smaller forces (typically 1–100 nN) and employs sharp microscopic tips for high-resolution profiling [4, 5].

Due to the fast development of STM and AFM in recent years, a large number of scanning tunneling and atomic force microscopes are built for operation in UHV, in air, and in liquids including electrochemical media. The STM and AFM images

Figure 3.1 Nanoscope III (Digital Instruments Inc.). The computer station and electronic unit are shown in the center, and the atomic force microscope combined with an optical microscope on the right.

presented in this book were obtained with commercial microscopes Nanoscope II and III (Digital Instruments Inc., Santa Barbara, California) (Fig. 3.1) in ambient conditions and in liquid. The operating principles of scanning tunneling and atomic force microscopes and their main components are discussed in this chapter. For a more detailed description of scanning probe microscopes, the reader is referred to recent reviews [6]. The practical aspects of STM and AFM measurements are summarized in Chapter 4.

3.1 Operating Principles and Main Components

In a scanning probe microscope, the sample surface is scanned with the sharp probe at a distance of less than a few nanometers, or in mechanical contact. For scanning, either the tip moves against the fixed sample, or the sample moves against the fixed tip. The common parts of the scanning probe microscope are the piezoceramic scanner, on which the moving element is mounted, and the coarse mechanism by which the tip and sample are brought close together so that the probing interactions can be measured with an appropriate detector. The detector signal is used for feedback control to adjust the tip–sample distance during the scanning. Since the nature of

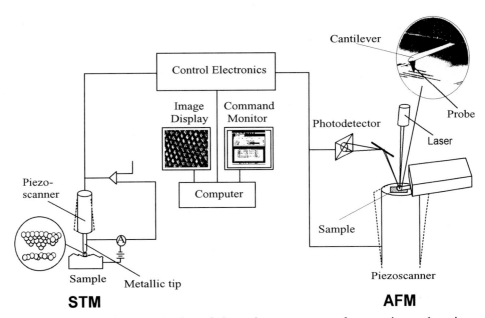

Figure 3.2 Schematic representation of the main components of a scanning probe microscope. This instrument incorporates an electronic unit and a computer station equipped with image and command monitors. The STM and AFM each includes a piezoscanner, probe, and detector.

the probing interactions is different in STM and AFM, they have different probes and detecting mechanisms. The main components of scanning tunneling and atomic force microscopes are shown in the schematic diagram of Fig. 3.2. The microscope head contains the scanner, the coarse mechanism, the probing tip, and the detector. Instrument operations are performed through the electronic unit controlled by the computer station. The software allows one to monitor the tip–sample approach, record the strength of the probing interaction as a function of the tip position, convert the collected information into the image on the screen, and store the data in the computer. Contemporary scanning probe microscopes are designed to perform both STM and AFM measurements by simply switching the heads.

3.1.1 Scanner

In scanning probe instruments the driving motion is performed by piezoceramic actuators, which change their dimensions under applied voltage. The most frequently used driver is a hollow-tube scanner, which can move the tip or sample in all three mutually perpendicular directions [7]. The outside surface of the tube is separated into four longitudinal segments, on which electrodes are attached. An additional electrode is placed on the inner surface of the tube. The application of different voltages to each pair of oppositely located outer electrodes induces a tube bending, which provides the lateral motion for the tip or sample attached to the tube. Application of different voltages to the outer and inner electrodes shrinks or extends the tube, which moves the tip or sample vertically. The detection of atomic-scale images requires a short-tube scanner with a piezomechanical coefficient in the 15–20 Å/V range, which allows the movement of the tip with ca. 0.01 Å accuracy. For large-scale scanning, which covers an area larger than 100 µm × 100 µm, it is necessary to use a long-tube scanner with a piezomechanical coefficient in the 200–300 nm/V range. The nonlinearity of the response of a long-tube scanner to low voltages may be corrected by the computer software, so that it can also be used for small-scale measurements. The calibration of scanners is carried out by scanning standard samples such as diffraction gratings for large-scale measurements, and mica and graphite for atomic-scale measurements. Because of the time-dependent properties of piezoceramic materials, the scanner parameters should be calibrated regularly.

3.1.2 Tip–Sample Approach and Electronic Feedback

The strength of the local probing interactions between the tip and the sample (i.e., the tunneling current in STM and the repulsive force in contact-mode AFM) becomes measurable only when the tip is positioned close enough to the sample surface. To prevent any possible damage of the sample by a tip–sample contact, the approach of the tip to the sample should be delicate. Initially the tip and sample are put close to each other by manually rotating the high-precision mechanical screws which are incorporated into the microscope stage. This procedure is facilitated by an optical control, when a scanning probe microscope is combined with an optical microscope or optical camera system. Closer approach is performed by the stepper motor, which brings the tip to the sample at separations that can be controlled by the scanner. The final adjustment of the tip–sample separation, to the value at which the probing interaction reaches the set-point level determined by the user, is performed by the scanner.

After this engagement, the lateral scanning is activated. The scanning tip can crash into the sample surface because of surface roughness and imperfect tip–sample alignment, unless the scanning is performed with a feedback mechanism. The difference Δ between the actual and the set-point values of the probing interaction is used for the feedback control. The electronic feedback system generates the response to the scanner, to adjust the vertical position to eliminate the difference Δ. Different types of feedback function (e.g., linear and logarithmic) can be used, depending upon the known or presumed relationship between the probing interaction and the tip–sample separation. The accuracy of imaging is strongly influenced by the feedback gain parameters. Use of very low gains keeps the tip position constant, whereas use of very high gains renders unstable the electronic circuit controlling the feedback, because the piezodrive and the current amplifier cannot respond instantaneously. Therefore the scanning should be performed with moderate gain parameters [8], which are commonly adjusted by the user.

3.1.3 Scanning Modes and Parameters

During scanning, the vertical movement of the tip is adjusted according to the chosen feedback gain, while the lateral motion proceeds independently. Conventionally, the fast and slow scanning directions are taken to be the x- and y-directions, respectively. The scanning from left to right is referred to as the "trace" scan, and that from right to left as the "retrace" scan. Along the slow-scanning direction, the tip moves from one border to another while performing the trace and retrace scans alternately. Thus, the tip covers a square area of the sample surface from the "bottom" to the "top" border (i.e., in the "up" direction) and then in the opposite direction (i.e., in the "down" direction). These directions are defined with respect to the image appearing on the computer display screen, where the horizontal and vertical axes correspond to the x- and y-axes of the sample surface, respectively. Commercial scanning probe microscopes provide the possibility of changing the fast-scanning direction away from the x-direction (this is known as the rotation of the scanning direction).

Two scanning modes are usually applied. In the "constant-height" mode, in which the feedback mechanism is turned off, one detects the changes in the probing interaction at a constant tip–sample separation. In the "constant-interaction" mode, in which the feedback mechanism is activated, one detects the variations of the local z-height of the tip with respect to the sample surface at a fixed interaction strength. For practical reasons, scanning probe microscopes have the feedback mechanism operating permanently. The constant-height mode is achieved by using very low feedback gain parameters, and the constant interaction mode by using high feedback gains.

Typically, in commercial scanning probe microscopes, each scan line consists of 512 points and scanning proceeds with line scanning frequencies in the 1–60 Hz range. Practically, the imaging of areas larger than 1 µm × 1 µm is performed with the scanning frequency in the 1–4 Hz range, which usually provides the optimal feedback response to the surface height corrugation. In general, zooming from a large to a smaller scanning area should be accompanied by an increase in the scanning speed to reduce the influence of the thermal drift on the images. For scanning flat areas of the order of 40 nm × 40 nm or smaller, where atomic- and molecular-scale images are collected, scan line frequencies in the 8–60 Hz range are commonly used. Ideally, the position of the sample is fixed in space when the tip scans an area of the sample from one border to another. However, at ambient temperature, the sample position drifts during the scanning and causes an image distortion, especially in small-scale measurements. The thermal drift is large just after the sample has been installed on the microscope stage and diminishes with time (in about 30 min) after thermal equilibrium is reached between the sample and the stage. It is also necessary to carry out small-area scanning with a high scanning speed to avoid any possible damage of the sample surface caused by the tip–sample force interaction. The probability of such damage increases as the tip spends more time at a given location. The user should select the optimal scanning speed for a given sample depending on the surface roughness and the feedback parameters.

3.1.4 Images and Filtering

An image typically consisting of 512 × 512 pixels can be generated by collecting the data for the trace scans only; this is called a trace image. Likewise, a retrace image is obtained by collecting the data for the retrace direction only. The use of trace and retrace images is most important for the analysis of lateral force measurements (see Section 3.3.1). In the images, the spatial variation of the tip-height or the strength of the probing interaction is represented by a gray or color-coded contrast. In measurements in the constant-interaction mode, the contrast refers to the spatial variation of the z-height of the tip, and such an image is called the height image. For operation in the constant-height mode, the image contrast refers to the spatial variation of the strength of the probing interaction. These images, known as the current images in STM and the force (or deflection) images in AFM, often exhibit a higher contrast of the morphological and atomic-scale details than do the height images. However, the latter provide more correct topographic information. In this book, grayscale images of both types are presented. Brighter spots correspond generally to elevated surface regions in height images, and to places with stronger probing interaction in current or force images.

In many cases, the as-received STM or AFM images are treated to give a better presentation. A planefit adjustment is necessary when the sample surface under examination is not exactly perpendicular to the scanner z-axis. This and other procedures of image modification (e. g., zooming, lowpass and highpass filtering, flattening, erasing scan lines) are typically included in the software package delivered with a commercial instrument. Some filtering procedures can be performed on-line during the scanning.

To emphasize the periodic nature of STM and AFM images of low signal-to-noise ratio, one can filter raw images by means of the two-dimensional fast Fourier transform (FFT) procedure. The original image is Fourier-transformed to obtain the power spectrum, in which the periodic patterns of the real-space image are represented by bright spots. One can construct the filtered image by selecting the most pronounced spots of the power spectrum for the reverse Fourier transformation (see Fig. 3.3). This filtering highlights the periodic features of the image but loses information about the nonperiodic features that might have resulted from local defects.

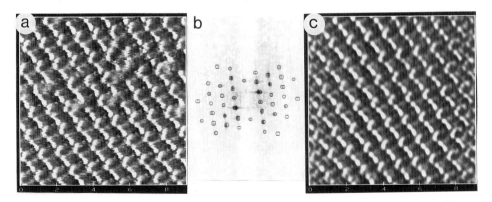

Figure 3.3 Actual example of FFT filtering for the STM height image of the organic salt α-(BEDT-TTF)$_2$IBr$_2$ (see Chapter 10): (a) unfiltered image; (b) power spectrum with the selected peaks; (c) filtered image. Scales are in nanometers.

3.1.5 Isolation of Vibrational Noise

To detect surface features such as atomic steps and corrugations in the sub-angstrom range by STM and AFM, it is necessary to avoid vibrational noises. For this purpose, the microscope head can be placed on the platform supported by rubber cords (a suspension-string system [9]), which has a rather low natural frequency (around 1 Hz). The suspension-string system is efficient in isolating the vertical and horizontal environmental vibrations. Within the STM head, one should prevent transmission of the external disturbance from the microscope stage, on which the sample is mounted, to the tip displacement with respect to the sample. For this purpose, construction of a rigid STM or AFM head is most suitable because of its high frequency (several kilohertz). The combination of the rubber cord isolation and rigid STM/AFM head makes it possible to avoid the interference of laboratory vibrational noises, whose frequency falls between several hertz and several kilohertz. In addition, acoustic noise might also disturb the measurements. This problem is more serious for AFM than for STM measurements. Quiet laboratory rooms or protective boxes are desirable for atomic force microscopes, especially for small-scale measurements.

3.2 Scanning Tunneling Microscope

The STM head contains the scanner and the microscope stage, which hold a metallic tip and a sample, respectively (Fig. 3.4). The scanner is placed on three high-precision screws. The two front screws are used for manual tip–sample approach, and the rear one is driven by a stepper motor. In an alternative construction, a sample can be placed on the scanner with the tip position fixed. In addition to the scanning speed and the feedback gains described above, the operation of a scanning tunneling microscope is controlled by the tunneling parameters, the set-point current (I_{set}) and the bias voltage (V_{bias}). In many cases, the gap resistance R_{gap} is used for the description of STM results.

3.2.1 STM Tips and Current Detection

Ideal tips for STM should have a monoatomically sharp apex; such tips can be prepared and characterized by means of a field-ion microscope [10a]. For most STM applications, however, such tips are less practical than are the tips prepared from a metallic wire (tungsten, gold, Pt/Ir, Rh/Ir) by mechanically cutting or by electrochemical etching [10b]. In general, the geometry of the tip apex is poorly defined and can change during the scanning. Although only the outermost atom, or few atoms, of the tip are expected to participate in the electron transfer, experimental images usually exhibit variations and artifacts caused by the nonideal tip geometry.

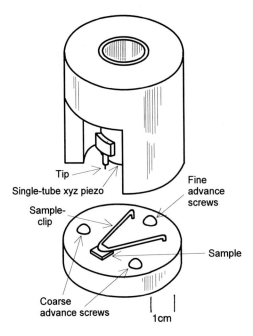

Figure 3.4 Typical STM head with a tip attached to the piezoscanner.

For ambient-condition measurements, Pt/Ir tips are preferred over tungsten (W) tips, because tungsten is easily oxidized in air [11]. Tungsten tips are commonly used in UHV, where oxidation is less likely to take place. Electrochemically etched W or Pt/Ir tips [10b] have a well-defined and sharp microscopic profile and are more suitable for measurements of corrugated surfaces. The profiling of complex surfaces of technological importance may demand further optimization of the tip shape (see Section 3.4). For studies of atomically flat samples, electrochemically etched tips have no advantages over mechanically cut tips, because on the submicron scales both have poorly defined and complex apex profiles [12].

For current detection in a scanning tunneling microscope, a combination of the logarithmic and standard amplifiers is applied because the current-vs.-distance dependence is expected to be exponential. Commonly the tunneling current employed in scanning tunneling microscopes is in the 0.1–40 nA range. The use of currents in the picoampere range extends the range of STM applications to samples with poor electrical conductivity [13].

3.2.2 Bias Voltage

The bias voltage V_{bias} determines which levels of the sample electronic states will participate in the tip–sample electron transfer and is therefore an important experimental parameter in STM. The V_{bias} values are typically in the 0.001–5 V range. In STM electrons flow from the tip to the sample when the bias voltage V_{bias} is positive, but from the sample to the tip when the bias voltage is negative (Fig. 3.5). Thus, tip–sample electron transfer involves the lowest unoccupied levels of the sample when $V_{bias} > 0$, but the highest occupied levels of the sample when $V_{bias} < 0$ (with the tip grounded). The highest occupied levels and the lowest unoccupied levels are similar in nature for metals and magnetic semiconductors, so the characteristics of their STM images do not depend on the polarity of the bias voltage.

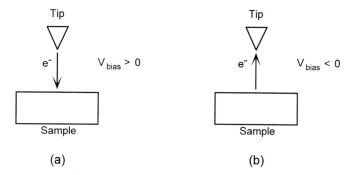

Figure 3.5 Electron transfer in STM: (a) tip-to-sample transfer at positive V_{bias}; (b) sample-to-tip transfer at negative V_{bias}.

In practice, the dependence of the image quality on I_{set} and V_{bias} is eventually related to the gap resistance R_{gap}, which is a qualitative measure of the tip–sample distance. For many metallic samples, well-resolved atomic-scale STM images are observed with relatively low R_{gap} values in the range 1–100 MΩ. Under such a condition, the tip–sample force interactions might contribute strongly to the image contrast.

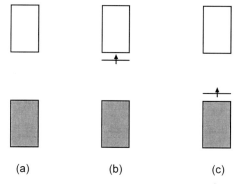

Figure 3.6 Energy levels of regular semiconductors in the vicinity of their band gaps: (a) defect-free semiconductor; (b) n-type semiconductor; (c) p-type semiconductor.

For normal semiconductors, the highest occupied levels (at the top of the valence band) generally differ in orbital character from the lowest unoccupied levels (at the bottom of the conduction band). Consequently, the STM images of semiconductors exhibit patterns which depend on the polarity and magnitude of the bias voltage. In addition, normal semiconductors invariably possess point defects (e. g., atom vacancies and atom substitutions), which create defect states in their band gaps (Fig. 3.6). These defect states strongly influence the nature of the images, depending on the polarity and magnitude of the bias voltage (see Chapter 7).

3.2.3 Scanning Tunneling Spectroscopy

The study of the tunneling current as a function of the bias voltage at a constant tip–sample separation is an important option of STM. For this purpose, the feedback is switched off, and the $I-V$ curve is determined while linearly ramping the voltage across the tunneling gap. The $I-V$ curves of ambient-condition measurements have a limited use because of the poorly defined experimental conditions. Nevertheless, one can deduce useful information about the sample's electrical conductivity (Fig. 3.7) and the range of the bias voltages that can be applied for the imaging. Most spectroscopic studies have been performed on clean surfaces in UHV. It is possible to carry out the scanning and spectroscopic measurements simultaneously (i. e.,

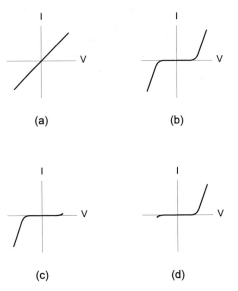

Figure 3.7 Schematic representations of the $I-V$ curves expected for metals and regular semiconductors: (a) metals; (b) defect-free semiconductors; (c) n-type semiconductors; (d) p-type semiconductors.

obtaining the tunneling spectra at a set of mesh-points on the scanned area), as demonstrated by Hamers et al. [14].

The results of STS measurements are usually presented by plotting the current I, the conductance I/V, or more commonly the normalized conductance $(dI/dV)/(I/V)$ [15] as a function of the voltage V. The plot of the "tunneling density of states" $(dI/dV)/(I/V)$ vs. V [i.e., $d(\log I)/d(\log V)$ vs. V] can resemble that of the electronic density of states versus energy [i.e., $n(e)$ vs. e] [15]. However, it is incorrect to assume that $d(\log I)/d(\log V)$ is proportional to $n(e)$ [16]. In another spectroscopic approach, STM images can be collected at bias voltages of different magnitude and polarity. A classical example is the observation of the empty and occupied states of GaAs(110), which are localized at Ga and As atoms, respectively. These states appear selectively in the images obtained at bias voltages of opposite polarities [17]. In most cases, however, the interpretation of the spectroscopic results is a difficult problem [18].

3.3 Atomic Force Microscope

The head of an atomic force microscope consists of the optical block and the base (Fig. 3.8). The stepper-motor, the alignment screws and the scanner are installed within the base, and a sample is mounted on the top of the scanner. The optical block (Figs. 3.9) contains a viewing-window at the top, a laser diode, mirrors, and

Figure 3.8 AFM head with the optical block and the base.

a

b

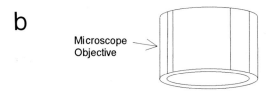

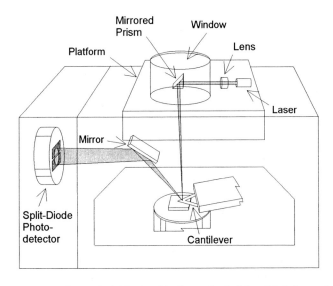

Figure 3.9 (a) The optical block with the probe holder. (b) Schematic diagram of the optical block incorporating the optical deflection system.

a four-quadrant positional photodetector. The cantilever holder is fixed on top of the alignment screws, with the tip positioned over the sample. The optical microscope or the optical camera attached to the viewing-window allows one to observe the cantilever for alignment of the laser beam and for positioning the tip over the sample areas of interest. In contemporary designs, an AFM head is multifunctional and allows measurements in different modes. The background of AFM instrumentation has been reviewed by Sarid [6c].

3.3.1 Contact Mode and Force Detection

In the contact mode the repulsive force experienced by the tip is measured by recording the cantilever deflection. The latter is controlled by the cantilever spring constant (k) and the magnitude of the force. The deflection can be measured by several methods (e.g., optical deflection, interferometric control, piezoresistivity) [6], but the optical deflection scheme [19] is most commonly used (Fig. 3.9 (b)). In this scheme, a beam from the laser diode is reflected from the cantilever surface to the photodetector. It should be noted that the cantilever in the rest position is slightly inclined (by about 12°) with respect to the horizontal plane so as to reflect the probing laser beam towards the detector. The vertical displacement of the cantilever is proportional to the differential signal from the vertical segments of the photodetector (i.e., $\Delta_{AB} = A - B$, where A and B are the signals in volts from the upper and lower segments, respectively) (see Fig. 3.10 (a)). For feedback control of the scanner movement in the z-direction, the reduced differential signal $\Delta_{AB}/(A + B)$ is used to avoid the influence of the laser beam instability. For cantilevers 100–200 µm long, the photodetector magnifies the surface features by a factor of 750–1500, which enables the detection of cantilever deflections in the sub-angstrom range. The optical detection system has a definite advantage over other methods because of the higher signal-to-noise ratio [20].

To engage the tip with the sample in contact-mode AFM, the user should choose the set-point Δ_{AB} value to be higher than that for the cantilever in the rest position. In this process, the sample comes in contact with the tip and bends the cantilever up until the actual Δ_{AB} reaches the set-point value. During the scanning, the user can change the set-point thereby adjusting the force applied to the sample. Depending upon the environment and the cantilever spring constant, the operational force in contact-mode AFM varies from one to several hundred nanonewtons. Practical examples of force optimization are further discussed in Chapter 4. Other scanning parameters applied in AFM measurements are similar to those used in STM, and the AFM results are presented in the form of height or deflection (force) images.

The detection of the lateral force is a useful option of contact-mode AFM. The lateral force applied to the tip causes buckling or torsion of the cantilever

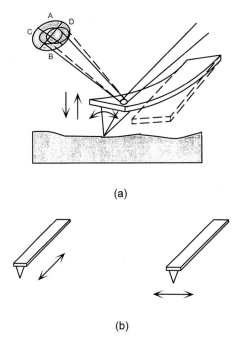

Figure 3.10 (a) Detection of the vertical and torsional deflections of the cantilever by the differential signals Δ_{AB} and Δ_{CD} on the vertical and horizontal photodetectors, respectively. (b) Depending on the orientation of the cantilever with respect to the fast-scanning direction (indicated by the arrows), the cantilever undergoes a buckling (left) or a torsion (right).

(Fig. 3.10 (b)) depending upon whether the fast-scanning direction is along or perpendicular to the cantilever beam, respectively. This cantilever twisting gives rise to the differential signal from the horizontal segments of the photodetector (i.e., $\Delta_{CD} = C - D$, where C and D are the signals from the left and right segments, respectively) (Fig. 3.10 (a)). The spatial variation of this signal is converted into a lateral force microscopy (LFM) image. On flat surfaces the lateral force is caused by the friction between the tip and the sample, so LFM is also referred to as frictional force microscopy. Because the deformation of the cantilever caused by the lateral force is complex, it is difficult to determine the corresponding spring constant. The lateral spring constant of a triangular cantilever is estimated to be high (about 135 N/m) [21].

For samples with flat surfaces, regions with different friction can be distinguished with LFM. This is clearly seen in Fig. 3.11, in which the lateral force profiles are recorded in the trace and retrace directions along the domains with different friction. If the domain of lower friction has a negative step in the trace profile, then it has a positive step in the retrace profile. This situation also occurs even when the height of the lower-friction area is different from that of the surrounding area. Conse-

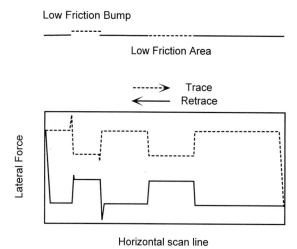

Figure 3.11 Trace and retrace scan profiles of the friction on a sample surface with two low-friction areas.

quently, in LFM images, regions with different friction will exhibit different contrasts, and a bright pattern in the trace image becomes a dark pattern in the retrace image, and vice versa. To accentuate this effect, one should find an optimum direction for fast scanning with respect to the cantilever. In practice, by rotating the scanning direction with respect to the cantilever, one can find a direction for which the trace and retrace profiles are similar. (Ideally, this happens when the fast-scanning direction is along the cantilever beam.) Then the direction perpendicular to it is optimal for the recording of LFM images.

The above discussion of lateral force detection was concerned with nearly flat, rigid surfaces. The surface corrugation, the geometry of tip–sample contact, and the mechanical response of the sample surface to the tip force makes the analysis of LFM images rather difficult. For example, for corrugated surfaces, the relationship between the Δ_{CD} signal and the lateral force is complicated or unknown. However, for practical purposes, simultaneous measurements of the height and LFM images can also be useful for samples with rough surfaces. In this imaging the surface corrugation is correctly reproduced by the height image, and the lateral force variations provide a better resolution of nanoscale details because steps and slopes are readily identified. In providing morphological information, LFM images are similar to force images.

3.3.2 AFM Probes

The probes commonly used for AFM measurements are cantilevers with integrated tips of Si_3N_4 or silicon (Fig. 3.12) [22]. The cantilevers are prepared with different

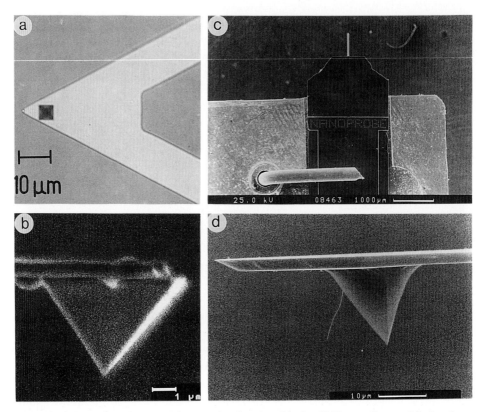

Figure 3.12 Scanning electron micrographs showing (a) the Si_3N_4 cantilever with the tip at the end, (b) the shape of the Si_3N_4 tip, (c) the silicon cantilever, and (d) the shape of the silicon tip at the end of the silicon cantilever.

lengths, thicknesses and shapes. Their elastic spring constants vary in the 0.01–50 N/m range. The shape of the tip apex and the elastic spring constant k of the cantilever are important parameters of AFM measurements.

The spring constant k of a rectangular cantilever is expressed as [22a]

$$k = Et^3 w / 4L^3$$

where E is the elasticity modulus, and w, L, and t are the width, length, and thickness of the cantilever, respectively. The spring constant strongly depends on the thickness of the cantilever, which is more difficult to measure than the width and length and varies within a set of cantilevers microfabricated on a silicon wafer. Practically, the thickness of any particular cantilever can be determined by measuring its resonance frequency F_r, which is written as [22a]

$$F_r = 0.162 (E/\rho)^{1/2} t/L^2$$

where ρ is the density of the cantilever material. There are several practical methods of determining this frequency in air and under liquid [22a, 23] (see also Chapter 4). Cantilevers of small spring force constant are most suitable for imaging soft materials, while rigid cantilevers are needed for probing the nanomechanical properties of sample surfaces and also for dynamic-mode AFM measurements.

The first commercial probe that satisfied the main requirements of AFM imaging was the Si_3N_4 microcantilever with an integrated pyramidal tip [22a]. The process of making this probe begins with an Si(100) wafer with a thermally grown SiO_2 layer. Using a photolithographic method, a square opening is etched on the SiO_2 film. The chemical etching of the part of the Si(100) wafer exposed through the square opening selfterminates at the Si(111) planes leading to a pyramidal pit. After removing the SiO_2 protection layer, Si_3N_4 is deposited on the wafer to form the shape of the cantilever using a lithographic method. Then, all remaining silicon is etched away, and the back of the cantilever is coated with gold to reflect the laser beam.

Si_3N_4 probes can be used for flat surfaces in large- as well as atomic-scale imaging. However, due to the relatively dull apex (i.e., the dihedral angle of about 70° between the *trans* faces of the pyramidal tip), these probes are less suitable for profiling highly corrugated surfaces, because the real topography can be hidden by the convolution of the tip geometry and the sharp surface features. Sharper tips are microfabricated by chemically etching an Si(100) wafer [22b]. These silicon probes are more suitable for the morphological study of samples. According to the control measurements of the (305) surface of $SrTiO_3$ [24], most silicon tips are reliable for profiling ridged surfaces with height corrugations on the ten-nanometer scale and also for atomic-scale imaging. More specialized probes than described are necessary for the profiling of surfaces with corrugations in the hundred-nanometer range (see Section 3.4).

3.3.3 Dynamic AFM Measurements

The development of modulation techniques in AFM was driven by attempts (a) to employ attractive forces for surface imaging (noncontact-mode AFM) [25a]; (b) to overcome the limitations of the contact mode (e.g., tapping-mode AFM and related techniques [25b]) in studying soft materials; and (c) to examine surface mechanical properties (force-modulation techniques) [25c] (see Fig. 3.13). In these methods either the cantilever is vibrated at its resonance frequency or the vertical sample position is modulated. The tip–sample force interaction causes a change in the amplitude, the phase, and the resonance frequency of the vibrating cantilever. For example, a tip–sample attractive interaction lowers the frequency, while a repulsive interaction raises it. The spatial variations of the changes can be presented in height or interaction (amplitude, phase, or frequency shift) images. Typically, height and interaction

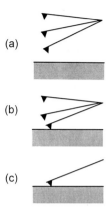

Figure 3.13 Tip–sample arrangements in the three modulation modes of AFM: (a) noncontact mode; (b) tapping mode; (c) force-modulation mode.

images are collected simultaneously. In general, the surface topography is better represented by height images, whereas the nanostructures and the fine morphological features are better distinguished in amplitude or phase images.

3.3.3.1 AFM Operation in the Attractive Force Regime

It is appealing to employ attractive interactions for probing soft surfaces [25a] and provide images with atomic resolution. Since the long-range attractive forces do not exhibit sharp force-vs.-distance behavior, the theoretical estimates relevant to AFM imaging indicate that the spatial variation of the VDW forces is very weak [26]. Girard et al. [27] proposed a discrete model for calculating the VDW forces between the NaCl(100) surface and a probe tip. Their calculations, based on pairwise potentials between all atoms in the solid, show that the difference in the forces on the Na and Cl sites decreases below 0.001 nN [27] when the tip–sample distance increases beyond 0.6 nm.

The AFM operation in the attractive force regime can be carried out in UHV by using the modulation technique with frequency shift detection, because the Q-factor of an oscillating cantilever is extremely high in UHV. The results obtained for Si(111) 7×7 and InP(100) demonstrate the possibility of observing surface atomic lattices and point defects [28]. In such experiments, the spatial variation of the negative frequency shift of the cantilever is used for the feedback [29]. Although this means that the total force exercised by the tip-cantilever macroscopic system is attractive, it is most likely that a few outermost atoms on the tip apex are in contact with the sample surface, and the spatial variation of the associated repulsive force component is responsible for the atomic-scale imaging. A truly noncontact mode imaging is unlikely to provide a sub-nanometer resolution because the long-range attractive forces are insensitive to a small change in the tip–sample separation.

3.3.3.2 Tapping Mode

The tip is vertically oscillated at its resonance frequency in the tapping mode [25b]. When the sample approaches the vibrating tip, they come into intermittent contact ("tapping"), thereby lowering the vibrational amplitude. The amplitude drop is used for the feedback. In this mode, the tip–sample lateral force is greatly reduced and the short tip–sample contact time prevents inelastic surface modification. As expected, comparison of the imaging in the contact and tapping modes shows that soft surfaces are less modified in the tapping mode [23b].

The tapping mode was originally introduced for ambient-condition experiments. For the tip to "penetrate" through the contamination overlayer, application of rigid cantilevers (resonance frequencies in the 300–400 kHz range) and high operating amplitudes (10–100 nm) is required. In the tapping mode, the energy delivered to the sample from the vibrating tip is determined by the amplitude of the free vibration (A_0) and the set-point drop (ΔA) in the amplitude. For high-resolution imaging and studies of soft materials, small values should be chosen for the A_0 and ΔA parameters. In ambient conditions, the reduction of these parameters is limited because of the contamination overlayer. Under liquid, however, one can operate the tapping mode with much smaller A_0 and ΔA values.

3.3.3.3 Force-Modulation Techniques

When the tip and sample are in contact, modulation of the tip (or the sample) makes it feasible to probe the nanomechanical properties of the sample surface [30]. In contrast to the noncontact and tapping modes, in which the amplitude of the vibrating cantilever is in the 20–100 nm range, the force-modulation technique uses a much smaller amplitude to prevent tip–sample disconnection. By measuring the amplitude and phase shift of the cantilever, one can detect spatial variation of the elasticity and viscoelasticity of a sample surface. For direct determination of the sample stiffness in the tip–sample contact area, it can be more convenient to apply the force to the tip rather than modulating the vertical position of the tip or sample. This can be achieved by coating the tip with magnetic material and applying an external magnetic field to produce a modulation of the force on the sample surface [31]. Using this technique with a force up to 1 nN, it was possible to measure directly the local compliance of ordered liquid layers at solid–liquid interfaces. The frequency of the force modulation employed varies from 10 Hz to 20 kHz [30, 31].

3.3.3.4 Magnetic Force Microscopy

In MFM the forces between a ferromagnetic tip and a ferromagnetic or paramagnetic sample are probed. Magnetic force microscopes are useful for measuring magnetic information for storage media (e.g., disks, tape), magnets, and soft magnetic materials such as garnet films. The performance (e.g., lateral resolution, sensitivity, etc.) of MFM is enhanced by detecting changes in the phase or frequency of the cantilever vibration. Such measurements are carried out by using the tapping mode with a magnetic tip (e.g., a silicon tip coated with a magnetic material such as cobalt). Since the magnetic force variation is weak, the two-pass technique (known as the lift mode) is applied. On the first pass over each scan, the sample's surface topography is measured and recorded. On the second pass, the tip is lifted to a user-selected distance above the recorded surface topography and the force measurement is made. MFM has emerged as an indispenable tool for magnetic recording technology [32b].

3.4 STM and AFM as Metrology Tools

The development of scanning probe techniques has opened new possibilities in metrology [32], because sub-angstrom resolution is routinely achieved on flat surfaces with STM and AFM. For metrological purposes, these methods are also attractive because they can be used in different environments and provide three-dimensional imaging of the surface features from macroscopic to atomic scales.

3.4.1 Resolution in STM and AFM

An important issue concerning microscopic methods is their resolution. In general, the as-obtained image refers to an "ideal" image containing the desired information, convoluted with an instrumental resolution function or line shape. This situation is rather complex in scanning probe techniques because in many cases the convolution

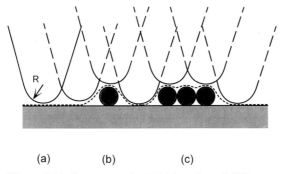

(a)   (b)   (c)

Figure 3.14 Contact mode AFM imaging of different surfaces: (a) flat surface; (b) single cylinder lying on the flat surface; (c) close-packed array of cylinders lying on the flat surface. The broken line refers to the height variation of the tip position during the imaging of the cylinders

is nonlinear and depends on the tip shape and the sample. In STM images sub-angstrom features are detected, but it is difficult to establish the accuracy and the physical meaning of these features from the viewpoint of metrology [32].

The dependence of the AFM resolution on the sample geometry is illustrated in Fig. 3.14. Suppose that a single cylinder (10 nm in diameter) and an array of close-packed cylinders lying on a surface are imaged by an AFM tip with 20 nm apex diameter, and that the diameter of the tip–sample contact diameter is 3 nm. An image with atomic resolution can be achieved for the flat surface (Fig. 3.14 (a)). Scanning over the single cylinder gives an image with overestimated width (Fig. 3.14 (b)). For the array of cylinders (Fig. 3.14 (c)), the probe touches only their topmost parts so that the repeat distance of the periodic image pattern is close to the diameter of a cylinder.

A situation similar to that described in Fig. 3.14 (b) occurs in studies of single macromolecules such as DNA on mica [33]. The image of Fig. 3.15, obtained by contact-mode AFM under water, shows several DNA strands, and their apparent dimensions are given by the cross-section profile. The apparent "width" of the strands is in the 15–20 nm range (as measured by the width at the baseline of the profile), which far exceeds the real value (ca. 2.5 nm). The apparent width obtained from the images is about 5 nm by tapping-mode AFM under water [34], and about 3.5 nm by low-current STM in humid air [13]. In all DNA images the apparent height is observed to be about 1 nm, which is much smaller than the real height (ca. 2.5 nm). This apparent flattening is most likely caused by tip-induced deformation of the DNA strands.

As already pointed out, STM and AFM differ in their capability of detecting atomic-scale defects. For a large number of inorganic layered materials and organic molecular crystals, AFM provides atomic-scale images but these images show perfect lattice patterns without atomic-size defects (e.g., atom vacancies). The smallest defects observed by AFM are linear dislocations [35] whose lengths exceed the estimated diameter of the tip–sample contact. Atomic-size defects are routinely detected in STM, which implies that just one or only a few atoms of the tip apex collect the tunneling current. In contact-mode AFM it is most probable that a multiatom contact takes place [36]. However, multiatom contact does not prevent the recording of atomic- and molecular-scale periodic features. Molecular-dynamics calculations for the scanning of a multiatom tip in contact with a crystalline sample show that the lateral and vertical forces experienced by the tip exhibit periodic variations with the repeat distance of the surface being examined [37a]. This finding is also consistent with the experimental observation of surface lattices in LFM images [37b].

It is challenging to find the experimental conditions appropriate for detecting atomic-size defects in contact-mode AFM. One approach is to diminish the force that the tip exerts on the sample surface and consequently the tip–sample contact area, for example by operating under liquid [38]. (It should be noted that the tip–sample contact area does not completely vanish even when the load is zero, due

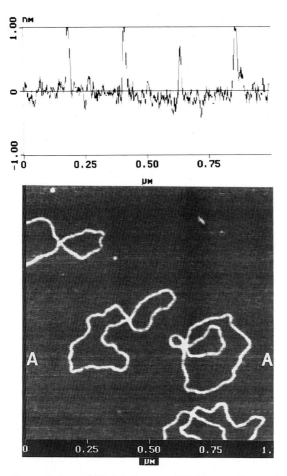

Figure 3.15 AFM height image of DNA strands on mica. The image was obtained in the contact mode under water. The image contrast covers height variations in the 0–2 nm range. The cross-section profile along the A–A line is shown above the image.

to the tip–sample adhesion.) In this approach the tip–sample interactions are reduced, which demands improvement of the signal-to-noise ratio (i.e., the sensitivity of detection).

It is surprising that contact-mode AFM operation with high forces (several hundred nanonewtons) leads to well-resolved atomic-scale images for many inorganic layered materials [39]. This finding is tentatively explained by the effective sharpening of the tip: under a high applied force, the tip–surface contact area would certainly be large, but not all the atoms of the tip in the contact area contribute equally to the total repulsive force between the tip and the sample. Since the repulsive force curve is very steep for small atom–atom distances, the total repulsive force may be dominated by the contributions of the few outermost atoms of the tip apex in high-force imaging. This tip sharpening effect might lead to the detection of atomic-size defects (see Chapter 4 for further details).

3.4.2 Metrological Applications

For measuring surface roughness with spatial variations in the submicron range and profiling nanoscale surface structures, STM and AFM are currently the only tools [32]. Nevertheless, improvements of the sensors, position control and probes are required to achieve metrological accuracy. With STM one can profile conducting samples and insulators coated with a metallic overlayer. In ambient-condition measurements, however, the tip–sample force interaction might introduce undesirable changes to the tip and sample geometry. Therefore, development of low-current STM instrumentation might be important for metrological applications. For these applications, AFM has a broader applicability due to its use of small and better-controlled forces and the rapid improvements in force sensors. Dynamic-mode detection leads to a high resolution in force (ca. 10^{-4} nN), and the lateral resolution of tapping- or noncontact-mode AFM is close to that of contact-mode AFM. To increase the force sensitivity of the contact mode, it is necessary to overcome the jump-into-contact instability of soft cantilevers. One way to solve this problem might be to use a rocking-beam force sensor whose pivoting motion is controlled by two capacitors [40].

The common actuator (i.e., piezoceramic cylinder) transports the sample or the tip laterally by bending. Because of tilting of the tube, hysteresis, and time-dependent properties of the piezoceramic material, a more precise scanning can be achieved with the use of independent mechanisms for monitoring the lateral and vertical positions of the piezoceramic cylinder [41].

Commercial silicon probes can be applied for roughness measurements of relatively flat surfaces such as silicon wafers. When the examined surface is more corrugated, the geometry of the tip apex plays an important role in topography profiling. The convolution of the tip apex with rough surface features was treated theoretically by Stedman [42]. In practice, the use of a corrugation standard such as the specially-prepared (305) surface of $SrTiO_3$ [24] helps to select sharp tips. As shown in the AFM height image of Fig. 3.16 (a), the surface of this standard consists of linear grooves with a sawtooth profile precisely characterized by transmission electron microscopy. The relative sharpness of the tips can be evaluated by their ability to reproduce the sawtooth profile, as shown in Figs. 3.16 (b) and (c). It is important to recall that, in addition to the sharpness of the tip, the applied force also affects the size of the tip–sample contact area and can be crucial in measurements of the roughness of soft materials.

Commercial silicon and Si_3N_4 tips cannot be applied for profiling surfaces with complex corrugations. Specially prepared sharp tips are needed to profile deep tren-

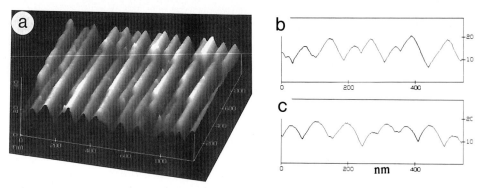

Figure 3.16 (a) Three-dimensional representation of the AFM height image of the (305) surface of SrTiO$_3$, which serves as a standard for selecting sharp AFM probes. (b) and (c) show the cross-section profiles of the (305) surface determined by sharp and dull tips in the direction perpendicular to the surface grooves.

ches accurately. For example, sharp tips can be prepared on the apexes of Si$_3$N$_4$ probes by the use of an electron beam in a scanning electron microscope [43]. However, these tips are not sufficiently hard because the tip material resembles a cross-linked polymer formed from organic traces. For profiling trenches with steep sides, a special probe shape is required. The three-point probe, with one tip pointed downward and two others oriented in opposite horizontal directions, was designed for this purpose [44]. The up–down and side-to-side scanning allows one to profile complex trenches and holes.

The application of scanning probe techniques in metrology is still in the initial stage. So far, most efforts have been aimed at identifying problems associated with such experiments. It is also evident that the spectrum of metrological measurements with AFM will not be limited to topographical studies and will include the evaluation of nanomechanical surface properties [45].

References

[1] R. Young, J. Ward, F. Scire, *Rev. Sci. Instrum.* **1972**, *43*, 275.
[2] J. N. Israelachvili, D. Tabor, *Proc. R. Soc. London Ser. A* **1972**, *331*, 19.
[3] Alpha-Step 250, Tencor Instruments Inc., Mountain View, CA.
[4] G. Binnig, C. F. Quate, Ch. Gerber, *Phys. Rev. Lett.* **1986**, *56*, 930.
[5] C. M. Mate, R. Erlandsson, G. M. McClelland, S. Chiang, *Surf. Sci.* **1989**, *208*, 473.
[6] (a) P. K. Hansma, J. Tersoff, *J. Appl. Phys.* **1987**, *61*, R1. (b) S. Park, C. F. Quate, *Rev. Sci. Instrum.* **1987**, *58*, 2010. (c) D. Sarid, *Scanning Force Microscopy*, Oxford University Press, New York, **1991**. (d) C. J. Chen, *Introduction to Scanning Tunneling Microscopy*, Oxford University Press, New York, **1993**. (e) D. A. Bonnell (Ed.), *Scanning Tunneling Microscopy and Spectroscopy*, VCH Publishers, New York, **1993**.

[7] G. Binnig, D. P. E. Smith, *Rev. Sci. Instrum.* **1986**, *57*, 1688.
[8] Y. Kuk, P. J. Silverman, *Rev. Sci. Instrum.* **1989**, *60*, 165.
[9] P. K. Hansma, V. Elings, O. Marti, C. E. Bracker, *Science* **1988**, *242*, 29.
[10] (a) V. T. Binh, *Surf. Sci.* **1988**, *202*, L539. (b) I. H. Muselman, P. E. Russell, *J. Vac. Sci. Technol. A* **1990**, *8*, 3358.
[11] J. Garnaes, F. Kragh, K. A. Morch, A. R. Thölen, *J. Vac. Sci. Technol. A* **1990**, *8*, 441.
[12] S. N. Magonov, *Appl. Spectrosc. Rev.*, **1993**, *28*, 1.
[13] R. Guckenberger, M. Heim, G. Cevc, H. P. Knapp, W. Wiegräbe W., A. Hillebrand, *Science* **1994**, *266*, 1538.
[14] R. J. Hamers, R. M. Trompt, J. E. Demuth, *Phys. Rev. B* **1986**, *34*, 5343.
[15] R. M. Feenstra, W. A. Thompson, A. P. Fein, *Phys. Rev. Lett.* **1986**, *56*, 608.
[16] R. J. Hamers, in D. A. Bonnell (Ed.), *Scanning Tunneling Microscopy and Spectroscopy*, VCH Publishers, New York, **1993**, Ch. 4.
[17] R. M. Feenstra, J. A. Stroscio, J. Tersoff, A. P. Fein, *Phys. Rev. Lett.* **1987**, *58*, 1192.
[18] R. M. Trompt, *J. Phys. Condens. Matter* **1989**, *1*, 10211.
[19] (a) N. H. Amer, G. Meyer, *Appl. Phys. Lett.* **1988**, *53*, 1045. (b) S. Alexander, L. Hellemans, O. Marti, J. Schneir, V. Elings, P. K. Hansma, M. Longmire, J. Gurley, *J. Appl. Phys.* **1989**, *65*, 164.
[20] M. G. L. Gustafsson, J. Clarke, *J. Appl. Phys.* **1994**, *76*, 172.
[21] J. M. Neimeister, W. A. Ducker, *Rev. Sci. Instrum.* **1994**, *65*, 2527.
[22] (a) T. R. Albrecht, S. Akamine, T. E. Carver, C. F. Quate, *J. Vac. Sci. Technol. A* **1990**, *8*, 3386. (b) O. Wolter, Th. Bayer, J. Greschner, *J. Vac. Sci. Technol. B* **1991**, *9*, 1353.
[23] (a) J. P. Cleveland, S. Manne, D. Bocek, P. K. Hansma, *Rev. Sci. Instrum.* **1993**, *64*, 1. (b) A. Wawkuschewski, K. Crämer, H.-J. Cantow, S. N. Magonov, *Ultramicroscopy* **1995**, *58*, 185.
[24] S. S. Sheiko, M. Möller, E. M. C. M. Reuvekamp, H. W. Zandbergen, *Phys. Rev. B* **1993**, *48*, 5675.
[25] (a) Y. Martin, C.C. Williams, H. K. Wickramasinghe, *J. Appl. Phys.* **1989**, *61*, 4723. (b) Q. Zhong, D. Innis, K. Kjoller, V. B. Elings, *Surf. Sci. Lett.* **1993**, *290*, L688. (c) M. Salmeron, *Mater. Res. Bull.* **1993**, *18*, 20.
[26] C. Ciraci, in *Scanning Tunneling Microscopy III*, R. Wiesendanger, H.-J. Güntherodt (Eds.), Springer-Verlag, Heidelberg, **1993**, p. 179.
[27] C. Girard, D. van Labeke, J. M. Vigoureux, *Phys. Rev. B* **1989**, *40*, 12133.
[28] (a) F. J. Giessibl, *Science* **1995**, *267*, 68. (b) Y. Sugawara, M. Ohta, H. Ueyama, S. Morita, F. Osaka, S. Ohkouchi, M. Suzuki, S. Mishima, in Abstracts of STM '95, Eighth International Conference on Scanning Tunneling Microscopy/Spectroscopy and Related Techniques (July 23–28, 1995, Snowmass, Colorado, USA), p. 228.
[29] T. R. Albrecht, P. Grütter, D. Horne, D. Rugar, *J. Appl. Phys.* **1991**, *69*, 668.
[30] (a) P. Maivald, H.-J. Butt, S. A. C. Gould, C. B. Prater, B. Drake, J. A. Gurley, V. B. Elings, P. K. Hansma, *Nanotechnology* **1991**, *2*, 13. (b) M. Radmacher, R. W. Tillmann, M. Fritz, H. E. Gaub, *Science* **1992**, *257*, 1900. (c) M. Radmacher, R. W. Tillmann, H. E. Gaub, *Biophys. J.* **1993**, *64*, 735.
[31] S. J. O'Shea, M. E. Welland, J. B. Pethica, *Chem. Phys. Lett.* **1994**, *223*, 336.
[32] (a) J. E. Griffith, D. A. Grigg, *J. Appl. Phys.* **1993**, *74*, R83. (b) D. A. Grigg, J. E. Griffith, G. P. Kochanski, M. J. Vasile, P. E. Russell, *Proc. SPIE* **1992**, *1673*, 557. (c) K. Babcock, M. Dugas, S. Manalis, V. Elings, *Mat. Res. Soc. Symp. Proc.* **1995**, *355*, 311.

[33] (a) C. Bustamante, J. Vesenka, C. L. Tang, *Biochemistry* **1992**, *31*, 22. (b) H. G. Hansma, J. Vesenka, C. Siegerist, G. Kelderman, H. Morret, R. L. Sinsheimer, V. Elings, C. Bustamante, P. K. Hansma, *Science* **1992**, *256*, 1180. (c) H. G. Hansma, M. Bezanilla, F. Zenhausern, M. Adrian, R. L. Sisheimer, *Nucl. Acids Res.* **1993**, *21*, 505. (d) Y. L. Lybchenko, L. S. Shlyaktenko, R. E. Harrington, P. I. Oden, M. Lindsay, *Proc. Natl. Acad. Sci.* **1993**, *90*, 2137. (e) T. Thundat, D. P. Allison, R. J. Warmack, T. C. Ferrel, *Ultramicroscopy* **1992**, *42-44*, 1101.

[34] P. K. Hansma, J. P. Cleveland, M. Radmacher, D. A. Walters, P. Hillner, M. Bezanilla, M. Fritz, D. Vie, H. G. Hansma, C. B. Prater, J. Massie, L. Fukunaga, J. Gurley, V. B. Elings, *Appl. Phys. Lett.* **1994**, *64*, 1738.

[35] (a) J. Garnaes, D. K. Scwarz, R. Viswanathan, J. A. N. Zasadzinski, *Nature* **1992**, *255*, 64. (b) L. Bourdieu, O. Ronsin, D. Chatenay, *Science* **1993**, *259*, 798.

[36] T. P. Weihs, Z. Nawaz, S. P. Jarvis, J. B. Pethica, *Appl. Phys. Lett.* **1991**, *59*, 3536.

[37] (a) U. Landman, W. D. Luedke, A. Nitzan, *Surf. Sci. Lett.* **1989**, *10*, L177. (b) C. M. Mate, G. M. McClelland, R. Erlandsson, S. Chiang, *Phys. Rev. Lett.* **1987**, *59*, 1942.

[38] (a) F. Ohnesorge, G. Binnig, *Science* **1993**, *260*, 1451. (b) H.-J. Butt, E. Seifert, E. Bamberg, *J. Phys. Chem.* **1993**, *97*, 7316.

[39] (a) H. Bengel, H.-J. Cantow, S. N. Magonov, L. Monconduit, M. Evain, W. Liang, M.-H. Whangbo, *Adv. Mater.* **1994**, *6*, 649. (b) H. Bengel, H.-J. Cantow, S. N. Magonov, L. Monconduit, M. Evain, M.-H. Whangbo, *Surf. Sci.* **1994**, *321*, L170.

[40] G. L. Miller, J. E. Griffith, E. R. Wagner, D. A. Grigg, *Rev. Sci. Instrum.* **1991**, *62*, 75.

[41] R. C. Barrett, C. F. Quate, *Rev. Sci. Instrum.* **1991**, *62*, 1393.

[42] M. Stedman *J. Microsc.* **1988**, *152*, 611.

[43] (a) Y. Akama, E. Nishimura, A. Sakai, H. Murakami, *J. Vac. Sci. Technol. A* **1990**, *8*, 429. (b) H. Ximen, P. Russel, *Ultramicroscopy* **1992**, *42-44*, 1526.

[44] D. Nyyssonen, L. Landstein, E. Coombs, *J. Vac. Sci. Technol. B* **1991**, *9*, 3612.

[45] B. Bhushan, V. N. Koinkar, *Appl. Phys. Lett.* **1994**, *64*, 1653.

4 Practical Aspects of STM and AFM Measurements

At present the STM and AFM imaging of samples with relatively flat surfaces is a routine task. However, more elaborate efforts are required to obtain the reliable images needed for thorough characterization of the surfaces. Therefore, in carrying out STM and AFM measurements, one should optimize the experimental conditions and consider the possibility of image artifacts having various different origins. Several practical aspects of STM and AFM experiments are discussed in this chapter.

4.1 Samples

The application of electron and optical microscopy methods often requires elaborate sample preparations and measurements in special environments. In this respect, the scanning probe techniques (especially AFM) are more universal and can be applied to a broad range of samples in different environments. The use of atomic force microscopes is mostly limited by the roughness of samples. Because commercial AFM probes have a height of several microns, samples with surface corrugations below 1 µm can be investigated with atomic force microscopes. With scanning tunneling microscopes, one can study surfaces with larger corrugations using longer tips. However, very long STM tips cannot be employed because of their mechanical instability during scanning. Also, surfaces with extremely sharp features cannot be properly examined with STM and AFM because they lead to image artifacts (see Section 4.3.3).

As for the size of the surface area to be imaged, there is no practical limit. Silicon wafers ca. 10 cm in diameter and microcrystals with dimensions in the micron range can be examined with scanning tunneling and atomic force microscopes. Using an optical microscope or an optical camera, each of which is commonly combined with a scanning probe microscope, one can position the probe in the area of interest and examine, for example, the same area of a conducting sample, successively with STM and AFM. This is better realized by using a large-scale scanner, which can cover areas greater than 15 µm × 15 µm and provide atomic-scale imaging.

STM and AFM experiments can be conducted in different environments (e.g., UHV, ambient conditions, under liquid, electrochemical medium) and at temperatures ranging from the liquid helium temperature up to several hundreds of degrees Celsius. To study samples under liquid or in an electrochemical medium one can use appropriate cells supplied with commercial STM and AFM instruments. In air, many metal and semiconductor surfaces are covered with an oxide layer, which prevents their examination in ambient conditions. Usually these materials are examined in UHV, and before measurement their surfaces are subjected to cleaning procedures. Recently, it was demonstrated that the oxide overlayers of semiconductors GaAs and silicon can be etched away in an electrochemical environment. Therefore, in-situ STM measurements in electrochemical cells make it possible to examine the clean surfaces of these semiconductors on an atomic scale [1]. Metallic and semiconducting materials stable to oxidation can be studied in ambient conditions.

It is natural to ask what minimum electrical conductivity a sample must have, to be examined by STM. Samples with bulk electrical conductivity as low as 10^{-9} S/cm can be examined with conventional STM instruments. If a sample has low conductivity, the electrical contact between the sample surface and its holder should be improved by applying silver or gold paste, and the tip should be positioned close to the contact. The development of low-current STM instrumentation allows the measurements of poorly conducting materials with I_{set} in the picoamp (pA) range and R_{gap} up to teraohms (TΩ).

4.2 Optimization of Experiments

To extract comprehensive physico-chemical information about the surface under examination, the experimental conditions need to be optimized. It is desirable first to image the sample with minimal tip-sample force interactions and then to vary the experimental parameters systematically to examine their influence on the image. This approach provides information that is useful for the image interpretation.

4.2.1 Optimization of STM Experiments

When the sample's electrical conductivity characteristics are known or determined by measuring the I-V dependence, one can estimate the tunneling parameters appropriate for the STM imaging. The choice of the bias voltage polarity is not crucial for metallic samples, but it is essential for semiconducting materials. To understand properly the nature of the STM images of semiconductors, it is necessary to examine their dependence on the magnitude and polarity of the bias voltage. In addition, one should consider a possible complication related to the bending of the valence and conduction bands due to the bias voltage (see Chapter 5). This band bending can alter electron occupation at the top of the valence band and at the bottom of the conduction band. This strongly influences the STM images, depending on the magnitude and the sign of the bias voltage.

To minimize the tip-sample force interaction in the imaging of metallic samples, an initial scan should be performed with high gap resistance (e.g., by using a small set-point current below 1 nA and a relatively high bias voltage around 1-2 V). Scanning with high gap resistance is likely to produce images with low signal-to-noise ratio. Then a gradual decrease in the gap resistance, which is equivalent to decreasing the tip–sample distance, can lead to better-resolved images. However, the increase of the tip–sample force interactions at small tip–sample distances might lead to image variation and even to surface etching.

In STM and AFM of layered compounds (e.g., transition-metal dichalcogenides and organic conducting salts), molecular adsorbates, and polymers [2–5], etching of the surface layers is frequently observed. Figure 4.1 shows the image changes associ-

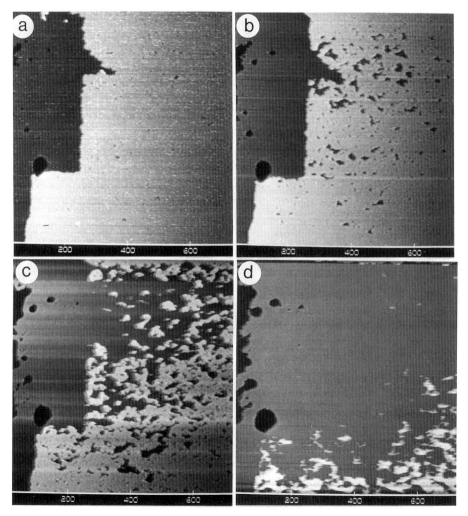

Figure 4.1 (a)–(d) Four successive STM height images of the organic conducting salt α-(BEDT-TTF)$_2$I$_3$ (I_{set} = 1 nA, V_{bias} = 0.1 V). The contrast covers height variations in the 0–10 nm range.

ated with etching on the surface of an organic conductor, which were recorded in successive scans. These images reveal a gradual removal of the topmost layer and a growth of surface holes in the lower-lying layer. Crystalline surfaces with a large number of steps, holes, and other structural imperfections are extensively etched during STM and AFM measurements. The etching proceeds faster at high applied force in AFM, and at small gap resistance in STM. The etching speed also increases when the tip spends more time in a particular area of the surface and when the surface steps are oriented perpendicular to the fast-scanning direction. In STM of organic conducting salts, not only surface etching but also surface growth can occasionally be observed (see Chapter 10). In atomic-scale STM and AFM imaging of layered crystals, layer removal by etching is seen as the successive disappearance and appearance of the periodic image patterns. The results of etching can be recognized by zooming out of the scanned area. In general, the etching mechanism in STM and AFM is not well understood, but its essential features can be rationalized in terms of the tip–sample force interactions. In STM it is also possible that electric field-related effects are involved in the etching process [2].

4.2.2 Optimization of Contact-mode AFM Experiments

In optimizing AFM experiments, it is important to adjust the level of the set-point force, which determines the force applied to the sample (i.e., the operational force, F_{opr}). This can be done by using the force-vs.-distance curve. As an example, Figs. 4.2 (a)–4.2 (c) show such curves recorded on a polyethylene tape in air and in water. These curves are obtained by plotting the cantilever deflection (ΔD) from its rest position ($\Delta D = 0$) as a function of the z-position of the sample. The force-vs.-distance curve is obtained by converting the cantilever displacement into the force unit according to the relationship $F = k_{cl} \Delta D$, where k_{cl} is the spring constant of the free cantilever. The total force experienced by the cantilever is repulsive in the region of $\Delta D > 0$ ($F > 0$), and attractive in the region of $\Delta D < 0$ ($F < 0$).

A strong hysteresis is commonly observed in the ambient-condition experiment due to the capillary force (F_{cap}) associated with the surface contamination layer (Figs. 4.2 (a) and 4.2 (b)). For converting the cantilever displacement ΔD to the force involved, it is convenient to redefine the pull-out force ($F_{pull-out}$) as the force corresponding to the minimum point of the retrieval curve, so that $F_{pull-out}$ is represented as a negative number in Figs. 4.2 (a) and 4.2 (b). When the cantilever is bent up to reach the set-point deflection, ΔD_{set} (or equivalently the set-point force, $F_{set} = k_{cl} \Delta D_{set}$), the operational force is given by $F_{opr} = F_{set} - F_{pull-out}$. When the tip–sample adhesion is negligible, $F_{pull-out}$ is approximately equal to the capillary force F_{cap}, which is typically higher than 10 nN in magnitude [6]. It is estimated that forces from tens to hundreds of nanonewtons are applied to the sample surface in ambient-condition experiments [7]. As a result, the tip with an apex radius of several nanometers

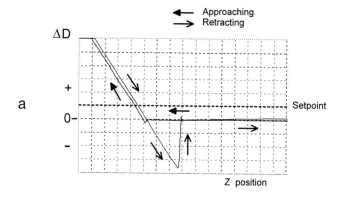
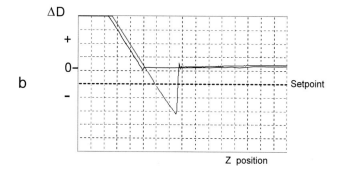
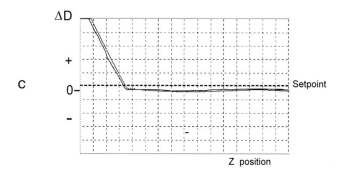

Figure 4.2 Force-vs.-distance curves determined for a polyethylene tape: (a), (b) in air; (c) under water. The arrows indicate the direction of the sample's approach to and retrieval from the tip. The cantilever deflection ΔD, plotted as the vertical axis, is converted into the force unit by the relationship $F = k_{cl} \Delta D$, where k_{cl} is the spring constant of the free cantilever. The set-point deflection ΔD_{set} is indicated by the broken line.

will produce a load of several gigapascals (GPa), which might lead to irreversible surface modification of soft materials such as polymers. A nondestructive imaging of polymers requires the application of smaller forces (a few nanonewtons or less).

It is clear from the above that by adjusting the set-point level, the user can image sample surfaces with different applied forces. By raising the set-point level, high-force imaging can be carried out. To minimize the operational force, one can adjust the set-point level close to the jump-out point (Fig. 4.2 (b)). In this case, the net force acting on the cantilever is attractive (i.e., the cantilever is bent down), but the image contrast is determined by the variation of the repulsive force between the sample and tip. Due to the cantilever instability near the jump-out point, however, a stable operation is achieved only when the set-point force is not close to $F_{\text{pull-out}}$ [7a]. Therefore, in ambient-condition AFM measurements, a part of the capillary force should be overcome by the repulsive force. A more substantial drop in the applied force is achieved in subliquid measurements. In the absence of the capillary force and the related hysteresis in the force-vs.-distance curve, the operational force can be diminished to subnanonewton range by adjusting the set-point close to the tip–sample disconnection point (Fig. 4.2 (c)). In such a case, the operational force can be reduced to the subnanonewton range by using soft cantilevers with spring constants smaller than ca. 0.06 N/m [7a].

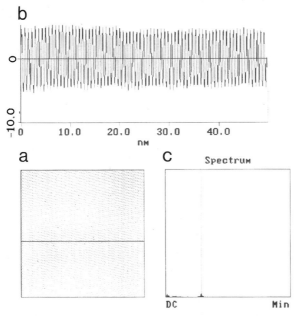

Figure 4.3 (a) Contact-mode AFM image recorded under water with a feedback gain high enough to cause the cantilever to vibrate. (b) Cross-section profile along the x-direction of the image in (a). (c) FFT power spectrum of the cross-section profile in (b).

For studies of soft materials, cantilevers with small spring constants are needed. To find soft ones among a batch of commercial cantilevers with different spring constants, one can determine their resonance frequencies by performing AFM experiments under water [8]. Scanning with a very high feedback gain causes the cantilever to vibrate, thereby leading to a characteristic oscillation image (Fig. 4.3 (a)). The FFT power spectrum of the horizontal cross-section profile of this image exhibits a pronounced peak at the frequency of the cantilever (Figs. 4.3 (b) and 4.3 (c)). This frequency is very close to that of the free cantilever, due to the weak tip–sample coupling under water.

The effect of the applied force on the resolution of AFM images is illustrated in Fig. 4.4, which shows the images of a stretched polyethylene tape recorded in air and under water with the same tip. In subwater measurements, where the operational force is smaller than that in air, the tip–sample contact area is smaller. This improves the lateral resolution significantly; the width of the nanofibrils measured in the experiments under water (15–25 nm) is smaller than that determined from the image recorded in air by a factor of at least two [9].

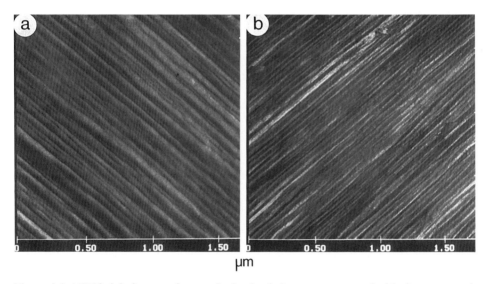

Figure 4.4 AFM height images of a stretched polyethylene tape measured with the same cantilever (a) in air and (b) under water. The contrast covers height variations in the 0–10 nm range.

4.2.3 Optimization of Tapping-Mode AFM Experiments

For soft materials such as polymers and biological compounds, the application of even the smallest force in the contact-mode AFM mode might damage the weak surface structures. One of the factors causing this problem is the lateral force that the tip applies to the sample surface. To avoid this lateral force, one can employ tapping-mode AFM in air or under liquid. Currently, this mode is more widely applied in AFM studies of soft materials than other modulation techniques. However, several aspects of the tapping mode are yet to be clarified. The most important questions are the range of the compressive forces experienced by the sample in this experiment, and the image resolution that can be achieved.

By using the tapping mode under water, one can substantially diminish the amplitude of the cantilever vibration in the free position compared with the corresponding amplitude in air. In this way, tip–sample force interaction can be reduced to a level lower than that reached in the contact mode under water [8]. In subwater imaging with the tapping mode, surface nanofibrils of 2–3 nm in diameter were recorded on a stretched polyethylene tape. The skin layer made up of these nanofibrils was less damaged in the tapping-mode operation than in the contact-mode measurements with ca. 1 nN force. Consequently, the compressive forces applied to the sample in the tapping-mode measurements were effectively smaller than 1 nN. Other estimates suggest that this force is around 0.4 nN [10].

The applicability of the tapping mode to atomic-scale imaging is not clear. In tapping-mode imaging of layered semiconductor WSe_2, the atomic-scale pattern of the surface lattice was detected (Fig. 4.5). The signal-to-noise ratio of this image is inferior to that of the contact-mode image obtained with the same tip in the same place on the sample. In the tapping mode, atomic-scale images are observed occasionally,

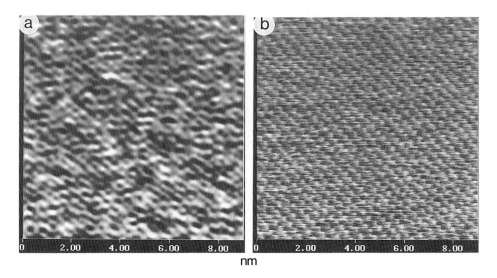

Figure 4.5 (a) Tapping-mode AFM height image of layered compound WSe_2. (b) Contact-mode AFM height image obtained with the same tip and in the same place on the WSe_2 crystal. The contrast covers height variations in the 0.0–0.3 nm range in (a), and in the 0.0–0.5 nm range in (b).

so it is difficult to establish their origin. Such imaging may be caused by a small lateral-force component that the tip experiences during the tapping. The latter is possible because the cantilever in the rest position is slightly inclined to the sample surface, as pointed out in Section 3.3.1.

4.3 STM and AFM Measurements

Several aspects of STM and AFM imaging on different scales are summarized in this section. The specific features of imaging that depend on the nature of the compounds under consideration will be described in later chapters.

4.3.1 Large-Scale Imaging

To examine a sample by STM and AFM, it is preferable to start with large-scale imaging. In general, images of areas larger than 100 nm × 100 nm do not provide information about the contrast variation within a unit cell. However, they contain invaluable data about the topography of the samples and other surface properties (e.g., hardness, friction). The images of amorphous materials such as polymers exhibit different types of surface morphology (for details, see Chapter 12). Flat terraces separated by single or multiple steps are typically found on the surfaces of crystalline samples and thin organic layers adsorbed on substrates. The step heights in the nanometer and angstrom range are rather difficult to measure by physical methods other than STM and AFM. Knowledge of the surface topography obtained with large-scale imaging also allows one to select a flatter and less defective region of the surface for atomic-scale imaging. In some cases, perfect surface layers can be prepared by deliberately etching away the defective topmost layers with the scanning tip.

When surfaces contain chemical and structural defects such as vacancies, dopants, and noncrystalline domains, large-scale STM images should be analyzed with care because they can be dominated by the electronic structures associated with the defects. When defective features are found in the images, it is important to check how the image contrast is influenced by the polarity of the bias voltage. This check can help identify the nature of the defects because, depending on the bias voltage polarity, different electronic states of the sample may become involved in the tunneling process (see Chapter 7). The STM images of defects may be influenced by tip–sample force interactions, and these effects can be observed in the large-scale as well in the atomic-scale images. For instance, in the STM study of defects in layered semiconductor WSe_2, the nanometer-scale bright spots at high R_{gap} are reversibly converted to bright rings at low R_{gap} (Fig. 4.6) [11]. The nanometer-scale bright spot is assigned to the trapped electron state around a donor dopant on the surface Se-atom

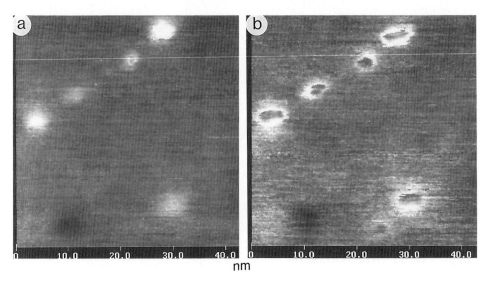

Figure 4.6 STM height images of WSe$_2$ (a) $I_{set} = 1$ nA and $V_{bias} = -0.73$ V; (b) $I_{set} = 1$ nA, $V_{bias} = -0.5$ V. The contrast covers height variations in the 0–1 nm range.

sheet, and this spot becomes a ring because the trapped electron around the dopant atom becomes freed when the ring center is compressed by the tip. Thus, it is also important to examine the image variations at different R_{gap} values.

The interpretation of large-scale AFM images can be complicated when chemically or structurally different domains are present on the surface, because they might modify the local hardness and friction of the surface. This leads to more complicated image patterns than those expected from the surface topography. To confirm the surface-hardness contribution to the images, it is necessary to conduct force-dependent measurements by varying the set-point force level, since the image contrast between the hard and soft regions of the surface increases with increasing load. The softer region of a surface is expected to be more depressed by the tip force, which increases the tip–sample contact area. This will enhance the lateral force (i.e., frictional force) that acts on the tip in the softer region of the surface, which will be manifest in the LFM images.

In principle, it should be possible to distinguish chemically different surface domains by utilizing the response of the AFM probe to the hardness, friction and different surface forces of a sample surface. In most cases, however, interpretation of the associated image contrast is not straightforward, because the surface hardness and friction are affected both by chemical structures and by their packing arrangements. In addition, hydrophilic and hydrophobic surface domains can be different in their mechanical properties. To deconvolute the different factors contributing to the images, it would be useful to employ various chemically modified tips to emphasize, and thus identify, specific types of tip–sample interactions [12].

4.3.2 Atomic-Scale Imaging

STM and AFM images obtained for small areas (below 100 nm × 100 nm) of crystalline surfaces exhibit molecular- and atomic-scale features. Currently, one can obtain atomic-scale AFM images with a signal-to-noise ratio comparable with that found for atomic-scale STM images. For compounds with known crystal structures, the images of the crystal faces are compared with the lattices expected from the bulk crystal data. Therefore, it is important to determine reliably the surface unit cells of the images. To minimize the effect of the thermal drift on the geometrical parameters of the images, it is necessary to carry out imaging after thermal equilibrium has been reached between the sample and microscope stage, and to conduct measurements with high scanning speed. Further, the images should be collected in the "up" and "down" scanning directions and at different rotation angles of the fast-scanning direction. Finally, the geometrical parameters of these measurements should be averaged. This procedure gives rise to more reliable surface unit-cell parameters from the images. For layered compounds the image and crystallographic lattice parameters are in agreement, typically to an accuracy of within 5%.

In STM and AFM studies it is common to observe variations of the image details in atomic-scale patterns depending on the experimental conditions. Image variation may occur spontaneously due the instability of the tip and the surface, even when the scanning is carried out without changing the experimental conditions. Alternatively, image variation may reflect a change in the tip–sample distance and applied force, so it can help to characterize the sample surface. For example, the STM images of some organic conducting salts exhibit an image variation as a function of R_{gap} (see Chapter 10). Such a variation may originate from tip–sample force interaction, the effect of which on the surface structure can be directly observed in the AFM images. To understand the dependence of the image features on the applied force, it is necessary to carry out systematic force-dependent measurements. Figure 4.7 shows the results of force-dependent AFM experiments for the layered telluride $NbGe_{3/7}Te_2$. These images show bright spots corresponding to surface Te atoms. The low-force AFM image exhibits the hexagonal pattern expected from the atomic arrangement of the Te-atom surface layer. In the high-force image, periodic rows of depressed surface atoms appear; upon reducing the applied force, this feature disappears. These reversible image changes show the occurrence of reversible, atomic-scale surface relaxation. As will be discussed in Chapter 9, the depressed atoms are associated with those Te atoms on the surface whose local hardness is small [13]. Thus, the high-force image reveals the variation of the surface local hardness, and the force-dependent AFM studies provide an experimental basis for nanomechanics.

As already mentioned, the images of inorganic layered compounds obtained in high-force experiments exhibit a high signal-to-noise ratio. Furthermore, atomic-size imperfections were found in high-force images of $TaGe_{0.355}Te_2$. Figure 4.8 shows four successive images of $TaGe_{0.355}Te_2$, with two imperfections indicated with ar-

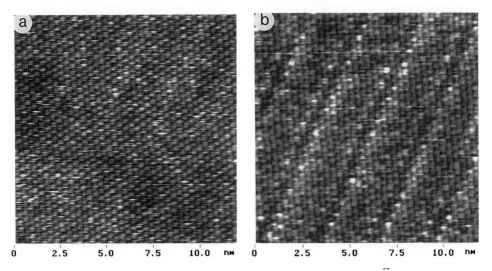

Figure 4.7 AFM height images of NbGe$_{3/7}$Te$_2$ recorded with (a) 170 nN and (b) 270 nN. The contrast covers height variations in the 0–1 nm range.

rows. The defects have brighter and darker contrasts compared with their neighbors in the rows. The cross-section profiles along these rows show these contrast differences in all four successive images. The atomic-size defects change their positions in the images due to thermal drift, but the distance between them remains constant. These findings imply that under high applied forces the tip–sample interactions may lead to an "effective sharpening" of the tip.

In AFM studies of crystalline surfaces, one can obtain LFM images of well-resolved atomic-scale patterns. As measurements of such patterns are useful in answering several questions concerning the atomic origin of friction, the examination of atomic-scale LFM images is necessary. Currently, various aspects of atomic-scale friction are being studied intensively [14].

4.3.3 Image Artifacts

The occurrence of artifacts in STM and AFM images has been a bothersome problem since the invention of scanning probe techniques [15, 16]. Striking image artifacts are often observed for a hard surface with steep corrugations. If a sample surface has structural features sharper than the tip apex, the imaging roles of the tip and surface are reversed so that the tip shape appears in the image. For example, the AFM imaging of a fractured surface of a carbon fiber (about 10 µm in diameter), which contains numerous needles and sharp edges of graphite planes, might result in an image showing the shape of the Si$_3$N$_4$ tip (Fig. 4.9) [17]. This example shows

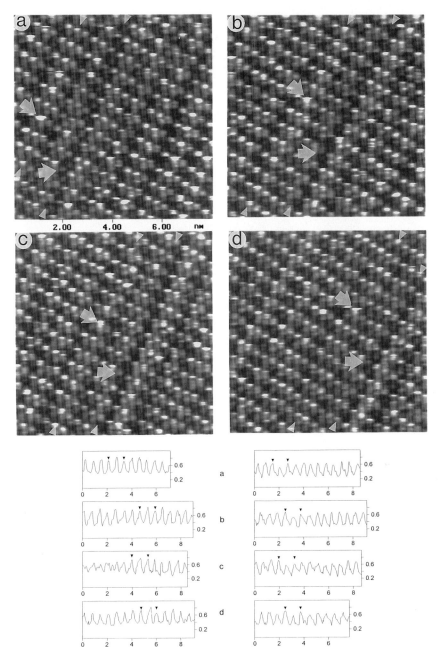

Figure 4.8 (a)–(d) Sequence of four contact-mode AFM height images of TaGe$_{0.355}$Te$_2$ recorded at 10 s intervals with high applied force. The contrast covers height variations in the 0.0–0.7 nm range. The large arrows point to the atomic-size defects. The rows containing the defects are indicated by small wedges, and the height profiles along these rows are shown below the images.

a definite limitation of STM and AFM in studying such surfaces. For flat surfaces, image imperfections might occur due to a multiple-tip effect and tip asymmetry. The image features associated with double-tip imaging are characterized by "ghost" patterns, in which the surface features are repeated. As an example, Fig. 4.9 (b) shows the AFM image of the supramolecular structure of an organic amphiphile (see Chapter 12).

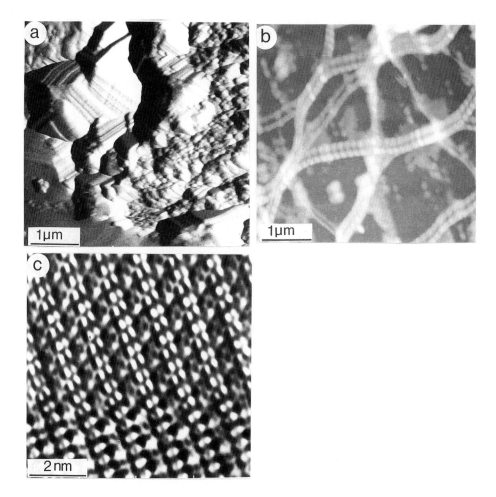

Figure 4.9 (a) Contact-mode AFM force image recorded on a fractured surface of a carbon fiber. The pyramidal-shape pattern in the upper left-hand corner shows the shape of the probing tip. The contrast covers force variations in relative units. (b) AFM height image of the adsorbate of organic amphiphile 10MS (see Chapter 12). The contrast covers height variations in the 0–90 nm range. (c) STM height image of organic conducting salt α-(BEDT-TTF)$_2$I$_3$ (I_{set} = 3.5 nA, V_{bias} = 88.5 mV). The contrast covers height variations in the 0–1 nm range.

It is worthwhile to comment on the surface imperfections of highly ordered pyrolytic graphite (HOPG), samples of which are used as a substrate for depositing organic molecules for their STM imaging. Despite its flat terraces, STM images of HOPG occasionally exhibit disordered and partially ordered features of different shapes, which occur at the surface steps and grain boundaries [18]. Therefore, in experiments with biological and organic molecules adsorbed on graphite, it is necessary to distinguish between the features of defective graphite and those of adsorbed molecules.

Typical artifacts that occur in atomic-scale STM images are shown in Fig. 4.9 (c). A sudden change happens in the center of the image, and the periodic features in the lower and upper parts have different contrasts. Such variations might occur during scanning, due to a spontaneous change in the tip and in the tip–sample gap. In some cases, such changes are accompanied by a small shift of the periodic features in different parts of the image. This can happen when the atoms located in different places on the tip apex become involved in tunneling over different places on the sample surface. The drastic pattern changes observed during an STM study of an Si(111) 7×7 surface in UHV [19] were explained by the sudden change in the electronic states of the tip participating in the tunneling.

Another kind of multiple-tip effect is observed in STM images of HOPG [20]. For example, Fig. 4.10 (a) shows an image of HOPG containing two regions (I and II) with different crystallographic orientations. The angle of misalignment is about $40°$. The border region III shows a pattern (Fig. 4.10 (b)) which resembles the superposition of the images of the domains I and II. Such images can be explained if the current is collected simultaneously from the two regions by two microtips [20]. It should be noted that the occurrence of STM images with superstructure patterns is not always caused by tip-related artifacts. For example, such images can arise from misalignment of a surface layer of graphite with respect to the underlying graphite lattice (see Chapter 9).

The atomic-scale image imperfections resulting from nonideal tip geometry are best illustrated by those found for HOPG. Mizes et al [21] simulated the anomalous images of HOPG by the superposition of three sine waves, whose amplitude and phases were adjusted to match the experimental data. A multiple-tip effect in the HOPG image has been also considered by Colton et al. [22]. Their results suggest that the image distortion is caused more strongly by the tip asymmetry than by the number of tip atoms. The same conclusion was reached by Ren et al. [23], who calculated the partial electron density plots for layered compounds β-Nb_3I_8 and WTe_2 with two- and three-atom tips of different geometry.

Many image artifacts are not self-evident, so one should carry out repeated (though tedious) measurements to characterize properly the sample under examination. For large-scale measurements of corrugated surfaces, sharp tips are desirable; it is therefore useful to test the tip shape with standard samples of known profiles (see Section 3.4.2). However, this does not totally eliminate image artifacts, because

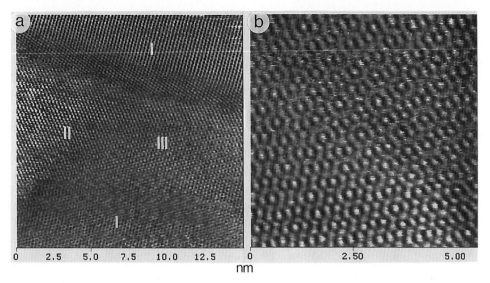

Figure 4.10 (a) STM current image of a defective region of HOPG (I_{set} = 1 nA, V_{bias} = 1 mV), where surface domains with different patterns are indicated with labels I–III. (b) STM current image recorded within the domain III. The contrast is proportional to current variations in relative units. (Courtesy of C. Wang)

a selected tip can be modified during the scanning. Therefore, in the absence of a reliable control of the tip apex geometry, it is common to employ several different tips for the examination of a given surface and to choose the most reproducible images as the genuine representations of the sample under consideration. In this process, one can take the advantage of the expected symmetry of the surface. For highly symmetrical surfaces, the most characteristic atomic-scale images are likely to be highly symmetrical ones.

References

[1] K. Itaya, R. Sugawara, Y. Morita, H. Tokumoto, *Appl. Phys. Lett.* **1992**, *60*, 2534.
[2] (a) B. A. Parkinson, *J. Am. Chem. Soc.* **1990**, *112*, 7498. (b) E. Delawski, B. A. Parkinson, *J. Am. Chem. Soc.* **1992**, *114*, 1661. (c) Y. Kim, C. M. Lieber, *Science* **1992**, *257*, 375.
[3] S. N. Magonov, A. Ya. Gorenberg, G. Bar, H.-J. Cantow, E. B. Yagubskii, *Adv. Mater.* **1993**, *5*, 453.
[4] A. Wawkuschewski, H.-J. Cantow, S. N. Magonov, *Langmuir* **1993**, *9*, 2778.
[5] O. M. Leung, M. C. Goh, *Science* **1992**, *255*, 64.
[6] C. M. Mate, M. R. Lorenz, V. I. Novotny, *J. Chem. Phys.* **1989**, *90*, 7550.

References

[7] (a) A. L. Weisenhorn, P. K. Hansma, T. R. Albrecht, C. F. Quate, *Appl. Phys. Lett.* **1989**, *54*, 2651. (b) A. L. Weisenhorn, P. Maivald, H.-J. Butt, P. K. Hansma, *Phys. Rev. B* **1992**, *45*, 11226.

[8] A. Wawkuschewski, K. Crämer, H.-J. Cantow, S. N. Magonov, *Ultramicroscopy* **1995**, *58*, 185.

[9] A. Wawkuschewski, H.-J. Cantow, S. N. Magonov, S. Sheiko, M. Möller, *Polym. Bull.* **1993**, *31*, 693.

[10] M. Radmacher, M. Fritz, P. K. Hansma, *Science* **1994**, *265*, 1577.

[11] S. N. Magonov, H.-J. Cantow, M.-H. Whangbo, *Surf. Sci. Lett.* **1994**, *318*, L1175.

[12] V. T. Moy, E.-L. Florin, H. E. Gaub, *Colloid. Surf.* **1994**, *93*, 343.

[13] (a) H. Bengel, H.-J. Cantow, S. N. Magonov, L. Monconduit, M. Evain, M.-H. Whangbo, *Surf. Sci. Lett.* **1994**, *321*, L170. (b) H. Bengel, H.-J. Cantow, S. N. Magonov, L. Monconduit, M. Evain, W. Liang, M.-H. Whangbo, *Adv. Mater.* **1994**, *6*, 649.

[14] H.-J. Günterodt, D. Anselmetti, E. Meyer (Eds.), *Forces in Scanning Probe Methods*, Kluwer, Dordrecht, **1995**.

[15] (a) D. Nyyssonen, L. Landstein, E. Coombs, *J. Vac. Sci. Technol. B* **1991**, *9*, 3612. (b) D. J. Keller, C. Chih-Chung, *Surf. Sci.* **1992**, *268*, 333.

[16] (a) P. Grütter, W. Zimmerman-Edling, O. Brodbeck, *Appl. Phys. Lett.* **1992**, *60*, 2741. (b) K. L. Westra, A. W. Mitchell, D. J. Thomson, *J. Appl. Phys.* **1993**, *74*, 368.

[17] S. N. Magonov, A. Ya. Gorenberg, H.-J. Cantow, *Polym. Bull.* **1992**, *28*, 577.

[18] H. Chang, A. J. Bard, *Langmuir* **1991**, *7*, 1143.

[19] J. P. Pelz, *Phys. Rev. B* **1991**, *43*, 6746.

[20] T. R. Albrecht, H. A. Mizes, J. Nogami, S. Park, C. F. Quate, *Appl. Phys. Lett.* **1994**, *64*, 1738.

[21] H. A. Mizes, S.-I. Park, W. A. Harrison, *Phys. Rev. B* **1987**, *36*, 4491.

[22] R. J. Colton, S. M. Baker, R. J. Driscoll, M. G. Youngquist, J. D. Baldschwieler, W. J. Kaiser, *J. Vac. Sci. Technol. A* **1988**, *6*, 349.

[23] J. Ren, M.-H. Whangbo, S. N. Magonov, unpublished results.

5 Simulations of STM and AFM Images

The STM and AFM studies of many crystalline compounds provide atomic-scale images. As already mentioned, these images are described by the partial and total electron density plots of the sample. The theoretical and computational aspects of electron density plot calculations are discussed in this chapter, without considering tip–sample interactions, which are described in Chapters 8 and 9.

5.1 Electronic Structures of Solids

The electronic structures of molecules are characterized by their discrete energy levels (Fig. 5.1 (a)), and those of crystalline solids by their energy bands (Fig. 5.1 (b)). The electronic band structures of solids can be calculated by employing the tight-binding electronic band structure method [1], in which the crystal band orbitals of a solid are represented as linear combinations of its atomic orbitals.

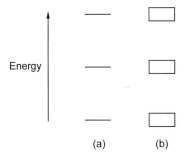

Figure 5.1. Characteristic features of electronic structures: (a) discrete levels of an isolated molecule; (b) energy bands of a solid.

Consider a crystalline solid with repeat vectors a, b, and c, and the corresponding unit cell lengths a, b, and c, respectively. Each unit cell has a set of M atomic orbitals $\{\chi_1, \chi_2, \chi_3, \ldots, \chi_M\}$. The positions of the unit cells are described by the lattice vectors $R = ma + nb + pc$, where m, n, and p are integers. Then, an atomic orbital χ_μ ($\mu = 1, 2, 3, \ldots, M$) located at the unit cell at R is written as $\chi_\mu(r - R)$. To satisfy the periodic boundary condition, the electronic band structure of the solid can be constructed in terms of the Bloch orbitals $\{\phi_1(r, k), \phi_2(r, k), \phi_3(r, k), \ldots, \phi_M(r, k)\}$.

Each Bloch orbital $\phi_\mu(r, k)$ ($\mu = 1, 2, 3, \ldots, M$) is defined by linearly combining the atomic orbitals χ_μ located at all the different unit cells as follows:

$$\phi_\mu(r, k) = \frac{1}{\sqrt{N}} \sum_R \exp(i k \cdot R) \chi_\mu(r - R) \tag{5.1}$$

Here N is the total number of unit cells in the solid ($N \to \infty$), and k is the wave vector. The solid with the repeat vectors a, b, and c has the corresponding reciprocal vectors a^*, b^*, and c^* [1]. For instance, for an orthorhombic solid, the directions of a^*, b^*, and c^* are identical with those of a, b, and c, respectively, and their lengths are $2\pi/a$, $2\pi/b$, and $2\pi/c$, respectively. In general, the wave vector k can be expressed as $k = xa^* + yb^* + zc^*$, where x, y and z are dimensionless numbers, so the term $\exp(ik \cdot R)$ is expressed as

$$\exp(i k \cdot R) = \exp(i k_a ma) \exp(i k_b nb) \exp(i k_c pc) \tag{5.2}$$

where $k_a = x(2\pi/a)$, $k_b = y(2\pi/b)$ and $k_c = z(2\pi/c)$. The exponential term of Eq. (5.2) is a periodic function of k. Thus the Bloch orbitals, which are linear combinations of all the atomic orbitals of the crystal, can be constructed by using only those wave vectors $k = (k_a, k_b, k_c)$ belonging to the region defined by $-\pi/a \leq k_a \leq \pi/a$, $-\pi/b \leq k_b \leq \pi/b$, and $-\pi/c \leq k_c \leq \pi/c$. This region of k values is known as the first primitive zone (FPZ) [1].

The band orbitals $\psi_i(r, k)$ ($i = 1, 2, 3, \ldots, M$) of the solid are expressed as linear combinations of the Bloch orbitals

$$\psi_i(r, k) = \sum_\mu C_{\mu i}(k) \phi_\mu(r, k) \tag{5.3}$$

The coefficients $C_{\mu i}(k)$ and the energies $e_i(k)$ of the crystal orbitals $\psi_i(r, k)$ ($i = 1, 2, 3, \ldots, M$) are determined by solving the eigenvalue problem associated with the effective Hamiltonian H^{eff}.

$$H^{\text{eff}} \psi_i(r, k) = e_i(k) \psi_i(r, k) \tag{5.4}$$

For a given k, the eigenvalue problem, Eq. (5.4) is solved by using the variation principle. The latter leads to the secular determinant

$$\begin{vmatrix} H_{11}(k) - e_i(k) S_{11}(k) & \ldots & H_{1M}(k) - e_i(k) S_{1M}(k) \\ \vdots & & \vdots \\ H_{M1}(k) - e_i(k) S_{M1}(k) & \ldots & H_{MM}(k) - e_i(k) S_{MM}(k) \end{vmatrix} = 0 \tag{5.5}$$

where the matrix elements $H_{\mu\nu}(k)$ and $S_{\mu\nu}(k)$ ($\mu, \nu = 1, 2, 3, \ldots, M$) are expressed as

$$H_{\mu\nu}(k) = \langle \phi_\mu(r, k) | H^{\text{eff}} | \phi_\nu(r, k) \rangle$$
$$= \langle \chi_\mu(r) | H^{\text{eff}} | \chi_\nu(r) \rangle + \sum_R \exp(-i k \cdot R) \langle \chi_\mu(r-R) | H^{\text{eff}} | \chi_\nu(r) \rangle +$$
$$+ \exp(i k \cdot R) \langle \chi_\mu(r) | H^{\text{eff}} | \chi_\nu(r-R) \rangle \tag{5.6}$$

and

$$S_{\mu\nu}(k) = \langle \phi_\mu(r, k) | \phi_\nu(r, k) \rangle$$
$$= \langle \chi_\mu(r) | \chi_\nu(r) \rangle + \sum_R \exp(-i k \cdot R) \langle \chi_\mu(r-R) | \chi_\nu(r) \rangle +$$
$$+ \exp(i k \cdot R) \langle \chi_\mu(r) | \chi_\nu(r-R) \rangle \tag{5.7}$$

The lattice sum in Eqs. (5.6) and (5.7) (i.e., the summation over R) can be limited to the terms involving the reference and nth nearest neighbors ($n = 1, 2, 3$, etc.) when the remaining terms are negligibly small.

One obtains the energy levels belonging to M different energy bands by solving the eigenvalue problem Eq. (5.4) for a fine mesh of k-points covering the FPZ. The electronic band structure of the solid thus obtained is presented by plotting $e_i(k)$ vs. k along certain directions of the FPZ or by plotting $n(e)$ vs. e, where $n(e)$ represents the electronic density of states (DOS) of the solid at a given energy e [1]. The $n(e)$ value is nonzero within the allowed energy region of a band and vanishes in the forbidden energy region.

Each energy band is made up of N energy levels. Since N is very large, the energy region of each band is continuously allowed from the bottom to the top. Thus, any given band can accommodate up to two electrons per unit cell. Whether a given solid is a metal or not depends upon the number of electrons per unit cell and the nature of the energy bands. The highest occupied band of a metal is not completely filled, so that there is no energy gap between the highest occupied and lowest unoccupied band levels (Fig. 5.2 (a)). A normal semiconductor or a normal insulator has no partially filled bands, so there is a band gap between the highest occupied (i.e., valence)

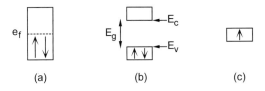

Figure 5.2 Band filling patterns: (a) half-filled band leading to a metal; (b) completely filled and completely empty bands leading to a normal semiconductor or insulator; (c) half-filled band leading to a magnetic semiconductor or insulator.

band and lowest unoccupied (i.e., conduction) band levels (Fig. 5.2 (b)). A magnetic semiconductor or a magnetic insulator has unpaired spin orbitals. In the electronic band picture, each is represented by energy bands whose levels are all singly filled (Fig. 5.2 (c)) [1].

In the extended Hückel method [2], only the valence atomic orbitals are used for electronic structure calculations, and the atomic orbitals $\chi_\mu(r, \theta, \phi)$ are approximated by Slater-type orbitals (STOs),

$$\chi_\mu(r, \theta, \phi) \propto r^{n-1} \exp(-\zeta r) \, \Phi(\theta, \phi) \tag{5.8}$$

where n is the principal quantum number of the atomic orbital, ζ is the orbital exponent, and $\Phi(\theta, \phi)$ is the spherical harmonics function determining the angular variation of the atomic orbital. The STOs are normalized so that $\langle \chi_\mu | \chi_\mu \rangle = 1$, and the integrals between two orbitals $\langle \chi_\mu | \chi_\nu \rangle$ ($\mu \neq \nu$) define the overlap integrals $S_{\mu\nu}$. The energy matrix elements $H_{\mu\nu}$ between two different orbitals χ_μ and χ_ν are approximated by the Wolfsberg–Helmholz formula $H_{\mu\nu} = (K/2)(H_{\mu\mu} + H_{\nu\nu}) S_{\mu\nu}$, where $H_{\mu\mu}$ and $H_{\nu\nu}$ are the valence-shell ionization potentials of the atomic orbitals χ_μ and χ_ν, respectively, and K is a constant.

5.2 Theoretical Aspects of STM

5.2.1 Tunneling Between Metals

Suppose that the tip and sample are both metals and form a metal–insulator–metal junction in the STM configuration (see Fig. 3.2). The gap between the tip and sample provides an insulating barrier (e.g., air, vacuum). When the bias voltage (V_{bias}) between the electrodes is zero, their Fermi levels become equal and there is no tunneling current between them (Fig. 5.3 (a)). When V_{bias} is positive (with the tip grounded), the energy levels of the sample are lowered by eV_{bias} so that the electrons in the occupied levels of the tip (between e_{f} and $e_{\text{f}} - eV_{\text{bias}}$) tunnel into the unoccupied levels of the sample (Fig. 5.3 (b)). When V_{bias} is negative (with the tip grounded), the energy levels of the sample are raised by $e|V_{\text{bias}}|$ so that the electrons in the occupied levels of the sample (between $e_{\text{f}} + e|V_{\text{bias}}|$ and e_{f}) tunnel into the unoccupied levels of the tip (Fig. 5.3 (c)).

The transmission probability of the tunneling is the largest for the electron at the Fermi level of the negatively biased electrode and steadily decreases as the energy is lowered from that level (shown by the lengths of the arrows in Figs. 5.3 (b) and 5.3 (c)). In general, most of tunneling electrons come from within 0.3 eV of the Fermi level of the negatively charged electrode [3].

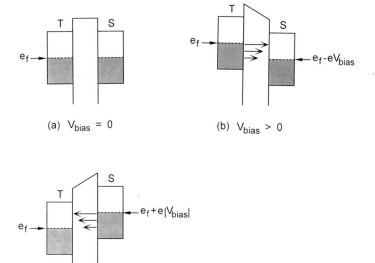

Figure 5.3 Energy bands associated with the metal–insulator–metal junction between a metallic tip and a metallic sample: (a) $V_{bias} = 0$; (b) $V_{bias} > 0$; (c) $V_{bias} < 0$.

5.2.2 Tunneling Between Metal and Semiconductor

Suppose that a pristine semiconductor sample forms a semiconductor–insulator–metal (SIM) junction with the metallic tip in the STM configuration. The bottom of the conduction band and the top of the valence band of the semiconductor may be referred to as E_c and E_v, respectively, so that the band gap is $E_g = E_c - E_v$. When V_{bias} is zero, the Fermi level of the tip lies at the midpoint of the semiconductor band gap [i.e., $e_f = (E_c + E_v)/2$] (Fig. 5.4 (a)) and there is no tunneling current between them. When the positive bias is greater than $E_g/2$ so that $E_c - eV_{bias} < e_f$, the electrons of the tip tunnel into the empty levels of the conduction band (between e_f and $E_c - eV_{bias}$) (Fig. 5.4 (b)). When the magnitude of the negative bias is greater than $E_g/2$ so that $E_v + e|V_{bias}| > e_f$, the electrons in the valence band of the sample (between $E_v + e|V_{bias}|$ and e_f) tunnel into the tip (Fig. 5.4 (c)).

A semiconductor with n-type dopants has donor levels near the bottom of the conduction band, and the thermal excitation of electrons from the donor levels populates the bottom of the conduction band. For an SIM junction with such an n-type semiconductor, the Fermi level is close to the donor level E_D of the semiconductor when $V_{bias} = 0$ (Fig. 5.5 (a)). A semiconductor with p-type dopants has acceptor levels near the top of the valence band. The thermal excitation of electrons from the valence band populates the acceptor level. For an SIM junction with such a p-type semiconductor, the Fermi level is close to the acceptor level E_A of the semiconductor when $V_{bias} = 0$ (Fig. 5.5 (b)). Depending upon the sign and magnitude of the bias voltage, the STM current of doped semiconductors may involve not only the bottom of the conduction band and the top of the valence band but also the donor and acceptor levels lying in the band gap. In addition, the wave functions of the donor and acceptor levels are spatially localized in the regions around the donor and acceptor dopants [4], unlike those of the valence and conduction band levels which are delocalized. Therefore, STM images of doped semiconductors or semiconductors with point defects can exhibit local imperfections. For more detailed discussion, see Chapter 7.

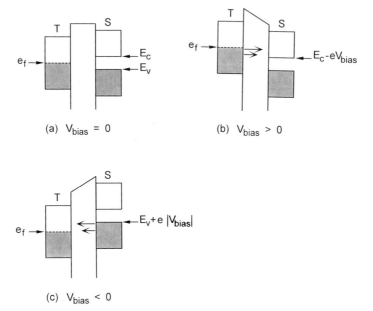

Figure 5.4 Energy bands associated with the metal–insulator–semiconductor junction between a metallic tip and a semiconductor sample: (a) $V_{bias} = 0$; (b) $V_{bias} > 0$; (c) $V_{bias} < 0$.

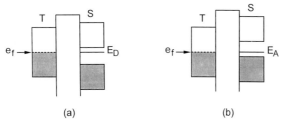

Figure 5.5 Energy bands associated with the metal–insulator–semiconductor junction between a metallic tip and a semiconductor sample: (a) for an n-type semiconductor at $V_{bias} = 0$; (b) for a p-type semiconductor at $V_{bias} = 0$.

When a bias voltage is applied through the SIM junction, the energy levels of the semiconductor near the junction surface become different from those further away from it. At positive V_{bias} the semiconductor bands bend downward so that the levels nearer the surface become higher in energy (Fig. 5.6 (a)), while at negative V_{bias} the bands bend upward so that the levels nearer the surface become lower in energy (Fig. 5.6 (b)) [5]. At a large positive bias, the band bending of a p-type semiconductor may empty the top portion of the valence band at the surface, so that the valence band top may also receive electrons from the tip (Fig. 5.7 (a)). At a large negative bias, the band bending of an n-type semiconductor may populate the bottom portion of the conduction band at the surface, so that electrons may also flow from the bottom of the conduction band to the tip (Fig. 5.7 (b)).

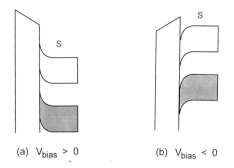

(a) $V_{bias} > 0$ (b) $V_{bias} < 0$

Figure 5.6 Bending of the valence and conduction bands of a semiconductor near the metal–insulator–semiconductor junction surface: (a) $V_{bias} > 0$; (b) $V_{bias} < 0$.

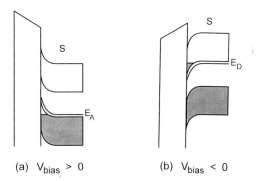

(a) $V_{bias} > 0$ (b) $V_{bias} < 0$

Figure 5.7 Possible effects of a strong band bending: (a) electron depletion from the valence band top near the junction surface for a p-type semiconductor under a strongly positive bias; (b) electron accumulation at the conduction band bottom near the junction surface for an n-type semiconductor under a strongly negative bias.

5.2.3 Tersoff–Hamman Theory and its Extension

The simplest and most practical theory of STM was developed by Tersoff and Hamman [6], who based their work on Bardeen's transfer Hamiltonian approach [7]. In Bardeen's perturbation treatment, the tunneling between two metals through a barrier is described in terms of the overlap between the tails of their wave functions. To describe the tunneling between the tip and sample using Bardeen's approach, it is necessary to specify their wave functions. In their theory, Tersoff and Hamman [6] introduced three simplifying assumptions: (1) the tip can be approximated by an atom with an s-orbital; (2) the tip–sample interactions are negligible; and (3) the bias voltage is small. Assumptions (1), (2), and (3) respectively allow one to treat the tip as a mathematical point, to neglect the possible modifications of the structures and wave functions of both electrodes, and to take into consideration only the sample electronic states at the Fermi level e_f for the tunneling. Under these assumptions, Tersoff and Hamman showed that the STM current I_{tun} is proportional to the local density of states (LDOS) of the sample at the position of the tip r_t,

$$I_{tun} \propto \sum_i \int d\mathbf{k}\, |\psi_i^s(r_t, \mathbf{k})|^2\, \delta(e_s - e_f) \tag{5.9}$$

where $\psi_i^s(r, \mathbf{k})$ is the ith electronic band orbital of the sample. For simplicity of notation, the band orbital energy $e_i^s(\mathbf{k})$ of the sample is written as e_s. The delta function insures that only the band orbitals of the sample at the Fermi level contribute to the LDOS. The equi-current contours observed in constant-height STM measurements correspond to the constant contours of the partial electron density $\rho(r_0, e_f)$ of the sample surface at the Fermi level (evaluated at the position of the tip-to-sample distance r_0). Therefore, for a given sample surface, the brightness pattern of the STM image is related to the high-density pattern of the $\rho(r_0, e_f)$ plot. It was reported that the s-orbital tip approximation fails to explain the observed atomic resolution for close-packed elemental metal surfaces [8], and Chen [9] explained the atomic resolution data on Al(111) by using the d_{z^2}-orbital tip model. Certainly, the s-orbital tip approximation oversimplifies the electronic structure of the tip. Nevertheless, Tersoff–Hamman theory has been successful in interpreting the STM images of a large variety of organic and inorganic materials [10]. This theory is most practical because the electron density plot calculations can be performed with various computational methods of electronic structure calculations.

Ou-Yang et al. developed a more general but still practical theory of STM [11], in which they kept assumptions (2) and (3) of Tersoff–Hamman theory. Using the time-dependent first-order perturbation theory, Ou-Yang et al. deduced an expression for the STM current as

$$I_{tun} \propto \sum_m \sum_i \int d\mathbf{k}\, |\langle \psi_m^t | H | \psi_i^s(r_t, \mathbf{k}) \rangle|^2\, \delta(e_s - e_f) \tag{5.10}$$

when the tip is treated as a semi-infinite linear chain of transition-metal atoms (perpendicular to the surface). The ψ_m^t denotes the mth d-orbital of the first tip atom (i.e., the atom closest to the sample), and H is the Hamiltonian appropriate for an electron moving in a system consisting of the tip and sample in STM. Equation (5.10) is reduced to Eq. (5.9) by introducing the following approximations [10a]:

$$\langle \psi_m^t | H | \psi_i^s(r_t, k) \rangle \propto \langle \psi_m^t | \psi_i^s(r_t, k) \rangle \propto \psi_i^s(r_t, k) \tag{5.11}$$

i.e., the energy matrix element $\langle \psi_m^t | H | \psi_i^s(r_t, k) \rangle$ is proportional to the overlap integral $\langle \psi_m^t | \psi_i^s(r_t, k) \rangle$, which is in turn proportional to the value of the orbital $\psi_i^s(r_t, k)$. These approximations are generally valid and form the basis of semi-empirical quantum mechanical computational methods.

An important aspect of Ou-Yang et al.'s theory is that for the STM measurements of insulating molecular species adsorbed on a metallic substrate, the $\psi_i^s(r_t, k)$ refer to the orbitals of the adsorbate/substrate system. From the viewpoint of Tersoff–Hamman theory, this means that the STM images of an adsorbate/substrate system can be simulated in terms of the partial electron density plot $\rho(r_0, e_f)$ calculated for the adsorbate/substrate system with the tip positioned above the adsorbate layer [10, 12].

In principle, the perturbation treatments of the electron transfer in STM by Tersoff and Hamman [6] and by Ou-Yang et al [11]. are not valid when the overlap between the wave functions of the tip and sample surface becomes significant. According to Ciraci's analysis [13] of an idealized tip–sample system, this electronic contact occurs before the mechanical contact as the tip approaches the sample surface. In practice, it is most likely that the tip–sample force interaction occurs through the contamination layer, thereby leading to macroscopic and microscopic deformations of the surface even before the tip and sample experience electronic contact (see Chapters 8 and 9) [14]. In such a case, the perturbation approaches are still useful, and the observed STM image is related to the partial electron density plot of the deformed surface.

5.2.4 Other Theories

In a general theory of STM, it is attempted to treat reliably the quantum properties of electrons with kinetic energy of a few electronvolts, because such electrons may interact with structures of dimensions of the order of their wavelength. For this purpose, one should not only use realistic potentials but also solve the associated three-dimensional scattering problem [15]. However, it is difficult to treat both parts of the question rigorously. In the studies based on realistic potentials, the STM current was calculated either in a rough approximation [16] or in perturbation theory [17]. Multiple scattering effects were studied using simple model potentials [18]. The STM theories in which the current is calculated beyond perturbation theory have been reviewed by Doyen [15].

5.3 Theoretical Aspects of AFM

The forces between closed-shell atoms and molecules are primarily determined by the outer regions of the atoms, in which the atomic electron densities overlap [19]. Therefore, if the AFM tip is approximated by an atom (by analogy with Tersoff–Hamman theory) and if the surface topography is negligibly affected by the tip–sample interactions, it is expected that the repulsive force F_{rep} experienced by the tip in contact-mode AFM is proportional to the LDOS of the sample at the position of the tip r_t,

$$F_{rep} \propto \sum_{e_i(k) \leq e_f} \int dk \, |\psi_i^s(r_t, k)|^2 \tag{5.12}$$

where the summation is over all the occupied band levels. Thus the equi-force contours in constant-height AFM measurements correspond to constant contours of the total electron density $\rho(r_0)$ of the sample surface at the position of the tip-to-sample distance r_0. Therefore, for a given sample surface, the brightness pattern of the AFM image is related to the high-density pattern of the $\rho(r_0)$ plot. It should be recalled that this approach is valid when the tip-force induced surface corrugation is negligible.

5.4 Image Simulation by Density Plot Calculations

5.4.1 STM Image Simulation

The partial density $\rho(r, e_f)$ necessary for the simulation of the STM image of a solid surface is obtained by summing the density contributions of the band orbitals $\psi_i(r, k)$ of the sample

$$\rho(r, e_f) = \sum_{|e_i(k) - e_f| \leq \Delta} \psi_i(r, k)^* \psi_i(r, k) \tag{5.13}$$

where the summation includes only those orbitals whose energies $e_i(k)$ belong to the energy window, $|e_i(k) - e_f| \leq \Delta$, appropriate for the problem. Here Δ is a small positive number. When the energy bands associated with tunneling are several electronvolts wide, the energy window Δ can be chosen to be about 0.25 eV. (In most cases, for metallic compounds, the variation of Δ in the 0.05–0.30 eV range does not qualitatively affect the results.) For metals, for which the highest occupied and lowest occupied levels are degenerate, the summation includes all the occupied band orbitals whose energies lie between e_f and $e_f - \Delta$ for the sample-to-tip tunneling (Fig. 5.8 (a)), and all the unoccupied band orbitals whose energies lie between $e_f + \Delta$ and e_f for the tip-to-sample tunneling (Fig. 5.8 (b)). Finally, the $\rho(r_0, e_f)$ plot for STM image simulation is obtained by calculating the values of the partial density $\rho(r, e_f)$ at the tip-to-surface distance r_0 and presenting the results in a two-dimensional (2D) contour plot or in a three-dimensional (3D) surface plot. Thus, three important steps of $\rho(r_0, e_f)$ plot calculations are (a) the electronic band structure calculations to obtain $\psi_i(r, k)$ and $e_i(k)$, (b) the selection of the band levels $\psi_i(r, k)$ appropriate for the partial density $\rho(r, e_f)$ with a certain energy window Δ, and (c) the evaluation of $\rho(r, e_f)$ at the tip-to-sample r_0.

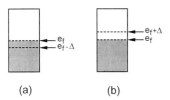

(a)   (b)

Figure 5.8 Energy windows appropriate for the partial density plot calculations of a metal: (a) for $V_{bias} < 0$; (b) for $V_{bias} > 0$.

For the calculations of $\rho(r_0, e_f)$ plots, the tip-to-surface distance r_0 value may be taken to be small (e.g., 0.5 Å from the atoms closest to the tip), because the orbital amplitudes of the wave functions of a surface decrease exponentially with distance from the surface so that the $\rho(r_0, e_f)$ values become too small for meaningful comparisons if r_0 is large (e.g., 4 Å). The $\rho(r_0, e_f)$ plots calculated for $r_0 = 0.5$ Å have been found to reflect the essential patterns of the observed STM images rather well [10]. Furthermore, a simultaneous STM/AFM study of 1T-TaS$_2$ at ambient conditions shows [20] that the STM image obtained with the tip in contact with the surface is essentially identical with that obtained in a traditional STM study. Therefore, use of the small r_0 value is justified.

For the sample-to-tip tunneling of an intrinsic semiconductor, the occupied band orbitals lying between E_v and $E_v - \Delta$ should be selected (Fig. 5.9 (a)). For the tip-to-sample tunneling of an intrinsic semiconductor, the unoccupied band orbitals lying between $E_c + \Delta$ and E_c should be selected (Fig. 5.9 (b)). A magnetic semiconductor has unpaired spin orbitals responsible for its magnetic properties. Within the framework of a band picture, a magnetic semiconductor is characterized by its highest occupied band(s) whose levels are singly filled (Fig. 5.2 (c)) [1, 21c]. The singly filled levels can either donate electrons to the tip or receive electrons from the tip. In addition, such bands representing localized electrons in each unit cell have a narrow bandwidth [1, 21]. Thus, to calculate the partial density plot of a magnetic semi-

76   5 Simulations of STM and AFM Images

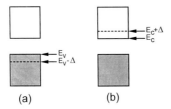

Figure 5.9 Energy windows appropriate for the partial density plot calculations of a normal semiconductor: (a) for $V_{bias} < 0$; (b) for $V_{bias} > 0$.

conductor, e_f should be set equal to the top of the partially filled band(s) and all the levels of the partially filled band(s) may be selected for both the sample-to-tip and tip-to-sample tunneling (Fig. 5.10).

The wave functions of the donor and acceptor states of a normal semiconductor are spatially localized around the defect sites [4] and hence are not described by electronic band structure calculations. Depending upon the polarity and magnitude of the applied bias voltage, the electron tunneling process may involve the defect states near the top of the valence band and the bottom of the conduction band. This complicates the STM image interpretation [5, 22], as will be discussed in Chapter 7.

Figure 5.10 Energy windows appropriate for the partial density plot calculations of a magnetic semiconductor sample for both $V_{bias} < 0$ and $V_{bias} > 0$.

5.4.2 AFM Image Simulation

The total density $\rho(r)$ necessary for the simulation of the AFM image of a solid surface is obtained by summing the density contributions of all the occupied band orbitals $\psi_i(r, k)$,

$$\rho(r) = \sum_{e_i(k) \le e_f} \psi_i(r, k)^* \psi_i(r, k). \tag{5.14}$$

The $\rho(r_0)$ plot for AFM image simulation is obtained by calculating the values of the total density $\rho(r)$ at the tip-to-sample distance r_0, which may be taken to be short (e.g., 0.5 Å) as in the case of the $\rho(r_0, e_f)$ plot calculations, and plotting the results in a 2D contour plot or in a 3D surface plot.

5.4.3 STM and AFM Images of Graphite

The use of density plot calculations can be illustrated by considering the STM and AFM images of highly ordered pyrolytic graphite (HOPG). Adjacent layers of HOPG are arranged in such a way that there occur two types of carbon atoms (A and B) on the surface layer (Fig. 5.11). The A-site carbon atoms lie directly above the carbon atoms of the underlying layer, and the B-site carbon atoms are located above the centers of the carbon hexagons of the underlying layer.

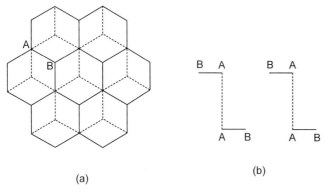

Figure 5.11 Schematic representation of the arrangement between two adjacent graphite sheets in HOPG: (a) top projection view, where the C–C bonds of the upper and lower sheets are indicated by solid and broken lines, respectively; (b) side projection view, where the interlayer C⋯C contacts are shown by dotted lines.

Most STM images of HOPG show a hexagonal pattern of bright spots, which are asigned to three nonadjacent carbon atoms for each carbon hexagon forming the surface graphite layer (Fig. 5.12 (a)) [14a, 23]. The "three-for-hexagon" patterns that are obtained for positive and negative bias voltages have an identical peak registry, i.e., the bright spots of a positive bias image are the same as those of a negative bias image [23]. Such images are more frequently observed than that shown in Fig. 5.12 (b). In the latter image, all six carbon atoms of each hexagon appear, but the brighter spots still form the three-for-hexagon patterns. The three-for-hexagon pattern is also observed in AFM images, as shown in Fig. 5.12 (c).

The three-for-hexagon STM pattern of HOPG is accounted for in terms of the interlayer interactions [14a, 23]. Simulation of the STM images of HOPG can be carried out by calculating the partial density plots $\rho(r_0, e_f)$ for a graphite bilayer (Figs. 5.13 (a) and 5.13 (b)). As shown in this figure, the high electron density spots of the $\rho(r_0, e_f)$ plots are located at the B-site carbon atoms for the sample-to-tip and tip-to-sample tunneling. This comes about because the interlayer C⋯C interactions through the A-site carbon atom, make the π-electron band levels around the

Fermi level more concentrated on the B-site than on the A-site carbon p_π-orbitals. The nature of the π-band levels of HOPG around the Fermi level can be deduced by analyzing that of the graphite bilayer. There are four p_π atomic orbitals per unit cell so there are four π bands. In each unit cell, the bonding and antibonding combinations of the atomic orbitals on the A-sites lead to the group orbitals ϕ_A^+ and ϕ_A^-, and those on the B-sites to the group orbitals ϕ_B^+ and ϕ_B^- (Fig. 5.14). Since the p_π-orbitals of the B-sites do not interact among themselves, the energies of ϕ_B^+ and ϕ_B^- are close to the p_π-orbital level. However, the sigma-type overlap through the C⋯C linkage makes the ϕ_A^+ level lower, and the ϕ_A^- level higher, in energy than the p_π-or-

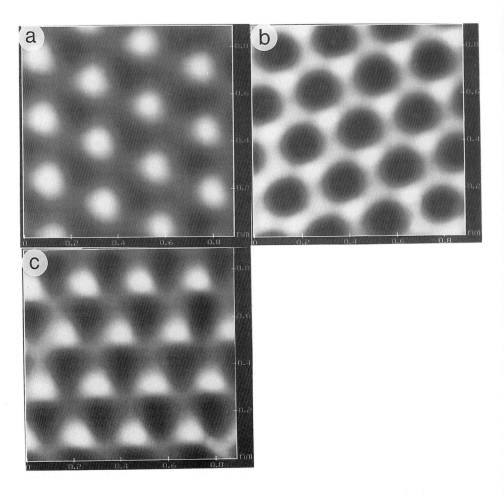

Figure 5.12 (a) STM current image of HOPG (I_{set} = 2 nA, V_{bias} = 28 mV). (b) STM current image of HOPG (I_{set} = 1 nA, V_{bias} = 30 mV). (c) AFM height image of HOPG. The contrast variation in (a) and (b) is proportional to the tunneling current, and that in (c) covers height variations in the 0.0–0.2 nm range.

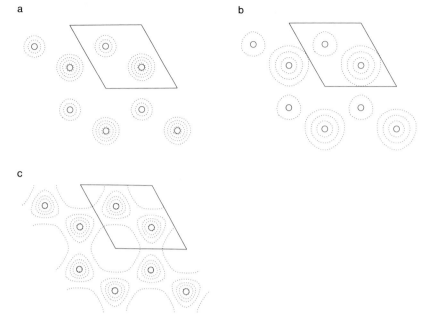

Figure 5.13 Partial and total density plots calculated for the graphite bilayer (a) $\rho(r_0, e_f)$ plot for $V_{bias} > 0$. The contour values used are 1.0×10^{-2}, 1.5×10^{-2}, 2.0×10^{-2} and 2.5×10^{-2} electrons/au^3. (b) $\rho(r_0, e_f)$ plot for $V_{bias} < 0$. The contour values used are 1.0×10^{-3}, 2.5×10^{-3}, 5.0×10^{-3} and 7.5×10^{-3} electrons/au^3. (c) $\rho(r_0)$ plot. The contour values used are 1.0×10^{-1}, 1.5×10^{-1}, 1.8×10^{-1}, 2.0×10^{-1} and 2.2×10^{-1} electrons/au^3.

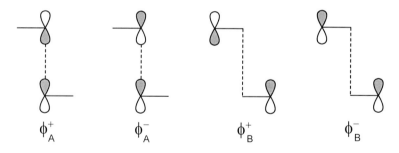

Figure 5.14 Four group orbitals resulting from each unit cell of the graphite bilayer.

bital level (Fig. 5.15). The π bands of the graphite bilayer can be constructed by first making Bloch orbitals from the group orbitals and then by combining the Bloch orbitals derived from ϕ_A^+ and ϕ_A^- with those derived from ϕ_B^+ and ϕ_B^-, respectively. Then, according to the orbital interaction theory [24], the π-band levels lying close to the Fermi level are more concentrated on the B-site atoms because they are derived mainly from the ϕ_B^+- and ϕ_B^--orbitals (Fig. 5.15). Likewise, the levels lying far removed from the Fermi level are more concentrated on the A-site atoms because they are mainly derived from the ϕ_A^+- and ϕ_A^--orbitals. Therefore, the partial electron density plots relevant for HOPG are concentrated on the B-sites. This in part explains why the bright spots of the STM images of HOPG are located on the B-site carbon atoms for the sample-to-tip and tip-to-sample tunneling. Another factor contributing to the three-for-hexagon pattern is the tip force induced surface corrugation (see below and Chapter 9).

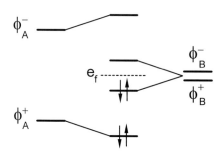

Figure 5.15 Orbital interactions between the Bloch orbitals derived from ϕ_A^+ and ϕ_A^- with those derived from ϕ_B^+ and ϕ_B^-, respectively, leading to the band orbitals of the graphite bilayer. The Fermi level reflects that there are four π-electrons per unit cell.

As expected, the total electron density plot $\rho(r_0)$ calculated for the graphite bilayer (Fig. 5.13 (c)) shows no difference in the electron density at the A- and B-sites, because the total electron density plot is based on all the occupied band levels. This plot is not in agreement with the observed AFM pattern (Fig. 5.12 (c)). It is important to note the implicit assumption behind the $\rho(r_0, e_f)$ and $\rho(r_0)$ calculations, i.e., that the z-heights of the A- and B-site carbon atoms are identical. This assumption is unlikely to be correct for the surface area under the tip because the tip force can induce a surface corrugation when the surface consists of sites with different local hardness, as will be discussed in Chapter 9.

References

[1] For a review, see: E. Canadell, M.-H. Whangbo, *Chem. Rev.* **1991**, *91*, 965.
[2] R. Hoffmann, *J. Chem. Phys.* **1963**, *39*, 1397.
[3] R. J. Hamers, in *Scanning Tunneling Microscopy and Spectroscopy: Theory, Techniques and Applications*, D. A. Bonnell (Ed.), VCH Publishers, New York, **1993**, Ch. 4.
[4] C. Kittel, *Introduction to Solid State Physics*, 2nd ed., Wiley, New York, **1956**, pp. 353–358.

[5] (a) Z. F. Zheng, M. B. Salmeron, E. R. Weber, *Appl. Phys. Lett.* **1994**, *64*, 1836. (b) J. F. Zheng, X. Liu, N. Newman, E. R. Weber, D. F. Ogletree, M. Salmeron, *Phys. Rev. Lett.* **1994**, *72*, 1490. (c) M. Maboudian, K. Pond, V. Bressler-Hill, M. Wassermeier, P. M. Petroff, G. A. D. Briggs, W. H. Weinberg, *Surf. Sci.* **1992**, *275*, L662. (d) R. M. Feenstra, J. A. Stroscio, *J. Vac. Tecnnol. B* **1987**, *5*, 923.

[6] J. Tersoff, D. R. Hamman, *Phys. Rev. B* **1985**, *31*, 85.

[7] J. Bardeen, *Phys. Rev. Lett.* **1961**, *6*, 57.

[8] (a) V. M. Hallmark, S. Chiang, J. F. Rabolt, J. D. Swalen, R. J. Wilson, *Phys. Rev. Lett.* **1987**, *59*, 2879. (b) J. Wintterlin, J. Wiechers, H. Brune, T. Gritsch, H. Höfer, R. J. Behm, *Phys. Rev. Lett.* **1989**, *62*, 59.

[9] (a) C. J. Chen, *Phys. Rev. Lett.* **1990**, *65*, 448. (b) C. J. Chen, *J. Vac. Sci. Technol. A* **1991**, *9*, 44.

[10] For a review, see: S. N. Magonov, M.-H. Whangbo, *Adv. Mater.* **1994**, *6*, 355.

[11] (a) H. Ou-Yang, B. Källebring, R. A. Marcus, *J. Chem. Phys.* **1993**, *98*, 7565. (b) H. Ou-Yang, R. A. Marcus, B. Källebring, *J. Chem. Phys.* **1994**, *100*, 7814.

[12] (a) W. Liang, M.-H. Whangbo, A. Wawkuschewski, H.-J. Cantow, S. Magonov, *Adv. Mater.* **1993**, *5*, 817. (b) A. Wawkuschewski, H.-J. Cantow, S. N. Magonov, M. Möller, W. Liang, M.-H. Whangbo, *Adv. Mater.* **1993**, *5*, 821.

[13] C. Ciraci, in *Scanning Tunneling Microscopy III*, R. Wiesendanger, H.-J. Güntherodt (Eds.), Springer-Verlag, Heidelberg, **1993**, p. 139.

[14] (a) M.-H. Whangbo, W. Liang, J. Ren, S. N. Magonov, A. Wawkuschewski, *J. Phys. Chem.* **1994**, *98*, 7602. (b) H. Bengel, H.-J. Cantow, S. N. Magonov, L. Monconduit, M. Evain, M.-H. Whangbo, *Surf. Sci. Lett.* **1994**, *321*, L170.

[15] G. Doyen, in *Scanning Tunneling Microscopy III*, R. Wiesendanger, H.-J. Güntherodt (Eds.), Springer, Berlin, **1993**, Ch. 3.

[16] S. Ciraci, *Phys. Rev. B* **1989**, *40*, 11969.

[17] M. Tsukada, K. Kobayashi, N. Ishiki, H. Kageshima, *Surf. Sci. Rep.* **1991**, *13*, 265.

[18] (a) W. S. Sacks, C. Noguera, *Phys. Rev. B* **1991**, *43*, 11612. (b) W. S. Sacks, C. Noguera, *J. Vac. Sci. Technol. B* **1991**, *9*, 488.

[19] R. G. Gordon, Y. S. Kim, *J. Chem. Phys.* **1972**, *56*, 3122.

[20] R. C. Barrett, J. Nogami, C. F. Quate, *Appl. Phys. Lett.* **1990**, *57*, 992.

[21] (a) N. F. Mott, *Metal-Insulator Transitions*, Barnes and Noble, New York, 1977. (b) B. H. Brandow, *Adv. Phys.* **1977**, *26*, 651. (c) M.-H. Whangbo, *J. Chem. Phys.* **1979**, *70*, 4963; **1980**, *73*, 3854.

[22] M.-H. Whangbo, J. Ren, S. N. Magonov, H. Bengel, B. A. Parkinson, A. Suna, *Surf. Sci.* **1995**, *326*, 311.

[23] (a) D. Tománek, S. G. Louie, H. J. Mamin, D. W. Abraham, R. E. Thompson, E. Granz, J. Clarke, *Phys. Rev. B* **1987**, *35*, 7790. (b) D. Tománek, S. G. Louie, *Phys. Rev. B* **1988**, *37*, 8327.

[24] T. A. Albright, J. K. Burdett, M.-H. Whangbo, *Orbital Interactions in Chemistry*, Wiley, New York, **1985**.

6 STM and AFM Images of Layered Inorganic Compounds

Many-layered inorganic materials possess flat surfaces, are stable in air, and exhibit high electrical conductivity. Therefore, they are suitable for ambient-condition STM and AFM studies. From the atomic-scale images routinely observed on the surfaces of these compounds, one can examine several fundamental problems of STM and AFM imaging, e.g., the origin of the images and the role of the tip–sample interactions. To interpret the atomic-scale features of the observed images, it is practical to begin with density plot calculations on the basis of the known bulk crystal structures. For this approach to succeed, it is necessary that

(a) the electronic state of the sample is a symmetry-adapted one which can be approximated by electronic band structure calculations,
(b) the sample has no point defects leading to localized states near the Fermi level, and
(c) the tip force induced surface corrugation is not significant.

In this chapter, the examples fulfilling all three conditions are discussed first. This is followed by the analysis of an example for which condition (a) is not met. Other cases requiring the use of an approach going beyond mere density plot calculations are discussed in Chapters 7 and 9.

6.1 Layers from MX_6 Trigonal Prisms and Octahedra

Many-layered transition-metal compounds can be regarded as made up of MX_6 trigonal prisms (Fig. 6.1) or MX_6 octahedra (Fig. 6.2), where M is a transition metal and X is a ligand atom (chalcogen or halogen). The X–M–X sandwich layers of

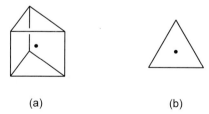

(a)   (b)

Figure 6.1 Views of an MX_6 trigonal prism: (a) perspective view; (b) projection view along the three-fold rotational axis.

84   6 STM and AFM Images of Layered Inorganic Compounds

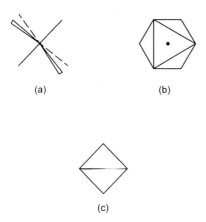

Figure 6.2 Views of an MX$_6$ octahedron: (a) perspective view; (b) projection view along one of the three-fold rotational axes; (c) projection view along one of the two-fold rotational axes.

composition MX$_2$ are obtained by sharing MX$_6$ trigonal prisms (Fig. 6.3 (a)) or by sharing MX$_6$ octahedra (Fig. 6.3 (b)). The MX$_2$ layers made up of MX$_6$ trigonal prisms may be represented by 2H-MX$_2$, and those made up of MX$_6$ octahedra by 1T-MX$_2$. In the undistorted MX$_2$ layers constructed from ideal MX$_6$ trigonal prisms or octahedra, both the metal and the ligand sheets have a hexagonal arrangement of atoms. The X–M–X sandwich layers of composition M$_3$X$_8$ are obtained by sharing MX$_6$ octahedra to form a hexagonal arrangement of M$_3$ cluster units in the metal sheet (Fig. 6.3 (c)), while those of composition MX$_3$ are obtained by sharing MX$_6$ octahedra to form a honeycomb pattern of M atoms (Fig. 6.3 (d)). In another layered structure to be discussed (Fig. 6.3 (e)), the MX$_6$ octahedra share their edges in one direction and corners along the perpendicular direction. This layer does not have an X–M–X sandwich structure. The various layers described above stack to form a three-dimensional structure with VDW interactions between them. The layers on the surface are practically identical in structure with those in the bulk. These compounds are easily cleaved along the layer plane to give clean and flat surfaces for studies with scanning probe techniques.

In the X–M–X sandwich layers mentioned above, the surface ligand-atom sheets possess an atomic-scale surface corrugation when there occurs a metal clustering in the metal-atom sheets. The metal atoms of an undistorted 1T-MX$_2$ layer form a hexagonal lattice (Fig. 6.4 (a)), which is found for a system containing d^0 metal ions. 1T-MX$_2$ phases with d^1 to d^3 exhibit various patterns of metal atom clustering, which originates from the formation of metal–metal bonding through the shared octahedral edges [1, 2]. For example, the 1T-MX$_2$ compounds with d^1 ions have $\sqrt{13} \times \sqrt{13}$ clusters (Fig. 6.4 (b)). The d-electron count of the 1T-MTe$_2$ (M = V, Nb, Ta) systems is formally d^1 but is actually close to d$^{4/3}$ because of a partial chalcogen-to-metal electron transfer [1], and their metal atoms form "ribbon-

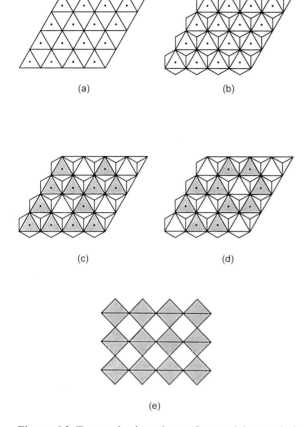

Figure 6.3 Top projection views of several layers derived from MX_6 trigonal prisms and octahedra: (a) 2H-MX_2 layer; (b) 1T-MX_2 layer; (c) M_3X_8 layer; (d) edge-sharing MX_3 layer; (e) edge- and corner-sharing MX_3 layer.

chains" (Fig. 6.4 (c)). The 1T-MX_2 phases with d^2 metal ions exhibit either "zigzag-chains" (Fig. 6.4 (d)) or $\sqrt{3} \times \sqrt{3}$ clusters (Fig. 6.4 (e)) [2], and those with d^3 metal ions show "diamond-chains" (Fig. 6.4 (f)). The metal-atom clustering raises the heights of the ligand atoms lying directly above the resulting metal clusters, thereby leading to several different ligand atoms of unequal z-heights on the surface (under the usual convention of having the layers in the xy-plane). It is noted that the metal–atom clustering patterns described above are often referred to as charge density waves (CDWs).

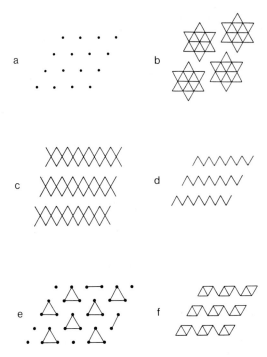

Figure 6.4 Metal atom arrangements in 1T-MX$_2$ layers: (a) hexagonal lattice of d^0 1T-MX$_2$ layer; (b) $\sqrt{13} \times \sqrt{13}$ clusters of d^1 1T-MX$_2$ layer; (c) ribbon chains of d$^{4/3}$ 1T-MX$_2$ layer; (d) zigzag chains of d^2 1T-MX$_2$ layer; (e) $\sqrt{3} \times \sqrt{3}$ clusters of d^2 1T-MX$_2$ layer; (f) diamond chains of d^3 1T-MX$_2$ layer.

6.2 Images of Layered Compounds

6.2.1 2H-MoS$_2$

The layered compound 2H-MoS$_2$ consists of 2H-MoS$_2$ layers (Fig. 6.3 (a)), both the metal- and S-atom sheets have a hexagonal arrangement (Fig. 6.4 (a)), and all the atoms of the surface sulfur sheet have the same z-height. The AFM and STM images of 2H-MoS$_2$ are shown in Figs. 6.5 (a) and 6.5 (b), respectively. Both images exhibit a hexagonal pattern in agreement with their layer structures. The $\rho(r_0)$ plot calculated for a single 2H-MoS$_2$ layer (Fig. 6.5 (c)) shows that the bright spots of the AFM image represent the surface sulfur atoms.

The Mo^{4+} ion of 2H-MoS$_2$ has the d-electron count d^2, and the lowest-lying d-block level of an isolated MoS$_6^{8-}$ trigonal prism (i.e., (Mo^{4+})(S^{2-})$_6$) is mainly represented by the d$_{z^2}$ orbital. Thus, in the earlier interpretation of the STM image of 2H-MoS$_2$, it was thought that the bright spots of the STM image are associated with the d$_{z^2}$-orbitals of Mo [3]. However, as shown in Fig. 6.5 (d), the $\rho(r_0, e_f)$ plot calculated for a single 2H-MoS$_2$ layer is dominated by the contribution of the surface sulfur atoms, not by the Mo d$_{z^2}$-orbitals [4]. Although the d-block bands of a single 2H-MoS$_2$ layer have small contributions from the sulfur orbitals, the Mo atom sheet is far removed from the surface sulfur atom sheet (by 1.59 Å). In general, the amplitude of an atomic orbital decreases exponentially with increasing distance from its center. As a consequence, the high electron density (HED) spots of the

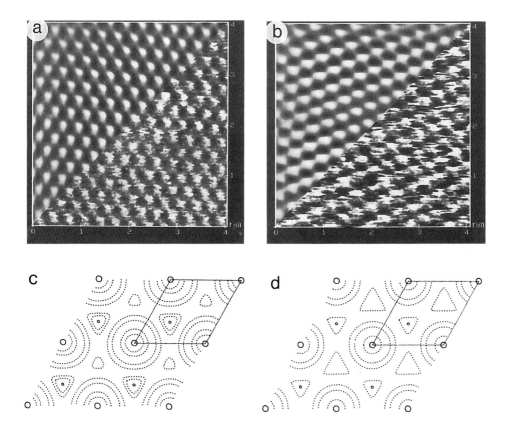

Figure 6.5 (a) AFM height image of 2H-MoS$_2$. The contrast variations cover the 0–3 Å range. (b) STM height image of 2H-MoS$_2$ (I_{set} = 2 nA, V_{bias} = 100 mV). The contrast covers height variations in the 0–2 Å range. (c) $\rho(r_0)$ plot calculated for a single 2H-MoS$_2$ layer, where the contour values used are 20×10^{-2}, 10×10^{-2}, 5×10^{-2}, 1×10^{-2}, and 0.5×10^{-2} electrons/au^3. (d) $\rho(r_0, e_f)$ plot calculated for the bottom of the conduction band of a single 2H-MoS$_2$ layer, where the contour values used are 20×10^{-4}, 10×10^{-4}, 5×10^{-4}, 1×10^{-4}, and 0.5×10^{-4} electrons/au^3. The upper left-hand parts of the images in (a) and (b) are shown after FFT filtering. In (c) and (d), the surface S and the Mo atoms are represented by large and small circles, respectively.

$p(r_0, e_f)$ plot are centered on the surface S atoms, with which the bright spots of the STM image are also associated. Thus, the surface S atoms are responsible for the bright spots of both the STM and the AFM images of 2H-MoS$_2$.

6.2.2 MoOCl$_2$

The MoOCl$_2$ layer is constructed from MoCl$_4$O$_2$ octahedra (with the oxygen atoms at the *trans* corners) by sharing the O corners in one direction and the Cl–Cl edges along the perpendicular direction [5], as shown in Fig. 6.3 (e). The arrangement of these layers in MoOCl$_2$ is presented in the perspective view in Fig. 6.6. In each MoOCl$_2$ layer, the Mo–Mo bonds are formed across the shared Cl–Cl edge to give rise to an Mo–Mo⋯Mo bond alternation along the crystallographic *c*-direction. However, there is no bond alternation along the Mo–O–Mo direction (i.e., the *b*-direction). Due to the Mo–Mo⋯Mo bond alternation, there are two kinds of Cl atoms on the MoOCl$_2$ layer (contained in the *bc*-plane). The Cl atoms above the Mo–Mo bonds lie 0.3 Å higher than those above the Mo⋯Mo linkage. The electrical resistivity measurements on single-crystal samples of MoOCl$_2$ show a metallic conductivity down to 4.2 K [6].

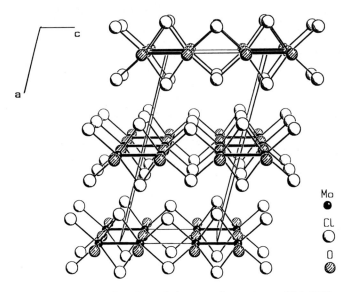

Figure 6.6 Perspective view of the crystal structure of MoOCl$_2$.

Figure 6.7 (a) presents the $p(r_0)$ plot calculated for the *bc*-plane surface of an MoOCl$_2$ layer. The HED spots are centered on the surface Cl atoms, and the higher-lying Cl atoms have the higher density. Figure 6.7 (b) shows the AFM image measured for the bc-plane surface, which consists of rows of bright and less bright spots as in the case of the $\rho(r_0)$ plot. The $\rho(r_0, e_f)$ plot of the bc-plane surface (Fig. 6.7 (c)) is strikingly different from the $\rho(r_0)$ plot. The HED spot on each Cl atom is not spherical but represents in-plane 3p-orbital density whose axis is perpendicular to the Mo–Mo⋯Mo direction. The higher-lying Cl atom has a stronger contribution to the $\rho(r_0, e_f)$ plot. These aspects are indeed found in the STM image of the bc-plane surface (Fig. 6.7 (d)).

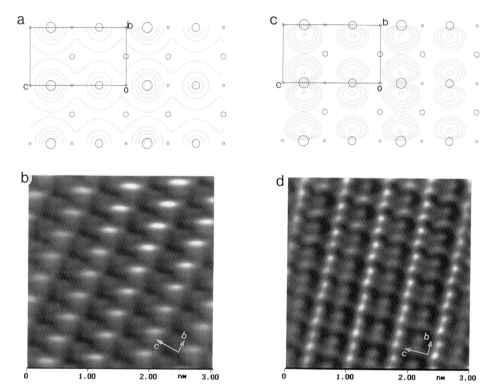

Figure 6.7 (a) $\rho(r_0)$ plot calculated for a single MoOCl$_2$ layer, where the contour values used are 15×10^{-2}, 10×10^{-2}, 5×10^{-2}, and 0.5×10^{-2} electrons/au^3. (b) AFM height image of MoOCl$_2$. The contrast covers height variations in the 0–0.5 nm range. (c) $\rho(r_0, e_f)$ plot calculated for a single MoOCl$_2$ layer, where the contour values used are 10×10^{-4}, 4×10^{-4}, 2×10^{-4}, and 0.5×10^{-4} electrons/au^3. (d) STM current image of MoOCl$_2$ ($I_{set} = 5$ nA, $V_{bias} = 25$ mV). The contrast covers current variations in relative units.

6.2.3 WTe$_2$

Due to the zigzag-chain formation in 1T-WTe$_2$ (hereafter referred to as WTe$_2$) (Fig. 6.4 (d)), there occur two W and two Te atoms with different z-heights [7]. Figure 6.8 (a) presents a top projection view (in the *ab*-plane) of a single WTe$_2$ layer. Figure 6.8 (b) shows a side projection view (in the *ac*-plane) of two adjacent WTe$_2$ layers along the zigzag chains, which run parallel to the *b*-direction. The Te(2) atom lies higher than the Te(1) atom by 0.62 Å. The atomic-scale AFM image of WTe$_2$ in Fig. 6.9 (a) is characterized by two kinds of rows consisting of bright spots with different contrast. The assignment of the brighter spots to the more protruding Te(2) atoms is supported by the $\rho(r_0)$ plot calculated for a single WTe$_2$ layer (Fig. 6.9 (b)). A similar pattern is observed in the STM images of this compound [8]. The STM image of Fig. 6.10 (a) exhibits two different rows of bright spots. The $\rho(r_0, e_f)$ plot calculated for a single WTe$_2$ layer (Fig. 6.10 (b)) reveals that the Te(1) and Te(2) atoms possess nearly identical density values despite their height difference of 0.62 Å. Nevertheless, the density pattern around the higher-lying Te(2) atoms is broader than that around the lower-lying Te(1) atoms. Thus, the brighter rows of the STM image are assigned to the Te(2) atoms [9].

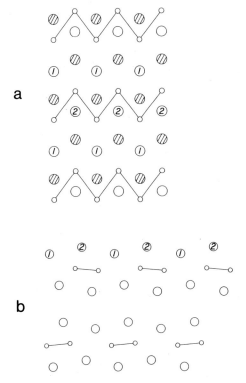

Figure 6.8 (a) Top projection view of a single WTe$_2$ layer, where the small and large circles represent W and Te atoms, respectively. The Te atoms of the lower surface are indicated by hatching, and the metal–metal bonds are connected by solid lines. (b) Side projection view of two adjacent WTe$_2$ layers along the zigzag chains. The numbers 1 and 2 represent the Te(1) and Te(2) atoms, respectively.

6.2 *Images of Layered Compounds* 91

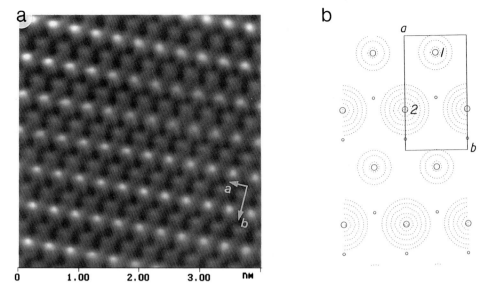

Figure 6.9 (a) AFM height image of WTe$_2$. The contrast covers height variations in the 0–0.5 nm range. (b) $\rho(r_0)$ plot calculated for a single WTe$_2$ layer. The contour lines correspond to 70×10^{-3}, 50×10^{-3}, and 30×10^{-3} electrons/au^3.

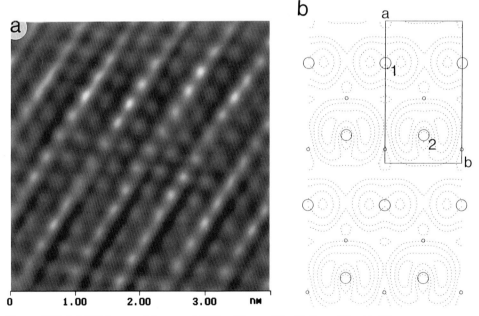

Figure 6.10 (a) STM current image of WTe$_2$ ($V_{\text{bias}} = 38$ mV, $I_{\text{set}} = 30$ nA). The contrast covers current variations in relative units; (b) $\rho(r_0, e_f)$ plot calculated for a single WTe$_2$ layer. The contour lines correspond to 6×10^{-4}, 3×10^{-4}, 2×10^{-4}, 1×10^{-4}, and 0.5×10^{-4} electrons/au^3.

6.2.4 NbTe$_2$

Due to the formation of ribbon-chains in 1T-NbTe$_2$ (hereafter referred to as NbTe$_2$) (Fig. 6.4 (c)), each NbTe$_2$ layer has two non-equivalent Nb atoms and three non-equivalent Te atoms [10]. Figure 6.11 (a) shows a top projection view (in the ab-plane) of a single NbTe$_2$ layer. Figure 6.11 (b) shows a side projection view (in the ac-plane) of two adjacent NbTe$_2$ layers along the ribbon-chains, which run parallel to the b-direction. The Te(3) atoms protrude most on the surface, and they lie higher than the Te(2) and Te(1) atoms by 0.585 and 0.082 Å, respectively.

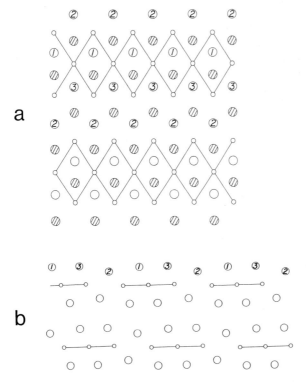

Figure 6.11 (a) Top projection view of a single NbTe$_2$ layer, where the small and large circles represent Nb and Te atoms, respectively. The Te atoms of the lower surface are indicated by shading, and the metal–metal bonds are connected by solid lines; (b) Side projection view of two adjacent NbTe$_2$ layers along the ribbon chains. The numbers 1, 2 and 3 represent the Te(1), Te(2), and Te(3) atoms, respectively.

The atomic-scale AFM image of NbTe$_2$ shown in Fig. 6.12 (a) is characterized by rows of bright spots, and there are three kinds of rows with different contrasts. The $\rho(r_0)$ plot calculated for a single NbTe$_2$ layer is given in Fig. 6.12 (b). As expected, the HED spots of this plot are represented by the most protruding atoms, i.e. Te(3),

6.2 Images of Layered Compounds   93

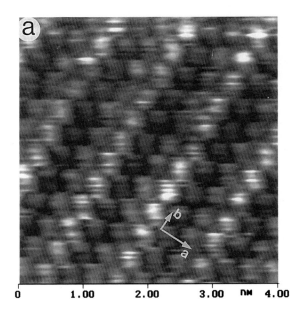

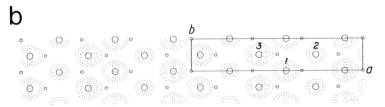

Figure 6.12 (a) AFM height image of NbTe$_2$. The contrast covers height variations in the 0–0.5 nm range; (b) $\rho(r_0)$ plot calculated for a single NbTe$_2$ layer. The contour lines correspond to 61×10^{-3}, 45×10^{-3}, and 30×10^{-3} electrons/au^3.

and the brightest rows of the AFM images are assigned to these atoms. The STM images of NbTe$_2$ (Figs. 6.13 (a) and 6.13 (b)) are also characterized by three kinds of rows with different contrast. However, the $\rho(r_0, e_f)$ plot calculated for a single NbTe$_2$ layer (Fig. 6.13 (c)) shows that the HED spots belong to the least protruding atoms Te(2), i.e., the latter atoms are seen as the brightest spots of the STM image.

To examine why the HED spots of the $\rho(r_0, e_f)$ plot are located at the least protruding atoms Te(2), it is necessary to calculate the electronic DOS for a single NbTe$_2$ layer. Figure 6.14 (a) shows the projected DOS (PDOS) plots calculated for the 4d-orbitals of the Nb atoms and the 5p-orbitals of the Te atoms. In the vicinity of the Fermi level, the Nb contribution is slightly larger than the Te contribution. The PDOS plots calculated for the 5p$_z$-orbitals of the Te(1), Te(2) and Te(3) atoms (Fig. 6.14 (b)) shows that around the Fermi level the Te(2) contribution is largest.

94   6 STM and AFM Images of Layered Inorganic Compounds

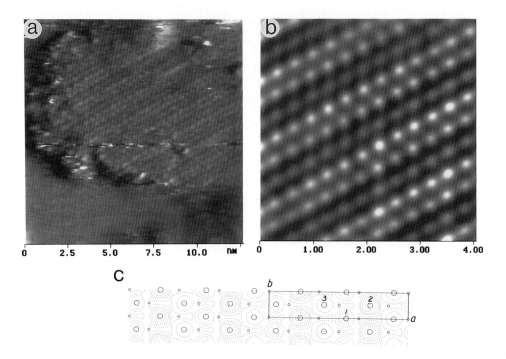

Figure 6.13 (a) STM current image of NbTe$_2$ (V_{bias} = −400 mV, I_{set} = 1 nA). The contrast covers current variations in relative units. (b) Zoomed-in part of the image in (a) after FFT filtering; (c) $\rho(r_0, e_f)$ plot calculated for a single NbTe$_2$ layer. The contour lines correspond to 40 × 10^{-4}, 20 × 10^{-4}, 10 × 10^{-4}, 5 × 10^{-4}, and 1 × 10^{-4} electrons/au^3.

This, combined with the diffuseness of the Te 5p$_z$-orbital, is responsible for the domination of the $\rho(r_0, e_f)$ plot by the electron densities of the least protruding atoms Te(2) [9a]. Similarly, the dominant contribution of the least protruding atoms to the STM pattern was found in the analysis of the STM images of another layered compound with corrugated surface, ReSe$_2$ [9b].

6.2.5 β-Nb$_3$I$_8$

Nb$_3$X$_8$ (X = Cl, Br, I) is made up of identical Nb$_3$X$_8$ layers (Fig. 6.3 (c)) [11]. In the Nb-atom sheet of each Nb$_3$X$_8$ layer, the Nb atoms form triangular Nb$_3$ clusters, so that each Nb atom is in a distorted octahedral environment. In each Nb$_3$X$_8$ layer, the top and bottom sheets of X atoms are not equivalent. On the bottom halogen sheet containing the X(2) and X(4) atoms in Fig. 6.15 (lower surface, A), the X(4) atoms lie farther away from the Nb-atom sheet than the X(2) atoms (by 0.60, 0.55, and 0.50 Å for X = I, Br, and Cl, respectively). On the top halogen sheet containing

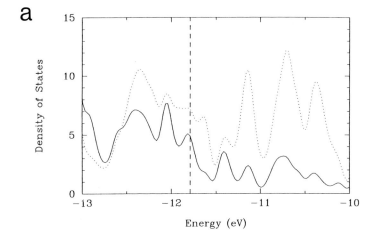

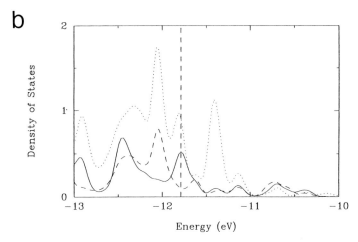

Figure 6.14 (a) PDOS plots for the Nb 4d orbitals (dotted line) and the Te 5p orbitals (solid line), where the vertical broken line refers to the Fermi level. (b) PDOS plots for the $5p_z$ orbitals of the Te(1) (solid line), Te(2) (dotted line) and Te(3) (broken line) atoms.

the X(1) and X(3) atoms in Fig. 6.15 (upper surface, B), the X(1) atoms lie farther away from the Nb-atom sheet than the X(3) atoms (by 0.44, 0.33, and 0.35 Å for X = I, Br, and Cl, respectively). In principle, the surfaces of Nb_3X_8 (X = Cl, Br, I) samples can be either surface A or surface B.

Figure 6.16 shows the atomic-resolution AFM image obtained for β-Nb_3I_8 [12]. This image exhibits the unit cell pattern consisting of one bright spot and three less bright ones, which resembles the atomic arrangement of surface A. The most representative STM images, recorded on the same crystal surface of β-Nb_3I_8, are shown in Fig. 6.17. STM images with one big bright spot per unit cell (Fig. 6.17 (a)) are ob-

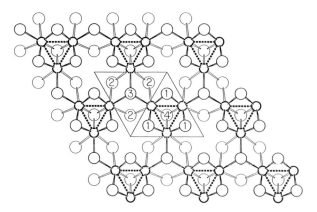

Figure 6.15 Schematic projection view, along the crystallographic c-axis direction, of a single Nb_3X_8 layer. The Nb and X atoms are represented by small and large circles, respectively. The Nb_3 clusters in the Nb-atom sheet are shown by connecting the Nb atoms with broken lines.

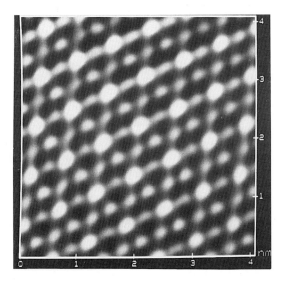

Figure 6.16 AFM height image of β-Nb_3I_8 crystal. The contrast covers height corrugations in the 0–0.5 nm range.

tained frequently. By changing the tunneling parameters to a smaller gap resistance, a more resolved STM image with four spots per unit cell is obtained (Fig. 6.17 (b)). Zooming and filtering part of such an image (Fig. 6.17 (c)) reveals three brighter triangular spots and a less pronounced one per unit cell [12].

Figure 6.18 (a) shows the dispersion relations of the energy bands calculated for a single Nb_3I_8 layer (nine t_{2g} block bands lying in the vicinity of the Fermi level) [12]. With the oxidation state I^-, there are seven electrons to fill the bottom four

Figure 6.17 STM current images of β-Nb_3I_8: (a) V_{bias} = 1000 mV, I_{set} = 10 nA; (b) V_{bias} = 180 mV, I_{set} = 0.75 nA; (c) zoomed-in part of the image in (b) after FFT filtering. The contrast covers current variations in relative units.

bands of Fig. 6.18. The highest occupied band, which is somewhat narrow, is half-filled. The $\rho(r_0)$ and $\rho(r_0, e_f)$ plots of a single Nb_3I_8 layer calculated for surface A are shown in Figs. 6.19 (a) and 6.19 (b), respectively. Within the unit cell of the $\rho(r_0)$ plot, one atom has a higher density than the other three atoms, in agreement with the AFM image of Fig. 6.16. The $\rho(r_0, e_f)$ plot, obtained by sampling all levels of the half-filled band of Fig. 6.18, is consistent with the STM pattern of Fig. 6.17 (c). In each unit cell of this plot, the three I(2) atoms have a greater density than do the I(4) atom, although the I(2) atoms lie farther away from the tip by 0.60 Å. As shown by the DOS analysis of Fig. 6.19 (c), this reflects the fact that the highest occupied band has a stronger contribution from the I(2) atoms than from

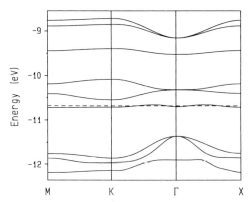

Figure 6.18 Dispersion relations of the nine bands, in the vicinity of the Fermi level (shown by the broken line), calculated for a single Nb_3I_8 layer of β-Nb_3I_8.

the I(4) atoms. In addition, in the $\rho(r_0, e_f)$ plot, the density peaks of the three I(2) atoms within a unit cell are closer to one another than expected on the basis of the crystal structure. This effect is caused by the hybridization of the I(2) atom p-orbitals in the half-filled band, and it explains the observation of triangular bright spots in the STM image in Fig. 6.17 (b). When these spots are not resolved, they appear as a big bright spot as shown in Fig. 6.17 (a).

6.2.6 1T-TaSe$_2$

The 1T-TaX$_2$ (X = S, Se) systems containing d^1 ions exhibit a $\sqrt{13} \times \sqrt{13}$ metal-atom clustering (Fig. 6.4 (b)) [13]. In a $\sqrt{13} \times \sqrt{13}$ cluster of 1T-TaX$_2$ (X = S, Se) (Fig. 6.20), the Ta(1) and Ta(2) atoms form a centered hexagon (i.e., a seven-atom cluster) and the Ta(3) atoms cap the edges of the hexagon. The clustering in the metal-atom sheets induces a corrugation in the chalcogen-atom sheets. In the crystal structure of 1T-TaSe$_2$, the Se(1) atoms are farthest away from the metal-atom sheet (at a distance of 1.78 Å). With respect to the Se(1) atoms, the Se(2), Se(3), Se(4), and Se(5) atoms are closer to the metal-atom sheet by 0.05, 0.28, 0.26, and 0.25 Å, respectively. Thus, the height corrugation in the Se-atom surface is less than 0.3 Å.

Figure 6.21 shows the dispersion relations of the bottom portion of the t_{2g}-block bands calculated for a single 1T-TaSe$_2$ layer with the $\sqrt{13} \times \sqrt{13}$ modulation. With 13 d-electrons per unit cell, the highest occupied band is half-filled [1, 14]. The highest occupied band is largely responsible for the $\rho(r_0, e_f)$ plot and hence the STM image. The PDOS plots calculated for the three different Ta atoms of 1T-TaSe$_2$ are given in Fig. 6.22 (a), and those for five different Se atoms in Fig. 6.22 (b) [14]. Fig. 6.22 (c) and 6.22 (d) compare the PDOS plots of the Se(1) and Se(2) atoms (solid line) with their $4p_z$-orbital contributions (broken line). Important observations to note from Figs. 6.22 (a)–6.22 (d) are:

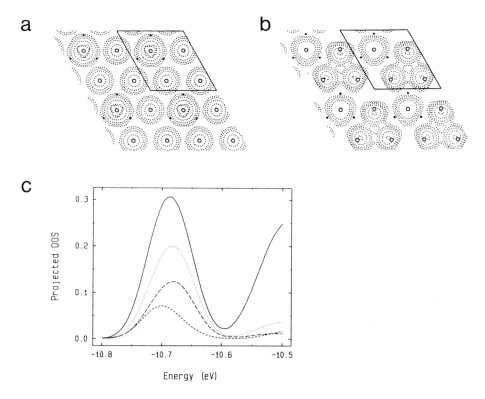

Figure 6.19 (a) $p(r_0)$ and (b) $p(r_0, e_f)$ plots calculated for surface A of a single Nb_3I_8 layer of β-Nb_3I_8. The plot area consists of four unit cells, and a unit cell is indicated by the rhombus. For clarity, the I atoms on the opposite surface are not shown. The Nb and I atoms are shown by small and large circles, respectively. The contour values used are 85×10^{-3}, 50×10^{-3}, 20×10^{-3}, 10×10^{-3}, and 5×10^{-3} electrons/au^3 in (a), and 100×10^{-5}, 50×10^{-5}, 20×10^{-5}, 10×10^{-5}, 5×10^{-5}, and 2×10^{-5} electrons/au^3 in (b). (c) PDOS values of the iodine atoms of surface A, in the energy region of the half-filled band, calculated for a single Nb_3I_8 layer of β-Nb_3I_8. The curves represent the p_z orbital of I(2) ($\cdots$), the p_x and p_y orbitals of I(2) (——), the p_z orbital of I(4) (----), and the p_x and p_y orbitals of I(4) (-- --).

(a) the highest occupied band has a larger contribution from the Ta atoms than from the Se atoms,
(b) the per-atom Ta contribution to the highest occupied band decreases in the order Ta(1) > Ta(2) > Ta(3),
(c) the per-atom Se contribution to the highest occupied band decreases in the order Se(1) > Se(2) ≫ Se(4), Se(3), Se(5), and
(d) the contributions of the Se(1) and Se(2) atoms to the highest occupied band are given almost exclusively by their $4p_z$-orbitals.

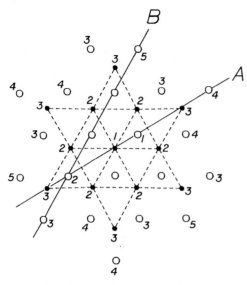

Figure 6.20 Schematic representation of the three different Ta atoms and five different Se atoms of a $\sqrt{13} \times \sqrt{13}$ cluster in 1T-TaSe$_2$. The Ta and Se atoms are presented by small filled and large empty circles, respectively. The Se atoms of the bottom Se-atom sheet are not shown for clarity. The lines A and B refer to the planes perpendicular to the layer of 1T-TaSe$_2$.

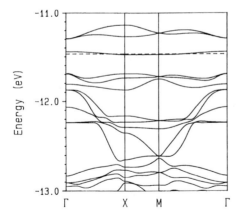

Figure 6.21 Dispersion relations of the bottom portion of the t_{2g}-block bands calculated for a single 1T-TaSe$_2$ layer with the $\sqrt{13} \times \sqrt{13}$ modulation. The broken line refers to the Fermi level.

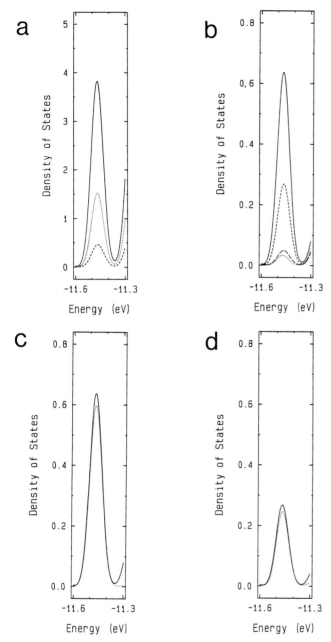

Figure 6.22 (a) PDOS values per atom calculated for the Ta(1) (——), Ta(2) (····), and Ta(3) (----) atoms of 1T-TaSe$_2$. (b) PDOS values per atom calculated for the Se(1) (——), Se(2) (----), Se(4) (– – –), and Se(5) (····) atoms of 1T-TaSe$_2$. The PDOS values per atom for Se(3) are practically identical with those for Se(5), and hence are not shown. (c) PDOS values per atom of Se(1) (——) and its 4p$_z$ orbital contribution (····). (d) PDOS values per atom of Se(2) (——) and its 4p$_z$ orbital contribution (····).

The $\rho(r_0)$ plot of a single 1T-TaSe$_2$ layer has density distributions exclusively on the surface Se atoms (Fig. 6.23 (a)). In each unit cell of the $\rho(r_0)$ plot, the electron density of the Se atom decreases in the order Se(1) ≥ Se(2) > Se(3), Se(4), Se(5). This is in agreement with the finding that the distances of the surface Se atoms to the tip increase in the order Se(1) > Se(2) ≫ Se(3), Se(4), Se(5). Consequently, the $\rho(r_0)$ plot

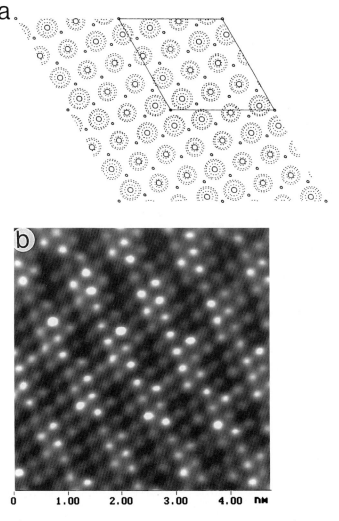

Figure 6.23 (a) $\rho(r_0)$ plot calculated for a single layer of 1T-TaSe$_2$. The plot area consists of four unit cells, and a unit cell is indicated by the rhombus. The contour values used are 10×10^{-2}, 5×10^{-2} and 3×10^{-2} electrons/au^3. For clarity, the Se atoms on the bottom surface are not shown. The Ta and Se atoms are shown by small and large circles, respectively. (b) AFM height image of 1T-TaSe$_2$. The contrast covers height variations in the 0–1 nm range.

of Fig. 6.23 (a) predicts that the bright spots of the AFM image are caused by the surface Se atoms, and that the AFM images should exhibit a pattern of six-chalcogen-atom triangles at the center of each $\sqrt{13} \times \sqrt{13}$ cluster unit. This is consistent with the AFM image of 1T-TaSe$_2$ in Fig. 6.23 (b).

The $\rho(r_0, e_f)$ plot of a single 1T-TaSe$_2$ layer presented in Fig. 6.24 (a) was obtained by sampling all levels of the highest occupied band of Fig. 6.21 [14]. The $\rho(r_0, e_f)$ plot shows that the electron density distribution has no contribution from the Ta atoms, and arises exclusively from the surface Se(1) and Se(2) atoms. Practi-

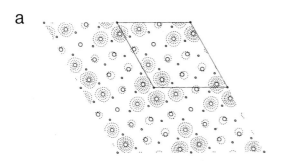

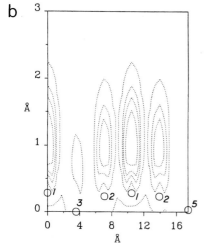

Figure 6.24 (a) $\rho(r_0, e_f)$ plots calculated for a single layer of 1T-TaSe$_2$. The plot area consists of four unit cells, and a unit cell is indicated by the rhombus. The contour values used are 25×10^{-4}, 20×10^{-4}, 10×10^{-4}, 5×10^{-4}, and 1×10^{-4} electrons/au^3. For clarity, the Se atoms on the bottom surface are not shown. The Ta and Se atoms are shown by small and large circles, respectively. (b) Cross-sectional view of the partial electron density distribution of a single 1T-TaSe$_2$ layer (resulting from the band levels at the Fermi level) in the plane B, perpendicular to the layer, defined in Fig. 6.20. The Se atoms are represented by empty circles. This cross-section contains the Se(1), Se(2), Se(3), and Se(5) atoms of a $\sqrt{13} \times \sqrt{13}$ cluster. The contour values used are 20×10^{-4}, 10×10^{-4}, 5×10^{-4}, and 1×10^{-4} electrons/au^3.

cally, there is no contribution from the Se(5) atoms, and the electron density decreases in the order Se(1) > Se(2) ≫ Se(3), Se(4). This is consistent with observation (c) and also with the fact that the Ta atoms are much farther away from the tip than the surface Se atoms. The peaks of the $\rho(r_0, e_f)$ plot for the Se(1) and Se(2) atoms (Fig. 6.24 (a)) are centered at the atomic positions; this arises from observation (d). It is clear from Fig. 6.24 (a) that the bright spots of the STM image are caused mainly by the surface Se(1) and Se(2) atoms, and that the STM image should exhibit a pattern of six-chalcogen-atom triangles for each $\sqrt{13} \times \sqrt{13}$ cluster unit, with the inner triangle of the Se(1) atoms much brighter than the outer triangle of the Se(2) atoms. Indeed, such a pattern is found in high-resolution STM images of 1T-TaSe$_2$ (Fig. 6.25).

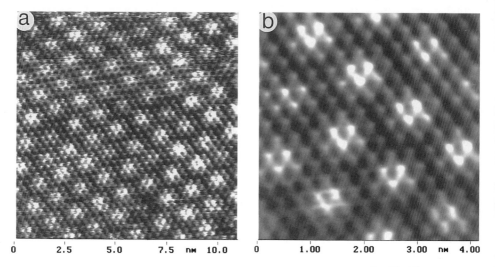

Figure 6.25 (a) STM current image of 1T-TaS$_2$ (I_{set} = 6 nA, V_{bias} = −10 mV). The contrast covers current variations in relative units. (b) Zoomed-in part of the image in (a) after FFT filtering.

The corrugation in the $\rho(r_0, e_f)$ plot is more pronounced than that in the $\rho(r_0)$ plot. This is consistent with the finding that the amplitude of the height modulation in the STM image is much stronger than that in the AFM image. Fig. 6.24 (b) shows the cross-sectional view of the partial electron density $\rho(r, e_f)$ in the plane B perpendicular to the layer (Fig. 6.20). This cross-section contains the Se(1), Se(2), Se(3), and Se(5) atoms of a $\sqrt{13} \times \sqrt{13}$ cluster. It is evident from Fig. 6.24 that the $\rho(r_0, e_f)$ plots are dominated by the 4p$_z$ orbitals of the surface Se(1) and Se(2) atoms. The STM height image is simulated by the contour plot of constant electron density in Fig. 6.24 (b). Compared with the Se(1) and Se(2) positions, the electron densities at the Se(3), Se(4), and Se(5) positions are very small. The z-height difference in the contour lines can easily reach 2 Å, which is consistent with the anomalously large surface corrugation deduced from constant-current STM measurements.

The bright spots of both the STM and AFM images of 1T-TaX$_2$ (X = S, Se) are associated with the surface chalcogen atoms. Certainly, the $\sqrt{13} \times \sqrt{13}$ clustering of 1T-TaX$_2$ (X = S, Se) are induced by the d-electrons of Ta, and their highest occupied bands are dominated by the Ta atoms [1, 14]. However, the surface chalcogen atoms determine the STM pattern because of their proximity to the scanning tip. It is noted that in the earlier studies of 1T-TaX$_2$ (S = S, Se), the metal d-orbitals were thought to dominate the STM pattern [15]. The large difference in the height amplitudes of the STM and AFM images simply reflects the difference in the density distributions of the $\rho(r_0)$ and $\rho(r_0, e_f)$ plots. Observations (c) and (d), together with the fact that the surface Se(1) and Se(2) atoms are closer to the scanning tip than any other atoms by about 0.25 Å, explain why the amplitude of the CDW modulation is so large in the STM image.

6.3 Charge Density Waves of MC$_8$ (M = K, Rb, Cs)

In interpreting the STM and AFM images of the layered compounds discussed in the previous section, the density plot calculations were satisfactory. As already mentioned, this approach is not always sufficient. For example, the superstructure STM patterns observed for the graphite intercalation compounds (GICs) MC$_8$ (M = K, Rb, Cs) are not reproduced by partial density plot calculations. The analysis given below indicates that the π-band electrons of the surface graphite monolayer of MC$_8$ are not in a metallic but in a CDW state.

6.3.1 Observations

GICs are termed stage-n compounds when every nth VDW gap is intercalated. Thus, stage-1 and stage-2 GICs consist of graphite monolayers and bilayers, respectively. In stage-1 GICs MC$_8$ (M = K, Rb, Cs), there are two kinds of carbon atoms in the surface graphite layer, i.e., those of the carbon rings lying above the K$^+$ ions and those joining such carbon rings (Fig. 6.26) [16]. The STM images of stage-1 GICs MC$_8$ (M = K,

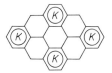

Figure 6.26 Schematic representation of the 2 × 2 arrangement of the K$^+$ ions in KC$_8$.

Rb, Cs) commonly exhibit a 2 × 2 superstructure superposed on a three-for-hexagon pattern, as shown in Fig. 6.27 [17, 18]. (The $\sqrt{3} \times 4$ and $\sqrt{3} \times \sqrt{13}$ superstructures are also found for stage-1 GICs RbC_8 and CsC_8 [17a,18c].) However, the STM images of stage-2 potassium GIC, KC_{24}, show the same three-for-hexagon pattern as is found for highly ordered pyrolytic graphite (HOPG) and do not exhibit a 2 × 2 superstructure [17c]. Consequently, the STM images of MC_8 with the 2 × 2 superstructure reflect the electron density distribution of its surface graphite monolayer. This implies that the surface graphite monolayer of MC_8 is in a certain CDW state [17b, c]. Angle-resolved photoelectron spectroscopy (ARPES) studies of KC_8 [19] and CsC_8 [20] reveal that the π-band as well as nondispersive states are present near e_f. It has been suggested [17, 19, 20] that the nondispersive states may originate from the "interlayer" state of bulk KC_8 found in some electronic structure calculations [21], and that the surface CDW state observed by STM may be associated with the nondispersive states [17b].

In an attempt to analyze the image of Fig. 6.27, the electronic structure of the surface of KC_8 was modeled by calculating that of a KC_8 bilayer (i.e., a graphite layer and a layer of K atoms), a $(KC_8)_2$ quadruple layer, etc. As expected, the partial density plots $\rho(r_0, e_f)$ calculated for such composite layers with the tip positioned on top of the surface graphite are represented by the p_π-orbitals of the surface graphite layer. In disagreement with the observed STM pattern, however, all eight carbon atoms of a surface unit cell have practically the same density in all these $\rho(r_0, e_f)$ plots [22]. It is important to recall that the electronic structures representing the metallic

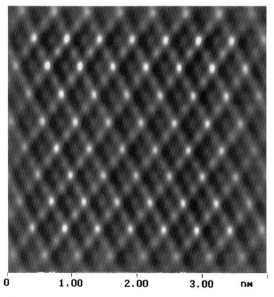

Figure 6.27 STM image of the KC_8 surface ($V_{bias} = -15$ mV, $I_{set} = 4$ nA). (Reproduced from Ref. 17b; courtesy of Dr. S. Kelty)

states are symmetry-adapted ones and hence cannot properly describe "low-symmetry" electronic states (e.g., CDWs) whose symmetry is lower than the symmetry of the lattice.

The likelihood that the CDW state is energetically more favorable than the metallic state for a low-dimensional metal occurs when the Fermi surface of the metallic state is nested [1, 23]. The Fermi surface of bulk KC_8 or a KC_8 bilayer does not possess any nesting [16, 24] that can cause the surface CDW observed by STM. The low-energy electron-diffraction pattern of KC_8 does not show any contrasted superstructure [25], so the surface graphite monolayer is probably almost flat. As described in Chapter 9, the surface topography can change under the tip force. The crystal structure of bulk KC_8 (Fig. 6.26) [16] suggests that under the tip force the carbons of the rings lying above the K^+ ions will be depressed less than are those joining the rings. Such an induced corrugation will lead to a 2×2 superstructure in which six of the eight carbons per surface unit cell are brighter than the remaining two. This pattern is different from the observed 2×2 superstructure superposed on a three-for-hexagon pattern (see Fig. 6.27) [17a–c]. Therefore, a tip force induced topography change alone cannot explain the surface CDW state of KC_8. (However, the $\sqrt{3} \times 4$ and $\sqrt{3} \times \sqrt{13}$ superstructures observed for stage-1 GIC RbC_8 and CsC_8 are explained in terms of the one-dimensional (1D) ordering of the alkali-metal atoms suggested by Anselmetti et al. [18c], if the surface graphite monolayer undergoes an appropriate tip force induced topography change.)

From the above discussion, it is clear that the STM image of KC_8 originates from the carbon p_π-orbitals of the surface graphite monolayer. The pattern within a unit cell is not explained by either the partial electron density calculations, the Fermi surface nesting, or the tip force induced topography change. What is clear though is that the additional electrons in the π band of the surface graphite monolayer (transferred from K) is not uniformly distributed.

6.3.2 Origin of Nonuniform Charge Distribution

The reason why the formation of the surface CDW is energetically favorable is found in the ionic interactions between the negatively charged surface graphite monolayer and the underlying potassium ions K^+ [26]. Since KC_8 behaves as a normal metal, the charge distribution in the graphite layer should be nonmagnetic, i.e., the up-spin density $n\uparrow$ and the down-spin density $n\downarrow$ on each carbon site are the same [27]. Table 6.1 summarizes the coulombic energies ($E_{coulomb}$) of a KC_8 bilayer calculated for several nonmagnetic charge distributions depicted in Fig. 6.28. The uniform charge distribution is given in Fig. 6.28 (a), the three-for-hexagon pattern in Fig. 6.28 (b), three-for-hexagon patterns with a 2×2 superstructure in Figs. 6.28 (c) and 6.28 (d), and a 2×2 superstructure with no three-for-hexagon pattern in Fig. 6.28 (e).

Table 6.1 Coulombic energy ($E_{coulomb}$) and on-site repulsion energy ($E_{on\text{-}site}$) of a KC_8 bilayer calculated for the point charge distributions of Figure 6.28 in the graphite monolayer.

Point charge distribution from Fig. 6.28	$E_{coulomb}$ (eV/KC_8)	$E_{on\text{-}site}$ (U/KC_8)
(a)	1.759	2.531
(b)	0.780	2.563
(c)	0.212	2.594
(d)	−0.689	2.648
(e)	−2.362	2.750

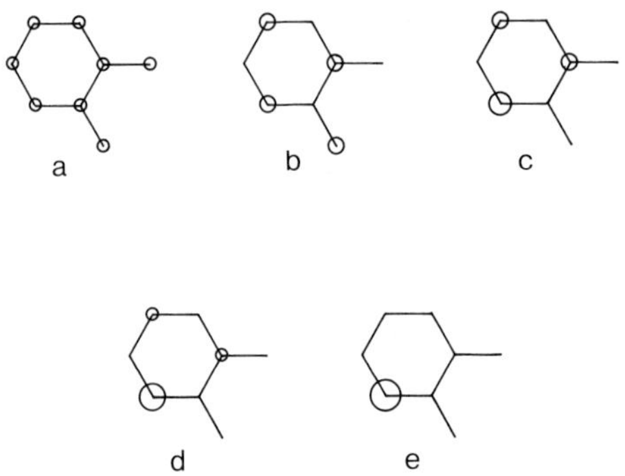

Figure 6.28 Model point charge distributions in the graphite monolayer carbon atoms of a KC_8 bilayer. Only the unit cell carbon atoms are shown, and the K^+ ion (not shown) is located underneath the center of the carbon hexagon. Only the carbon atoms with negative charge are shown by circles, with the larger circle representing the greater negative charges. The negative charges are as follows: (a) each carbon with $-1/8$; (b) four carbons with $-1/4$; (c) two carbons with $-1/4$ and one with $-1/2$; (d) two carbons with $-1/8$ and one with $-3/4$; (e) one carbon with -1.

Table 6.1 shows that the coulomb energy of a KC_8 bilayer becomes more favorable for a more nonuniform charge distribution and is attractive only for the strongly nonuniform charge distributions shown in Figs. 6.28 (d) and 6.28 (e) [26]. This reflects the fact that in the surface KC_8 bilayer, a nonuniform charge distribution in the graphite layer reduces the coulomb repulsion within the graphite layer. Also, the coulomb attraction between the graphite and the K^+ ion layers is less favorable when the charge is placed on the carbon atoms outside the carbon hexagon lying directly above each K^+. It should be noted that the charge distribution of Fig. 6.28 (d) resembles that of the surface graphite monolayer expected from the STM image of KC_8 (Fig. 6.27).

An energy term disfavoring nonuniform charge distribution is the on-site repulsion. At a site with $n\uparrow$ and $n\downarrow$ electrons, it is given by $n\uparrow n\downarrow U$, where U is the repulsion expected when two electrons reside on the atom [27, 28]. Table 6.1 lists the total on-site repulsion energies ($E_{\text{on-site}}$) calculated for the graphite monolayer with the point charge distributions given in Figs. 6.28 (a)–6.28 (e) [26]. Clearly, in terms of on-site repulsion, the more nonuniform charge distribution is less energetically favorable. In terms of $E_{\text{on-site}}$ and E_{coulomb}, the nonuniform charge distribution of Fig. 6.28 (d) or 6.28 (e) is predicted to be more stable than the uniform charge distribution of Fig. 6.28 (a), because the U value for carbon is much smaller than 11 eV [29].

To summarize, it is most likely that the surface CDW states of MC_8 are formed from the π-band electrons of the surface graphite monolayer to lower the coulombic interaction energy associated with the surface KC_8 bilayer. These states are electronic states in which the electron density is localized on a few nonadjacent carbon atoms of the unit cell (e.g., as in Fig. 6.28 (d)). Consequently, the electronic band levels describing them cannot be dispersive in the dispersion relations. Thus, the non-dispersive states of MC_8 observed by the ARPES studies most probably refer to the CDW states of the π-band electrons in the surface graphite monolayer.

6.4 Concluding Remarks

To a first approximation, the atomic-scale patterns of the AFM and STM images of a given sample surface are described by the HED patterns of the $\rho(r_0)$ and $\rho(r_0, e_f)$ plots, respectively. The $\rho(r_0)$ plot is determined by all the occupied energy levels of the sample. For a given surface containing one kind of atom, therefore, the HED spots of this plot are related to the most protruding atoms of the surface. However, the $\rho(r_0, e_f)$ plot is determined by the energy levels only in the vicinity of the Fermi level. Thus, the HED spots of the $\rho(r_0, e_f)$ plot may not be related to the most protruding atoms, the centers of these spots may not coincide with the surface atomic positions, and the lower-lying atoms of the surface can dominate the $\rho(r_0, e_f)$ plot and hence the STM image. For a rational interpretation of STM and AFM images, therefore, calculations of appropriate $\rho(r_0, e_f)$ and $\rho(r_0)$ plots are essential.

In all the X–M–X sandwich layer systems described above, the surfaces are formed by sheets of the ligand atoms, and the metal atoms lie lower than the surface ligand atoms by more than 1.5 Å. The $\rho(r_0, e_f)$ and $\rho(r_0)$ plots calculated for these systems are dominated by the surface ligand atoms, because the amplitudes of atomic orbitals decrease exponentially with increasing distance from their centers. Consequently, the STM and AFM images of these compounds are all represented by the surface ligand atoms.

The STM images of stage-1 GICs MC_8 (M = K, Rb, Cs) exhibit a 2 × 2 superstructure superposed on a three-for-hexagon pattern. The analysis in the previous section indicates that these images are associated with the surface CDW states of the π-band electrons in the surface graphite monolayer, which are formed to lower the coulombic interaction energy of the surface MC_8 bilayers. These CDW states are most likely to be the nondispersive states of MC_8 observed by ARPES studies.

References

[1] M.-H. Whangbo, E. Canadell, *J. Am. Chem. Soc.* **1992**, *114*, 9587.
[2] C. Rovira, M.-H. Whangbo, *Inorg. Chem.* **1993**, *32*, 4094.
[3] (a) G. W. Stupian, M. S. Leung, *Appl. Phys. Lett.* **1987**, *51*, 1560. (b) M. Weimer, J. Kramar, C. Bai, J. D. Baldeschwieler, *Phys. Rev. B* **1988**, *37*, 4292. (c) S. Akari, M. Stachel, H. Birk, S. Schreck, M. Lux, K. Dransfeld, *J. Microsc.* **1988**, *152*, 521.
[4] S. N. Magonov, M.-H. Whangbo, *Adv. Mater.* **1994**, *6*, 355.
[5] (a) H. Schäfer, J. Tillack, *J. Less-Common Metals* **1964**, *6*, 152. (b) H. G. von Schnering, H. Wöhrle, *Angew. Chem.* **1963**, *75*, 684. (c) H. Schäfer, H. G. von Schnering, *Angew. Chem.* **1964**, *76*, 833.
[6] P. Zönnchen, G. Thiele, H. Bengel, S. N. Magonov, C. Hess, C. Schlenker, D.-K. Seo, M.-H. Whangbo, *New J. Chem.*, in press.
[7] B. E. Brown, *Acta. Cryst.* **1966**, *20*, 268.
[8] (a) S. L. Tang, R. V. Kasowski, A. Suna, B. A. Parkinson, *Surf. Sci.* **1990**, *238*, 280. (b) A. Crossley, S. Myhra, C. J. Sofield, *Surf. Sci.* **1994**, *318*, 39.
[9] (a) H. Bengel, S. N. Magonov, D. Jung, J. Ren, M.-H. Whangbo, *New J. Chem.*, in press. (b) B. A. Parkinson, J. Ren, M.-H. Whangbo, *J. Am. Chem. Soc.* **1991**, *113*, 7833.
[10] B. E. Brown, *Acta. Cryst.* **1966**, *20*, 264.
[11] (a) H. Schäfer, H. G. von Schnering, *Angew. Chem.* **1964**, *76*, 833. (b) A. Simon, H. G. von Schnering, *J. Less Common Met.* **1966**, *11*, 31. (c) F. Hulliger, *Structural Chemistry of Layer-type Phases* (Ed.: F. Lévy) Reidel, Dordrecht, The Netherlands, **1976**, p. 319.
[12] S. N. Magonov, P. Zönnchen, H. Rotter, H.-J. Cantow, G. Thiele, J. Ren, M.-H. Whangbo, *J. Am. Chem. Soc.* **1993**, *115*, 2495.
[13] R. Brouwer, F. Jellinek, *Physica B* **1980**, *99*, 51.
[14] M.-H. Whangbo, J. Ren, E. Canadell, D. Louder, B. A. Parkinson, H. Bengel, S. N. Magonov, *J. Am. Chem. Soc.* **1993**, *115*, 3760.
[15] (a) R. V. Coleman, B. Giambattista, P. K. Hansma, W. W. McNairy, C. G. Slough, *Adv. Phys.* **1988**, *37*, 559. (b) R. V. Coleman, Z. Dai, W. W. McNairy, C. G. Slough, C. Wang, in *Scanning Tunneling Microscopy* (Ed.: J. A. Stroscio, W. J. Kaiser), Academic Press, New York, **1993**, p. 349.
[16] M. S. Dresselhaus, G. Dresselhaus, *Adv. Phys.* **1981**, *30*, 139.
[17] (a) S. P. Kelty, C. M. Lieber, *Crit. Rev. Surf. Sci.* **1992**, *1*, 217. (b) S. P. Kelty, C. M. Lieber, *J. Phys. Chem.* **1989**, *93*, 5983. (c) S. P. Kelty, C. M. Lieber, *Phys. Rev. B* **1989**, *40*, 5856. (d) S. P. Kelty, C. M. Lieber, *Phys. Rev. B* **1991**, *44*, 4064.
[18] (a) D. Anselmetti, R. Wiesendanger, V. Geiser, H. R. Hidber, H.-J. Güntherodt, *J. Microsc.* **1988**, *152*, 59. (b) D. Anselmetti, R. Wiesendanger, H.-J. Güntherodt, *Phys. Rev. B* **1989**, *39*, 11135. (c) D. Anselmetti, V. Geiser, G. Overney, R. Wiesendanger, H.-J.

Güntherodt, *Phys. Rev. B* **1990**, *42*, 1848. (d) R. Wiesendanger, D. Anselmetti, in *Scanning Tunneling Microscopy I* (Eds.: R. Wiesendanger, H.-J. Güntherodt) p. 131, Springer, Heidelberg, **1992**, p. 131.
[19] D. Marchand, C. Fretigny, N. Lecomte, M. Laguès, J. E. Fischer, *Synth. Met.* **1988**, *23*, 165.
[20] N. Gunasekara, T. Takahashi, F. Maeda, T. Sagawa, H. Suematsu, *J. Phys. Soc. Jpn.* **1987**, *56*, 2581, and references therein.
[21] H. Kamimura, *Ann. Phys. (Fr.)* **1986**, *11*, 39.
[22] J. Ren, M.-H. Whangbo, unpublished results.
[23] (a) M.-H. Whangbo, E. Canadell, P. Foury, J. P. Pouget, *Science* **1991**, *252*, 96. (b) E. Canadell, M.-H. Whangbo, *Chem. Rev.* **1991**, *91*, 965.
[24] N. A. W. Holzwarth, in *Graphite Intercalation Compounds II* (Eds.: H. Zabel, S. A. Solin) Springer, Berlin, **1992**, pp. 1–51.
[25] N. J. Wu, A. Ignatiev, *Phys. Rev. B* **1983**, *28*, 7288.
[26] M.-H. Whangbo, W. Liang, J. Ren, S. N. Magonov, A. Wawkuschewski, *J. Phys. Chem.* **1994**, *98*, 7602.
[27] M.-H. Whangbo, *J. Chem. Phys.* **1979**, *70*, 4963. (b) M.-H. Whangbo, *J. Chem. Phys.* **1980**, *73*, 3854.
[28] (a) N. F. Mott, *Metal-Insulator Transitions*, Barnes and Noble, New York, **1977**. (b) B. H. Brandow, *Adv. Phys.* **1977**, *26*, 651.
[29] A. L. Tchougreeff, R. Hoffmann, *J. Phys. Chem.* **1992**, *96*, 8993.

7 STM Images Associated with Point Defects of Layered Inorganic Compounds

The local character of STM provides one with the unique opportunity to detect different surface irregularities on the molecular and atomic scales. These include linear defects such as surface steps, edge dislocations and point defects (e.g., adatoms, substitutional defects and vacancies). The image features corresponding to linear surface defects are routinely found in STM and AFM studies of crystalline surfaces, whereas the atomic-size imperfections are observed only in STM images. It is not an easy problem to explain the structural and electronic origins of atomic-size local disturbances and larger nanoscale patterns observed in different regions of periodic images. As a first step toward solving this problem, it is of importance to study how point defects of layered inorganic compounds might manifest themselves in STM images. Theoretical studies in this direction have already been undertaken for several layered systems [1–7]. In this chapter, the STM image imperfections of two metal cluster compounds, β-Nb_3I_8 and 1T-$TaSe_2$, are analyzed first. Then, the effects of point defects on the STM images of semiconductors are examined in detail by considering 2H-MoS_2 [7]. As will be discussed below, the electronic band structure description is applicable to the defects which do not strongly perturb the surrounding lattice and electronic structures. In other cases, it is necessary to resort to qualitative analysis because the defects induce geometric and electronic structure perturbations of the host lattice around them, with the spatial extension reaching several nanometers.

7.1 Imperfections in Compounds with Metal Clusters

The STM images of β-Nb_3I_8 exhibit large defects of approximately the size of one Nb_3I_{13} cluster (Fig. 7.1 (a)). Likewise, the STM images of 1T-$TaSe_2$ show large defects, which appear as if whole $\sqrt{13} \times \sqrt{13}$ cluster units are missing (Fig. 7.1 (b)). The STM images of β-Nb_3I_8 and 1T-$TaSe_2$ are primarily determined by the partial electron density plots $\rho(r_0, e_f)$ associated with the highest occupied half-filled bands of their surface layers (Chapter 6). To examine a probable cause for the large STM defects, it is necessary to analyze the nature of these bands.

An Nb_3I_8 layer is derived from the $Nb_3I_{13}^{5-}$ clusters **1**, and a 1T-$TaSe_2$ layer from the $Ta_{13}Se_{42}^{32-}$ clusters **2**, by sharing their octahedral edges. There are 7 and 13 d-

114   7 STM Images Associated with Point Defects of Layered Inorganic Compounds

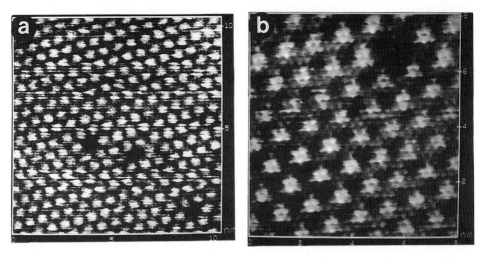

Figure 7.1 Large imperfections in the STM current images of (a) β-Nb$_3$I$_8$ ($V_{bias} = -400$ mV, $I_{set} = 1$ nA) and (b) 1T-TaSe$_2$ ($V_{bias} = -10$ mV, $I_{set} = 6$ nA). The contrast is proportional to current variations in relative units.

electrons in the Nb$_3$I$_{13}^{5-}$ and Ta$_{13}$Se$_{42}^{32-}$ clusters, respectively. The energy levels of the Nb$_3$I$_{13}^{5-}$ and Ta$_{13}$Se$_{42}^{32-}$ clusters calculated by the extended Hückel molecular orbital (EHMO) method are shown in Figs. 7.2 and 7.3, respectively. The HOMOs of these clusters are each singly filled. It is important to note that the highest occupied bands of the Nb$_3$I$_8$ and 1T-TaSe$_2$ layers are largely made up of the singly filled HOMOs from their cluster units **1** and **2**, respectively. This means that the bright spots in the STM images of β-Nb$_3$I$_8$ and 1T-TaSe$_2$, obtained with either positive or negative V_{bias}, represent the electron densities associated with the HOMOs of their cluster units.

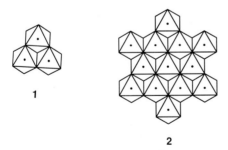

The tunneling current is sensitive to the partial electron density associated with the Fermi level. Thus, for β-Nb$_3$I$_8$ and 1T-TaSe$_2$, when one cluster unit is modified to have a lower-lying HOMO and a higher-lying LUMO than do other neighboring cluster units, it cannot contribute to a tunneling current and appears as dark in the STM image as if the whole cluster is missing. Such a situation is easily created when the metal cluster units have an atom vacancy. For example, Fig. 7.2 compares the energy

7.1 Imperfections in Compounds with Metal Clusters

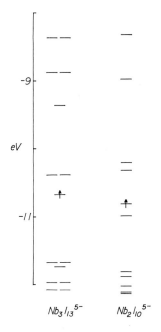

Figure 7.2 Low-lying energy levels of (a) $Nb_3I_{13}^{5-}$ and (b) $Nb_2I_{10}^{5-}$ clusters. For simplicity of presentation, only the highest occupied levels are shown to have electrons.

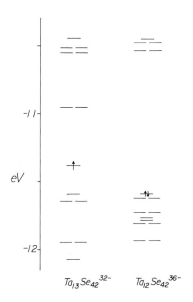

Figure 7.3 Low-lying energy levels of (a) $Ta_{13}Se_{42}^{32-}$ and (b) $Ta_{12}Se_{42}^{36-}$ clusters. For simplicity of presentation, only the highest occupied levels are shown to have electrons.

levels of an $Nb_3I_{13}^{5-}$ cluster with those of an $Nb_2I_{10}^{5-}$ cluster. (The latter is derived from an $Nb_3I_{13}^{5-}$ cluster by deleting one Nb atom and the resulting three dangling I atoms.) Figure 7.3 compares the energy levels of a $\sqrt{13} \times \sqrt{13}$ cluster $Ta_{13}Se_{42}^{32-}$ with those of the $Ta_{12}Se_{42}^{36-}$ cluster derived from it by deleting the central Ta atom (i.e., the Ta(1) atom in Fig. 6.20). In both figures, the HOMO and LUMO levels of the clusters with a metal-atom vacancy are quite different from those of the defect-free clusters. Therefore, the clusters with metal-atom vacancies appear as dark holes in the STM images (see Figs. 7.2 and 7.3).

7.2 Point Defects in Semiconductor 2H-MoS$_2$

As shown in Chapter 6, layered transition-metal dichalcogenides 2H-MX$_2$ are made up of face-sharing trigonal prisms, MX$_6$. Due to the oxidation state $M^{4+}(X^{2-})_2$ per formula unit MX$_2$, each MX$_6$ trigonal prism contains d^2 metal cations M^{4+}, and these dichalcogenides are regular semiconductors [8]. These compounds have been the subject of numerous STM studies [6, 9–12], and a number of interesting observations concerning STM image imperfections have been reported.

For the STM imaging of a semiconductor, the important energy region is from the top portion of the valence band to the bottom portion of the conduction band. To classify the types of STM images associated with its point defects (i.e., atom vacancy and substitution), it is necessary to examine how the electronic structure of this region is modified by the defects. Semiconductors with shallow donor and acceptor dopants (i.e., donor and acceptor substitution defects) give rise to donor and acceptor levels within their band gaps. The spatial extension of these defect states is localized in the vicinity of the dopant sites. The participation of these defect states in the tunneling process is strongly affected by the polarity and the magnitude of the bias voltage. A satisfactory description of the electronic structure change induced by a point defect should consider the local structure relaxation around the defect site. Since this problem is rather difficult to treat quantitatively, one may assume that the geometry of the host lattice is unaffected by the defect and then consider qualitatively, when necessary, how the results of such calculations might be affected by the local geometry relaxation.

For the discussion given below, it is important to recall the essential nature of bonding in transition metal compounds. In most cases, the interactions between the metal (M) d-orbitals and the ligand (L) p-orbitals lead to the electronic structure depicted in Fig. 7.4. The interaction between M and L is bonding in the p-block bands and is antibonding in the d-block bands. For a d^2 2H-MX$_2$ system such as 2H-MoS$_2$, the bottom d-block band is completely filled, and a band gap separates the valence band from the conduction band.

Within the approximation of a rigid band picture, in which the local geometry relaxation around a defect is neglected, Figs. 7.5 (a)–7.5 (e) present some possible

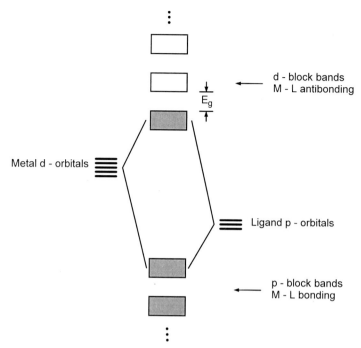

Figure 7.4 Schematic representation of the orbital interaction between the metal (M) d-orbitals and ligand (L) p-orbitals leading to the p-block and d-block bands. The interaction between M and L is bonding in the p-block bands, and antibonding in the d-block bands. Filled bands are indicated by shading.

cases of the states a point defect might bring about in the energy region of the valence and conduction bands of 2H-MoS$_2$ [7]. The valence and conduction bands of an ideal 2H-MoS$_2$ layer with no defects, depicted in Fig. 7.5 (a), are d-block bands. These bands are antibonding in nature between the metal and ligands [13]. The extent of this antibonding is reduced by the Mo and S atom vacancies (denoted by V_{Mo} and V_S, respectively), because the V_{Mo}–S and Mo–V_S linkages do not have antibonding interactions. Thus, the vacancies can give rise to states lying in the band gap, as will be presented below.

A few examples of substitutional point defects are presented in Figs. 7.5 (b)–7.5 (e). A substituting atom is a donor if it has more valence electrons than the replaced host atom, and an acceptor if it has less. If a donor metal M such as Mn substitutes for an Mo atom, the MS$_6$ trigonal prism has a d^3 cation M^{4+} and the singly filled d-block level of MS$_6$ lies within the band gap (Fig. 7.5 (b)). When an acceptor metal M, such as vanadium, substitutes for an Mo atom, the MS$_6$ trigonal prism has a d^1 cation M^{4+} and the singly filled d-block level of MS$_6$ lies below the valence band top (Fig. 7.5 (c)). Within a rigid band approximation, a donor atom D (e.g., Cl) substituting for an S atom releases an electron to the host lattice and creates an

118 7 STM Images Associated with Point Defects of Layered Inorganic Compounds

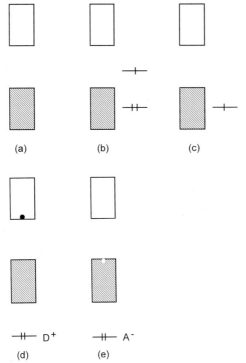

Figure 7.5 Some representative cases of the energy levels that point defects can bring to the energy region of the valence and conduction bands of a defect-free MoS_2 layer: (a) free of defects; (b) a donor metal producing a d^3 cation at the Mo site; (c) an acceptor metal producing a d^1 cation at the Mo site; (d) a donor ligand producing a D^+ cation at the S site; and (e) an acceptor ligand producing an A^- anion at the S site. The shaded and unshaded square boxes represent the valence and conduction bands, respectively.

electron at the bottom of the conduction band (Fig. 7.5 (d)). Likewise, an acceptor atom A (e.g., P) substituting for an S atom accepts an electron from the host lattice and creates a hole at the top of the valence band (Fig. 7.5 (e)). Another important effect to be considered in STM of semiconductors is band bending [14], discussed in Chapter 5.

7.3 Cases Tractable by Electronic Band Structure Calculations

To generate the electronic structures of MoS_2 layers containing point defects, one might introduce a point defect in a supercell, e.g., in an $n \times n$ supercell $(MoS_2)_{n^2}$, and then calculate the electronic structures by using an electronic band structure method. Such calculations are suitable for those defect states for which the spatial extension of their wave functions is strongly localized around the defect sites.

The PDOS plots calculated for a defect-free MoS$_2$ layer are presented in Fig. 7.6, which shows only the bottom portion of the d-block bands where the valence and conduction bands occur. For convenience, the partial density plot associated with the highest-lying occupied states will be represented by $\rho_{HO}(r_0, e_f)$, and that associated with the lowest-lying unoccupied states by $\rho_{LU}(r_0, e_f)$. For a defect-free MoS$_2$ layer, the $\rho_{HO}(r_0, e_f)$ and $\rho_{LU}(r_0, e_f)$ plots are nearly identical and consist of density spots centered on the surface sulfur atoms. Thus, as discussed in Chapter 6, the hexagonal pattern of bright spots found for atomic-scale STM images of MoS$_2$ at positive and negative V_{bias} represents the surface sulfur atoms. Point defects disturb this ideal pattern and produce image imperfections.

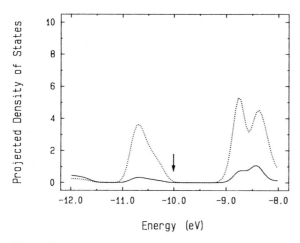

Figure 7.6 PDOS plot calculated for a defect-free MoS$_2$ layer. Only the energy region of the valence and conduction bands is shown, where the solid and dotted lines refer to the S and Mo contributions, respectively, and the top of the valence band is indicated by an arrow. The PDOS values for each atom are normalized to one atom.

7.3.1 Ligand-Atom Vacancy

The electronic structure of the MoS$_2$ layer containing S-atom vacancies (V$_S$) can be approximated by that calculated for the layer containing one vacancy in every 4 × 4 supercell, i.e., the layer with unit cell (MoSV$_S$)(MoS$_2$)$_{15}$. The $\rho_{HO}(r_0, e_f)$ plot calculated for this layer (Fig. 7.7) shows no density around the vacancy sites, nor does the $\rho_{LU}(r_0, e_f)$ plot. Thus, in STM images of both positive and negative V_{bias}, an S-atom vacancy of MoS$_2$ should appear as a dark spot covering the vacancy site.

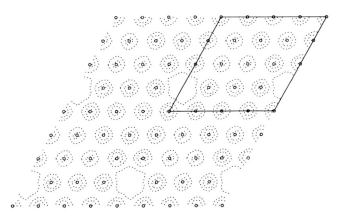

Figure 7.7 2D contour representation of the $\rho_{HO}(r_0, e_f)$ plot calculated for an MoS$_2$ layer with unit cell $(MoSV_S)(MoS_2)_{15}$, where the contour values used are 15×10^{-6}, 10×10^{-6}, 5×10^{-6}, and 1×10^{-6} electrons/au^3. The surface S atoms are represented by small circles.

7.3.2 Metal-Atom Vacancy

The electronic structure of the MoS$_2$ layer containing Mo-atom vacancies (V_{Mo}) can be approximated by that calculated for the layer with unit cell $(V_{Mo}S_2)(MoS_2)_{15}$. The PDOS, $\rho_{HO}(r_0, e_f)$ and $\rho_{LU}(r_0, e_f)$ plots of this layer are shown in Figs. 7.8 (a)–7.8 (c), respectively. In the $\rho_{HO}(r_0, e_f)$ plot, the density is slightly lower around the area of the three surface sulfur atoms of each $V_{Mo}S_6$ trigonal prism than in its surroundings. In the $\rho_{LU}(r_0, e_f)$ plot, the density is largely concentrated on the six sulfur atoms around each vacancy site (i.e., the three in the $V_{Mo}S_6$ trigonal prism, plus the three adjacent ones), and the central three are higher in density than the other three. The PDOS plot of Fig. 7.8 (a) shows that in the band gap of the defect-free MoS$_2$ layer, the V_{Mo} introduces the "empty d-block states" of, mainly, the six Mo atoms adjacent to the vacancy site. The $\rho_{LU}(r_0, e_f)$ plot represents the contributions of the surface sulfur atoms belonging to the MoS$_6$ trigonal prisms of these six Mo atoms. The most likely geometry relaxation around the Mo vacancy is the depression of the three surface S atoms of the $V_{Mo}S_6$ trigonal prism. This will reduce the brightness of the three atoms to some extent. Thus, each Mo-atom vacancy will produce a triangular STM pattern of six bright spots at positive V_{bias}, and this pattern will disappear at negative V_{bias}.

7.3.3 Donor Substitution at the Metal Site

The donor-substitutional defect considered in Fig. 7.5 (b) can be simulated by calculating the electronic band structure of the layer with unit cell $(MnS_2)(MoS_2)_{15}$. The

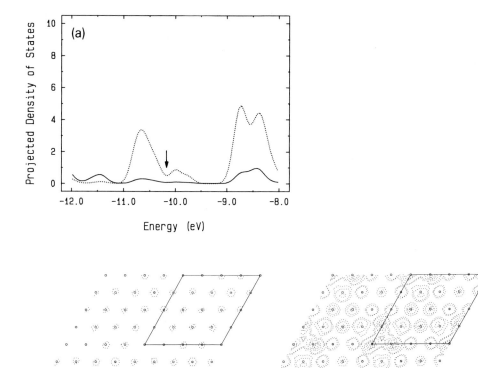

Figure 7.8 PDOS, $\rho_{HO}(r_0, e_f)$ and $\rho_{LU}(r_0, e_f)$ plots calculated for an MoS$_2$ layer with unit cell $(V_{Mo}S_2)(MoS_2)_{15}$. (a) PDOS plot, where the solid and dotted lines refer to the S and Mo contributions, respectively, and the Fermi level is indicated by an arrow. The PDOS values for each atom are normalized to one atom. (b) $\rho_{HO}(r_0, e_f)$ plot, where only one contour value $(15 \times 10^{-6}$ electrons/au$^3)$ was used to emphasize the small differences between the S atoms. (c) $\rho_{LU}(r_0, e_f)$ plot, where the contour values used are 15×10^{-6}, 10×10^{-6}, 5×10^{-6}, 1×10^{-6}, and 0.5×10^{-6} electrons/au^3. The surface S atoms are represented by small circles, and a unit cell is indicated by the parallelogram. The defect position is under the center of the sulfur triangle at the lower-left corner of each unit cell.

PDOS plot of this layer shows that the half-filled d-block level of the MnS$_6$ trigonal prism lies within the band gap (Fig. 7.9 (a)). The $\rho_{HO}(r_0, e_f)$ plot has the density concentrated on the three surface sulfur atoms of each MnS$_6$ trigonal prism (Fig. 7.9 (b)). $\rho_{LU}(r_0, e_f)$ is nearly the same as $\rho_{HO}(r_0, e_f)$, because the half-filled d-block level of each MnS$_6$ trigonal prism will participate in the tunneling process at both positive and negative V_{bias}. Thus, the STM image imperfection induced by the donor defect producing a d^3 cation at the Mo site will be given by three bright spots representing the three surface sulfur atoms of the defect-containing S_6 trigonal prism at both polarities of V_{bias}.

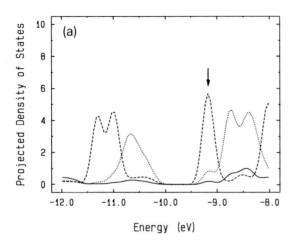

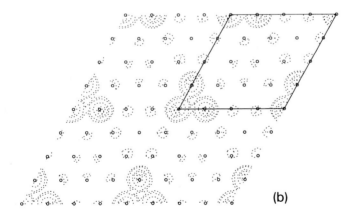

Figure 7.9 PDOS and $\rho_{HO}(r_0, e_f)$ plots calculated for an MoS$_2$ layer with unit cell (MnS$_2$)(MoS$_2$)$_{15}$. (a) PDOS plot, where the solid, dotted and broken lines refer to the S, Mo and Mn contributions, respectively, and the Fermi level is indicated by an arrow. The PDOS values for each atom are normalized to one atom. b) $\rho_{HO}(r_0, e_f)$ plot, where the contour values used are 90×10^{-5}, 50×10^{-5}, 30×10^{-5}, 10×10^{-5}, and 5×10^{-5} electrons/au^3. The surface S atoms are represented by small circles, and a unit cell is indicated by the parallelogram. The defect position is under the center of the sulfur triangle at the lower-left corner of each unit cell.

7.4 Cases Intractable by Electronic Band Structure Calculations

The electronic states associated with point defects can disturb the surrounding lattice and electronic structures with spatial extension on the nanometer scale [15]. These states destroy the translational symmetry of the lattice and hence are not described by the electronic band structure method. Such examples are found when the donor or acceptor substitution occurs at the ligand sites of the 2H-MoS$_2$ lattice (Figs. 7.5 (d) and 7.5 (e)). The electronic band structure description is also inapplicable to cases when the metal atom has a large on-site repulsion, so that its d-block levels are singly filled [16]. These cases arise when an acceptor metal substitutes the Mo site in the 2H-MoS$_2$ lattice to create a singly filled d-level in the middle of the valence band (Fig. 7.5 (c)). For all these cases, partial density plot calculations using the $n \times n$ supercell approach as described in the previous section are of no use in the interpretation of the associated image imperfections. Qualitative ideas should be developed to provide guidance in such cases.

7.4.1 Donor Substitution at the Ligand Site

A donor atom such as Cl at the S site (Cl$_S$) creates an electron at the bottom of the conduction band within a rigid band scheme (Fig. 7.5 (d)). It is energetically more favorable for this electron to be trapped around the donor cation (e.g., Cl$^+$) by relaxing the geometry around it (Fig. 7.10 (a)). The valence and conduction bands of MoS$_2$ are antibonding between the Mo and its ligand atoms L (i.e., L = S or Cl)

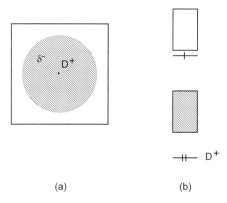

(a)   (b)

Figure 7.10 (a) Schematic representation of the electron-trapped area around a donor cation D$^+$, where the shading indicates that the trapped area has higher electron density than the surroundings. (b) Schematic representation of the donor level associated with the trapped-electron state.

[13]. In the local area around the donor defect where an electron is trapped (i.e., the electron-trapped area), there will be a geometry relaxation which slightly lengthens the Mo–L bonds to decrease the extent of the antibonding. This creates the singly filled donor level (i.e., one composed mainly of the orbitals of the electron-trapped area) below the bottom of the conduction band (Fig. 7.10 (b)), and the bottom of the conduction band loses the orbital character of the electron-trapped area. The radial distribution of the trapped electron state being analogous to that of a hydrogen 1s orbital [15], the extent of the geometry relaxation should decrease with the distance from the defect site. By analogy with the donor defects in the elemental semiconductors Ge and Si [15], one might expect the electron-trapped area to have a nanometer-scale size. The size of the electron-trapped area should decrease as the strength of the trapping increases.

7.4.1.1 The Case of Negative Bias

At negative V_{bias}, the donor level and the valence band top can donate electrons to the tip. In general, the transmission probability for the electrons to tunnel out from an electrode is highest at the Fermi level of the negatively biased electrode, and

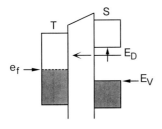

(a) $E_V < e_f < E_D$

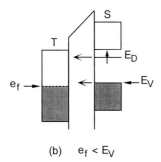

(b) $e_f < E_V$

Figure 7.11 Two extreme cases of sample-to-tip tunneling for an n-type semiconductor: (a) small bias, $E_V < e_f < E_D$; (b) large bias, $e_f < E_V$.

steadily decreases as their energy is lowered from the Fermi level [17]. At zero bias voltage, the Fermi level of the sample is closer to the donor level than to the valence band top (Fig. 5.5). Therefore, the donor level contributes more to the tunneling process than can the valence band top. Only the donor level can participate in the tunneling (Fig. 7.11 (a)) when the negative bias is not large; the donor levels and the valence band top can both participate if the bias is large (Fig. 7.11 (b)). In both cases, the electron-trapped area around the donor site should appear brighter than the surrounding area in the STM image [14c]. When the bias is large so that the band bending is significant, then the bottom of the conduction band can be filled near the interface (Fig. 5.7 (b)) and hence participate in the tunneling. This will lessen the brightness of the electron-trapped area.

7.4.1.2 The Case of Positive Bias

At positive V_{bias}, both the donor level and the bottom of the conduction band can receive electrons from the tip. Since the two levels are close to each other, their contributions to the tunneling current will be similar. The donor level makes the electron-trapped area brighter, while the bottom of the conduction band makes it darker, than the surroundings. Thus, the electron-trapped area at positive V_{bias} will be less bright than that at negative V_{bias}. In addition, the electron-trapped area would have a weaker tendency to receive electrons from the tip due to its electron-excess nature than do the surroundings. This further lowers the brightness of the electron-trapped area at positive V_{bias}, so the area may even appear darker than the surroundings.

7.4.2 Acceptor Substitution at the Ligand Site

An acceptor atom such as P at the S site (P_S) creates a hole at the top of the valence band within a rigid band scheme (Fig. 7.5 (e)), but it is energetically more favorable for this hole to be trapped around the acceptor anion (e.g., P^-) by relaxing the geometry around it (Fig. 7.12 (a)).

The filled levels of the acceptor anion lie below the valence band (Fig. 7.12 (b)), i.e., in the ligand p-block band region where the orbital character is bonding between the Mo and its surrounding ligand atoms L (L = S, A) [13]. To stabilize these filled levels, the local area around the defect where a hole is trapped (i.e., the hole-trapped area) should relax to have slightly shorter Mo–L bond lengths than the surroundings to enhance the extent of the bonding. In the valence band where the Mo–L bonds are antibonding between Mo and L, this relaxation of the hole-trapped area enhances the extent of antibonding and hence creates an acceptor level (i.e., one composed largely of the orbitals of the hole-trapped area) above the top of the valence band (Fig. 7.12 (b)), and the top of the valence band loses the orbital character of the hole-trapped area. The acceptor level is half-filled because of the loss of one electron to the acceptor. The situation just described is analogous to a two-orbital three-electron interaction in isolated molecules [13].

The size of the hole-trapped area should decrease as the strength of the trapping increases, i.e., as the acceptor level is raised farther from the valence band top. As in the case of the electron-trapped area, one might expect the hole-trapped area to have a nanometer-scale size.

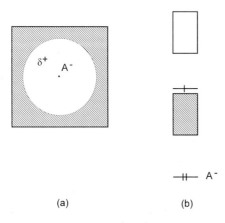

(a)   (b)

Figure 7.12 (a) Schematic representation of the hole-trapped area around an acceptor anion A^-, where the shading indicates that the trapped area has lower electron density than the surroundings. (b) Schematic representation of the acceptor level associated with the trapped-hole state.

7.4.2.1 The Case of Positive Bias

At positive V_{bias}, the acceptor level and the bottom portion of the conduction band can receive electrons from the tip. Only the acceptor level contributes to the tunneling process when the bias is not large (Fig. 7.13 (a)). The hole-trapped area will have a stronger tendency to receive electrons from the tip due to its electron-deficient nature. Both factors will make the hole-trapped area appear brighter than the surroundings at positive V_{bias}. If the bias voltage is large so that the tip Fermi level lies above the conduction band bottom, the latter can receive electrons from the tip (Fig. 7.13 (b)). This has the effect of reducing the brightness of the defect image. When the bias is large so that the band bending is significant, the valence band top becomes empty near the interface (Fig. 5.7 (a)). This makes holes accumulate around the acceptors (formally negative ions), enhances the tip-to-sample electron tunneling in the vicinities of the acceptors, and hence increases the contrast of the defect images [14b].

7.4 Cases Intractable by Electronic Band Structure Calculations 127

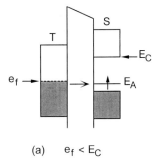

(a) $e_f < E_C$

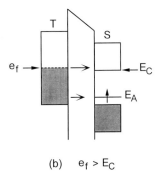

(b) $e_f > E_C$

Figure 7.13 Two extreme cases of tip-to-sample tunneling for a p-type semiconductor: (a) small bias, $e_f < E_C$. (b) large bias, $E_V < e_f$.

7.4.2.2 The Case of Negative Bias

At negative V_{bias}, both the acceptor level and the valence band top can donate electrons to the tip. Since the two levels are close to each other, the contributions of the valence band top and the acceptor level should be comparable. The latter makes the hole-trapped area brighter, while the former makes it darker, than the surrounding area. Thus, the hole-trapped area at negative V_{bias} will be less bright than that at positive V_{bias}. Furthermore, the hole-trapped area will have a weaker tendency to give electrons to the tip than do the surroundings, due to its electron-deficient nature. This further lowers the brightness of the hole-trapped area at negative V_{bias}, so that the contrast of this area may even become darker than the surroundings.

7.4.3 Acceptor Substitution at the Metal Site

For the case of the acceptor metal defect M at the Mo site (Fig. 7.5 (c)), there are two extreme situations to consider. When the on-site repulsion (i.e., pairing energy) at the defect site is negligible, the half-filled d-block level of the defect cation becomes doubly filled, and a hole is generated at the valence band top. This can induce a hole trapping around the MS$_6$ trigonal prism and leads to image imperfections similar to those discussed for an acceptor substitution at the S site. A transition-metal acceptor at the Mo site is less electronegative than a nontransition-metal acceptor at the S site. Thus, for the transition-metal acceptor, the hole trapping will be weaker, the hole-trapped area will be larger, and the image contrast of the area will be weaker. This situation would also be applicable to metal acceptors leading to d^0 cations M^{4+} at the Mo site, since their d-block levels are empty.

When the on-site repulsion at the defect site is large, it is energetically unfavorable for the half-filled level to accept an electron from the valence band [16]. Then, the defect site is locally magnetic and acts as a local insulating center. With the half-filled level lying below the valence band top (Fig. 7.5 (c)), it is less likely to participate in the tunneling process at negative V_{bias}. Because of the large on-site repulsion, it is energetically less favorable for an electron to enter the defect site MS$_6$ than the MoS$_6$ sites at positive V_{bias}. Therefore, the defect site will appear as a dark spot at both polarities of V_{bias}. If the electron of the singly filled level is truly localized at the defect center, the dark spot will have the approximate size of the three surface sulfur atoms belonging to the defect-containing S$_6$ trigonal prism. However, this situation is unlikely according to the nature of the valence band [18]. Although this band is often referred to as the d$_{z^2}$ band by analogy with the lowest-lying d-block level of an isolated MoS$_6$ trigonal prism, it has a strong in-plane d-orbital (i.e., d$_{x^2-y^2}$ and d$_{xy}$) character. The latter leads to metal–metal bonding along the three metal chain directions within the metal-atom sheet, [18b], and the Wannier orbitals (i.e., a solid-state analogue of localized molecular orbitals in isolated molecules) obtained from the valence band contain contributions from three adjacent metal atoms [18a]. Therefore, it is probable that the localized electron of the defect center spreads into its surrounding area on a nanometer scale, with a corresponding increase in the size of the associated dark STM spot.

Which of the two extreme cases discussed above is more appropriate depends on the nature of the defect atoms. In general, consideration of the on-site repulsion is more important for the 3d than for the 4d and 5d transition metals.

7.5 Survey of Image Imperfections Observed for d^2 2H-MX$_2$ Systems

The essential features of the STM image imperfections expected for various types of point defects in 2H-MX$_2$ compounds (M = Mo, W; X = S, Se) are summarized in Table 7.1. Several representative imperfections observed in STM images of 2H-MX$_2$ (M = Mo, W; X = S, Se) are analyzed below, in an attempt to assign the nature of the point defects that might have caused them.

7.5 Survey of Image Imperfections Observed for d^2 2H-MX$_2$ Systems

Table 7.1 Characteristic features of STM image imperfections expected for various point defects in 2H-MX$_2$ (M = Mo, W; X = S, Se) as a function of the polarity of the bias voltage $V^{(a)}$

Defect type	V_{bias}	Image characteristics
Vacancy at X	(+)	D; atom size
	(−)	D; atom size
Vacancy at M	(+)	B; triangle of six X
	(−)	NC
Donor at M	(+)	B; triangle of three or six X
	(−)	B; triangle of three or six X
Acceptor at M (strong U)	(+)	D; nm-scale circle
	(−)	D; nm-scale circle
Acceptor at M (weak U)$^{(b)}$	(+)	LB; nm-scale circle
	(−)	NC, D; nm-scale circle
Donor at X$^{(c)}$	(+)	D, NC, LB, B$^{(d)}$; nm-scale circle
	(−)	B; nm-scale circle
Acceptor at X	(+)	B; nm-scale circle
	(−)	D, NC, LB, B$^{(d)}$; nm-scale circle

(a) The symbols are defined as follows: B, bright; LB, light bright; NC, no contrast; D, dark; nm, nanometer; U, on-site repulsion.
(b) It may be difficult to observe due to the weakness of the contrast.
(c) The contrast and shape may change under strong tip forces.
(d) This can occur at large $|V_{bias}|$ due to a strong tip-induced band bending at the sample surface.

7.5.1 Atomic-Scale Images

A chalcogen atom vacancy on a surface MX$_2$ layer is expected to produce an image imperfection of an atom-size dark spot at positive and negative V_{bias} (Fig. 7.7). This is indeed the case, and a representative example from the images recorded on WSe$_2$ is shown in Fig. 7.14. The image imperfection expected for a donor defect producing a d^3 or d^4 cation M^{4+} at the Mo site is a triangular pattern of three bright spots (Fig. 7.9 (b)) at both polarities of V_{bias} according to the rigid band approximation. However, it is probable that the defect site undergoes a geometry relaxation because the Mo^{4+} and M^{4+} cations have different ionic sizes [19]. During ambient-condition STM measurements, the sample surface is likely to undergo a tip force induced surface corrugation (for details see Chapters 8 and 9). Since the defect site differs in local hardness from other sites, there will be a tip force induced surface relaxation around the defect site as well. Because of such a dynamic geometry relaxation, the image imperfection may expand its size slightly to appear as a triangular pattern of six bright spots. Figure 7.15 presents the STM image of MoS$_2$ recorded at negative

130   7 STM Images Associated with Point Defects of Layered Inorganic Compounds

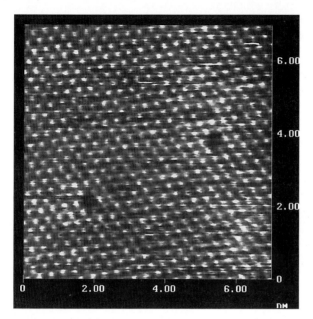

Figure 7.14 STM image of WSe$_2$ (V_{bias} = 727 mV, I_{set} = 3.2 nA). The contrast covers height variations in the 0.0–0.5 nm range.

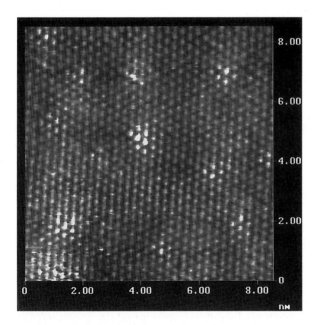

Figure 7.15 STM image of MoS$_2$ (V_{bias} = −1000 mV, I_{set} = 3.0 nA). The contrast covers height variations in the 0.0–0.5 nm range.

7.5 Survey of Image Imperfections Observed for d^2 2H-MX$_2$ Systems

V_{bias}, which shows triangular patterns of three as well as a larger number of atomic-size bright spots. These imperfections are consistent with the donor transition-atom defects leading to d^3 and d^4 cations (Fig. 7.9 (b)).

7.5.2 Nanometer-Scale Images

According to Table 7.1, there are several kinds of contrast dependence on the bias polarity which may help assign the types of point defects giving rise to nanometer-scale STM image imperfections. In this section, representative nanometer-scale images of the 2H-MX$_2$ compounds are analysed.

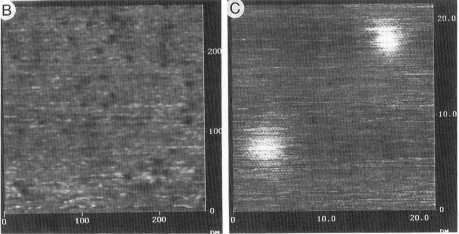

Figure 7.16 STM images of MoSe$_2$: (a) $V_{bias} = 1000$ mV, $I_{set} = 0.5$ nA; (b) $V_{bias} = -1000$ mV, $I_{set} = 0.5$ nA; (c) $V_{bias} = 1.333$ V, $I_{set} = 3.0$ nA. The contrast covers height variations in the 0.0–1.0 nm range for (a) and (c), and in the 0.0–1.2 nm range for (b).

Figure 7.16 (a) shows an STM image of MoSe$_2$ recorded at positive V_{bias}. The zoomed-in part of this image (Fig. 7.16 (c)) shows the bright spots to have a circular shape of about 5 nm in diameter. Figures 7.16 (a) and 7.16 (b) show that the bright nanometer-scale spots of MoSe$_2$ at positive V_{bias} become dark at negative V_{bias}. This bias-polarity dependence suggests that these imperfections originate from acceptor dopants on the surface chalcogen-atom sheets. Large-scale STM images of MoS$_2$ obtained at positive and negative V_{bias} are given in Fig. 7.17. The images show that

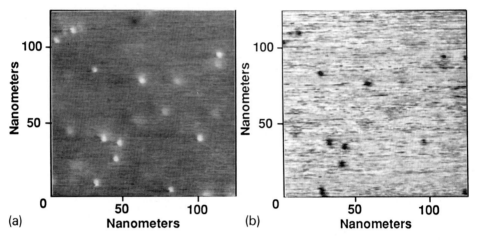

Figure 7.17 STM images obtained by sequential scans of the same area of an MoS$_2$ surface: (a) $V_{bias} = -800$ mV, $I_{set} = 1$ nA; (b) $V_{bias} = 800$ mV, $I_{set} = 1$ nA. Some thermal drift is evident between the two scans (Courtesy of Dr. B. Parkinson).

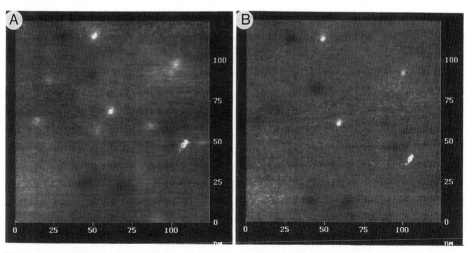

Figure 7.18 STM images of WSe$_2$: (a) $V_{bias} = -600$ mV, $I_{set} = 1.0$ nA; (b) $V_{bias} = 600$ mV, $I_{set} = 1.0$ nA. The contrast covers height variations in the 0.0–1.0 nm range.

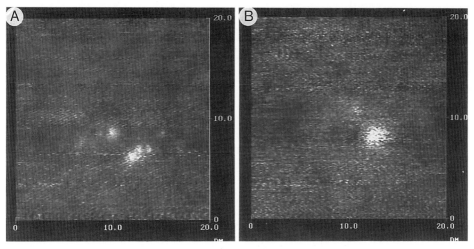

Figure 7.19 STM images of MoS$_2$: (a) $V_{bias} = -500$ mV, $I_{set} = 3.6$ nA; (b) $V_{bias} = 1400$ mV, $I_{set} = 3.6$ nA. The contrast covers height variations in the 0.0–0.3 nm range.

bright spots at negative V_{bias} become dark at positive V_{bias}. Large-scale STM images of WSe$_2$ (Fig. 7.18) show bright spots at negative V_{bias} that remain bright at positive V_{bias}. Figure 7.19 shows nanometer-scale bright spots of MoS$_2$ that are bright at both polarities of V_{bias}. Although these observations are apparently perplexing, the bright spots at negative V_{bias} can be assigned to the electron-trapped areas around donor dopants on the surface chalcogen-atom sheets. The brightness of these areas is affected by the nature of the donors and the tip–sample interactions. In addition, band-bending effects can contribute to lack of contrast change upon changing the bias polarity [14c]. The most likely donor dopants are Cl atoms for MoS$_2$, and I atoms for MoSe$_2$ and WSe$_2$ because, for the growth of the sulfide and selenides, Cl$_2$ and I$_2$ are used as the respective transport agents [20]. In the STM image of WSe$_2$ (Fig. 7.18), there are nanometer-scale spots that are dark at both polarities of V_{bias}. This fits the case of the transition-metal acceptor defect producing a magnetic d^1 cation in the Mo-atom sheet.

7.6 Concluding Remarks

The theoretical analysis presented above suggests a rational way of interpreting the STM image imperfections resulting from point defects of layered semiconducting inorganic materials. To develop this approach further, one needs to carry out controlled experiments, from synthesis to measurements. On the part of synthesis, it would be desirable to prepare samples with various types of dopants in differing amounts. On the part of STM measurements of such samples, it is necessary to examine their image imperfections as a function of the bias polarities as well as the gap resistance. A decrease in R_{gap} leads to stronger tip-sample force interactions, and imaging at different R_{gap} allows one to recognize the effects of tip force induced surface relaxation at defective locations (see Chapter 9).

References

[1] H. A. Mizes, W. A. Harrison, *J. Vac. Sci. Technol. A* **1988**, *6*, 300.
[2] M. R. Soto, *Surf. Sci.* **1990**, *225*, 190.
[3] G. P. E. M. van Bakel, J. T. M. De Hosson, *Phys. Rev. B* **1992**, *46*, 201.
[4] M.-H. Whangbo, J. Ren, E. Canadell, D. Louder, B. Parkinson, H. Bengel, S. N. Magonov, *J. Am. Chem. Soc.* **1993**, *115*, 3760.
[5] S. N. Magonov, P. Zönnchen, H. Rotter, G. Thiele, H.-J. Cantow, J. Ren, M.-H. Whangbo, *J. Am. Chem. Soc.* **1993**, *115*, 2495.
[6] S. N. Magonov, H.-J. Cantow, M.-H. Whangbo, *Surf. Sci. Lett.* **1994**, *318*, L1175.
[7] M.-H. Whangbo, J. Ren, S. N. Magonov, H. Bengel, B. A. Parkinson, A. Suna, *Surf. Sci.* **1995**, *326*, 311.
[8] (a) J. A. Wilson, A. D. Yoffe, *Adv. Phys.* **1969**, *18*, 193. (b) E. Doni, R. Girlanda, in *Electronic Structure and Electronic Transitions in Layered Materials* (Ed.: V. Grasso), Reidel, Dordrecht, The Netherlands, **1986**, p. 1.
[9] (a) G. W. Stupian, M. S. Leung, *Appl. Phys. Lett.* **1987**, *51*, 1560. (b) D. Sarid, T. D. Henson, N. R. Armstrong, L. S. Bell, *Appl. Phys. Lett.* **1988**, *52*, 2252. (c) M. Weimer, J. Kramar, C. Bai, J. D. Baldeschwieler, *Phys. Rev. B* **1988**, *7*, 4292. (d) M. Hara, Y. Iwakabe, K. Tochigi, H. Sasabe, A. F. Garito, Y. Yamada, *Nature (London)* **1990**, *346*, 228. (e) D. P. E. Smith, W. M. Heckl, *Nature (London)* **1990**, *346*, 616.
[10] R. V. Coleman, B. Giambattista, P. K. Hansma, W. W. McNairy, C. G. Slough, *Adv. Phys.* **1988**, *37*, 559.
[11] S. Hosoki, T. Hosaka, T. Hasagawa, *Appl. Surf. Sci.* **1992**, *60/61*, 643.
[12] (a) W. M. Heckl, F. Ohnesorge, G. Binnig, M. Specht, M. Hashimi, *J. Vac. Sci. Technol. B* **1991**, *9*, 1072. (b) Th. Schimmel, H. Fuchs, S. Akari, K. Dransfeld, *Appl. Phys. Lett.* **1991**, *58*, 1039.
[13] T. A. Albright, J. K. Burdett, M.-H. Whangbo, *Orbital Interactions in Chemistry*, Wiley, New York, **1985**.
[14] (a) R. M. Feenstra, J. A. Stroscio, *J. Vac. Sci. Technol. B* **1987**, *5*, 923. (b) Z. F. Zheng, M. B. Salmeron, E. R. Weber, *Appl. Phys. Lett.* **1994**, *64*, 1836. (c) J. F. Zheng, X. Lu, N. Newman, E. R. Weber, D. F. Ogletree, M. Salmeron, *Phys. Rev. Lett.* **1994**, *72*, 1490.
[15] C. Kittel, Introduction to Solid State Physics, 2nd ed., Wiley, New York, **1956**, pp. 353-358.
[16] (a) N. F. Mott, *Metal-Insulator Transitions*, Barnes and Noble, New York, **1977**. (b) B. H. Brandow, *Adv. Phys.* **1977**, *26*, 651. (c) M.-H. Whangbo, *J. Chem. Phys.* **1979**, *70*, 4963.
[17] R. J. Hamers, in *Scanning Tunneling Microscopy and Spectroscopy*, (Ed.: D. A. Bonnel), VCH, New York, **1993**, p. 88.
[18] (a) K. A. Yee, T. Hughbanks, *Inorg. Chem.* **1991**, *30*, 2321. (b) M.-H. Whangbo, E. Canadell, *J. Am. Chem. Soc.* **1992**, *114*, 9587.
[19] R. D. Shannon, *Acta Crystallogr. A* 1976, *32*, 751.
[20] F. Levy, *Nuovo Cimento B* **1977**, *38*, 359.

8 Tip–Sample Interactions

For a number of layered materials, the essential aspects of their STM and AFM images can be understood in terms of the partial and total electron density plots calculated without considering the possible effects of tip–sample force interactions (Chapter 6). However, many experimental results are not explained unless tip–sample interactions are taken into consideration. STM images can be influenced by switching the electronic state of the tip participating in the tunneling process, by perturbing the surface electronic structure of the sample, or by deforming the sample surface mechanically. The latter is also expected to affect AFM images. How much the tip–sample interactions influence the images depends on whether the experiments are conducted in air, under liquid or in UHV. This chapter provides a brief survey of experimental and theoretical studies concerning electronic and mechanical tip–sample interactions.

8.1 Electronic Interactions in STM

8.1.1 Tip Electronic States

According to Tersoff and Hamman, the nature of the tip orbitals participating in the tunneling process can have a significant effect on atomic-scale STM features [1]. The participation of the tip orbitals other than the s-orbital, and their effect on the images, have been the subject of several studies [2, 3]. For a tip of transition-metal atoms (e.g., W), the d_{z^2} orbitals of the tip-apex atoms play a significant role in the electron transfer [2]. First-principles calculations for the electronic states of several different W clusters show that the d_{z^2} orbitals of the apex atoms contribute to the states close to the Fermi level and are involved in the tunneling process [3]. STM studies of Au(111) and Al(111) surfaces show that the atomic-scale images have a much sharper height corrugation (0.3–0.8 Å) [4] than expected for elementary metals whose surface electron densities are dominated by s-type orbitals. Chen explained this finding in terms of the participation of the tip d_{z^2} orbital in probing these surfaces (Fig. 8.1) [5a]. (For an alternative explanation, see below.) Calculations of partial electron density plots with several different tip states indicate [5b] that the image features change by the "switching" from one electronic state of the tip to another. It was suggested that such a switching is responsible for the image changes observed for the reconstructed Au(111) and Si(111) surfaces [6a,b]. The STM images recorded for the (111) surface of a Pt/Ni alloy $Pt_{25}Ni_{75}$ occasionally exhibit two kinds of atomic-size spots with different contrast [6c]. This atom-selective imaging was attributed to the interaction of an adsorbate at the tip with the individual surface atoms. However, the origin of the observed contrast difference is not known, so it is not possible to establish which atoms (Pt or Ni) are represented by the brighter (or dimmer) spots.

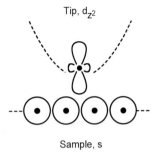

Figure 8.1 Participation of the tip d_{z^2} orbital in the STM imaging of an elemental metal whose partially filled band is largely represented by s orbitals. (Reproduced from Ref. 5a.)

8.1.2 Tip-Induced Local States

As mentioned earlier, the theoretical study by Ciraci and co-workers [7] suggests that as the tip–sample separation gradually decreases, the tip–sample interaction changes from the conventional tunneling regime to the electronic-contact regime and then to the mechanical-contact regime. Thus, the tip–sample force interaction changes from the attractive VDW type under the conventional tunneling regime to the repulsive one under the mechanical contact regime. It was suggested [7] that the charge density rearrangement within the unified tip–sample system (i.e., the tip-induced electronic state) plays an important role in the electron transfer. Therefore, the latter strongly depends on the local surface sites where the tip–sample interactions take place (e.g., the tops of the atoms or the hollow positions between the atoms on the surface). The tip-induced electronic states in the electronic-contact regime were analyzed in terms of first-principles calculations for an Al-atom tip interacting with an Al surface [7d]. With a separation of about 4.0 Å between the tip and the surface, the electronic states of the two electrodes start to interact. When the separation decreases, the electron density becomes concentrated within the atom–atom junction between the electrodes. This accounts for the enhanced corrugation observed on the surfaces of elemental metals whose electron densities are dominated by s-type orbitals. In such cases, the presence of the tip leads to a simple electron density increase and does not change the STM pattern. That the tip-induced local states can be site-dependent was also found in the study of tip–sample interactions between an Al atom tip and a graphite surface [8]. Figure 8.2 illustrates the charge accumulation in the tip–sample gap resulting from a tip-induced electronic state of the Al-atom tip/graphite system.

The primary objectives of the theoretical studies discussed above were the evaluation of the electronic effects of tip–sample interactions on STM images and the modeling of relevant electron-transport processes. These studies assumed that sample surfaces do not undergo a geometry relaxation under the tip force. This assumption is unlikely to be valid for many experimental situations. The tip–sample force interactions can induce macroscopic and microscopic surface structure relaxations, and the latter may strongly affect the patterns of atomic-scale images. It should be recalled that the microscopic deformation refers to a surface relaxation on nanometer and subnanometer scales, which are not treated by the continuum theory. As briefly mentioned in Section 5.2.3, the tip–sample force interaction occurring through the contamination layer causes a surface deformation even before the tip and sample experience electronic contact.

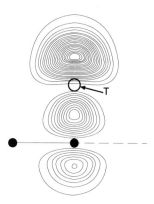

Figure 8.2 Cross-sectional view of the charge density contour plot of a tip-induced state for the Al-atom tip/graphite system. The cross-section is perpendicular to the surface graphite sheet and contains the Al atom lying on top of one graphite carbon atom (Reproduced from Ref. 8 b).

8.2 Force Interactions in STM

In STM the tip approaches the sample surface until the set-point current is reached (at a given bias voltage). This can bring the tip apex very close to the surface and hence lead to significant tip–sample force interactions. Due to the presence of the contamination layer which transmits the tip force, these interactions are typically stronger in ambient conditions than in UHV measurements, and can be responsible for the low tunneling barrier heights and anomalous atomic-scale image corrugations found for several compounds [9, 10]. Thus, it is necessary to consider tip–force effects in the analysis of the STM images obtained in air and in vacuum.

8.2.1 Force Interactions in Ambient Conditions

The influence of the tip–sample force interaction on the image contrast in ambient-condition measurements was first suggested in the explanation of giant corrugations in the STM images of HOPG [9], which exceed several times the corrugation of the total electron distribution (0.2 Å). Soler et al [9a]. attributed this effect to a force-induced amplification of the vertical displacement of the tip. In atomic-scale imaging of HOPG in air, Mamin et al [9b]. found that the image corrugations depend strongly on the gap resistance R_{gap} (in other words, the tip–sample distance), and that the corrugation increases from several ångströms to 24 Å with decreasing R_{gap}. They suggested that the tip force strongly deforms the graphite surface, and that the force is transmitted to the sample surface through the insulating contamination layer, which is always present on the surface in air and in a moderate vacuum (Fig. 8.3). To confirm this hypothesis, Mamin et al. performed additional experiments in UHV in which the contamination layer was practically absent. They found that the image corrugations were diminished, and the current vs. tip–sample distance (I-vs.-z) curve is steeper than that measured in air. These findings support their contamination-mediated deformation model of STM imaging in air [9b].

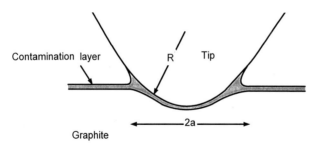

Figure 8.3 Contamination-mediated deformation model by Mamin et al [9b]. The tip force mediated by the contamination layer deforms the graphite surface. (Adapted from Ref. 9b.)

An alternative explanation for the large corrugation in the STM image of HOPG was given by Tekman and Ciraci in terms of tip-induced electronic states [11]. On the basis of pseudopotential calculations, they found that at small tip–sample separations the tunneling barrier gradually collapsed, and that the electronic states relevant for the tunneling process evolved into local resonance states which had relatively high weights in the tip–sample gap.

The importance of tip–sample force interactions became more evident from simultaneous STM/AFM measurements of HOPG [10, 12]. It was shown that under typical tunneling conditions in air, significant repulsive forces cause the compression of the sample surface. The study of Mate et al. [10b], motivated in part to understand the giant corrugations observed in the STM image of HOPG, strongly suggests that a large apparent corrugation of up to 20 Å is caused by a weak-spring element such as a whisker on the end of the tip. As the tip initially approaches the surface, the tip experiences repulsive forces before detecting a tunneling current. This important observation was confirmed by Salmeron et al. [12a], whose simultaneous STM/AFM imaging of HOPG showed that the current and force variations are in phase (Fig. 8.4). Both research groups found that the repulsive force that the tip experiences in STM experiments is in the 100–1000 nN range, which is considerably larger than the repulsive force typically employed in contact-mode AFM.

In the regime of strong tip–sample interactions, the nature of the electron transport can deviate from electron tunneling. In the conventional tunneling regime (at large tip–sample separations), it is expected that the I–V curve is linear, and the I-vs.-z relationship is exponential (see Chapter 2). As the tip approaches the sample

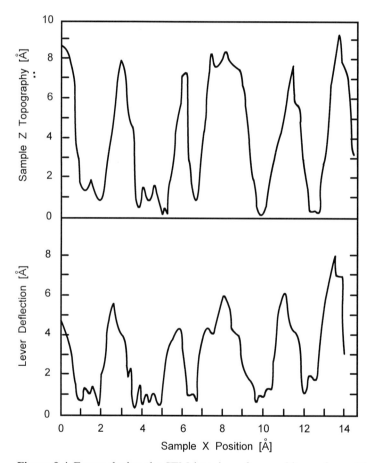

Figure 8.4 Forces during the STM imaging of a graphite surface with a Pt/Ir tip. The calibrated line profiles over 15 Å in the sample x-direction show several periods of the graphite lattice. The cantilever displacement and the z-piezo displacement were measured simultaneously (Reproduced from Ref. 12a).

further, the I-V curve is expected to become nonlinear, because the electron-transfer mechanism may deviate from the conventional tunneling one (see Section 2.1.1). However, in ambient-condition STM measurements on graphite and organic conductors, the opposite trend has been observed. For example, Fig. 8.5 shows a nonlinear dependence at R_{gap} higher than 20–30 MΩ. As R_{gap} is reduced to a few megaohms, the I-V curve becomes linear. The latter indicates the occurrence of an ohmic contact as suggested by Colton et al [13] (rather than that of a conventional tunneling). Several examples show that atomic-scale STM images can be obtained in the contact regime in air and in UHV. Smith et al. recorded atomic-scale images of HOPG when the tip was tentatively brought into contact with the sample [14a]. As mentioned in Chapter 5, Barrett et al. obtained atomic-scale STM images of 1T-TaS$_2$ from simultaneous STM/AFM experiments with the tip in contact with the surface [14b]. The image obtained in this mode is essentially identical with that recorded in normal STM experiments.

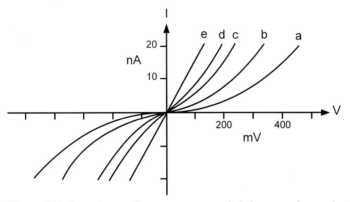

Figure 8.5 Current-vs.-voltage curves recorded for organic conductor α-(BEDT-TTF)$_2$I$_3$ at $I_{set} = 1$ nA and $V_{bias} =$ (a) 100, (b) 50, (c) 25, (d) 12.5, and (e) 6 mV.

8.2.2 Force Interactions in Ultra High Vacuum (UHV)

When STM experiments are conducted in UHV, the absence of the liquid contamination layer is expected to diminish the tip–sample interactions. Dürig et al. [15] carried out experimental studies of tip–sample force interactions in STM in UHV. They measured R_{gap} and the gradient (C_i) of the tip–sample force interaction using an iridium tip and an iridium sample, with the latter serving as a cantilever (Fig. 8.6 (a)). The gradient C_i was determined by monitoring the resonance frequency of the cantilever from the spectrum of tunneling current fluctuations (caused by the vibrations of the cantilever). The difference $\Delta\omega$ in the resonance frequencies of the interacting and free cantilevers is related to the force gradient by $\Delta\omega = 1.1 |C_i|$ ($\Delta\omega$ in Hz; C_i in N/m). The tip excursion z toward the sample was measured relative to the initial tip position (defined by $R_{gap} = 20$ MΩ), with positive z

corresponding to a decrease in the tip–sample distance. The results of such experiments show a nearly linear $\log(R_{gap})$-vs.-z relationship for $R_{gap} > 100$ kΩ (i.e., $z \leq 2.5$ Å) (Fig. 8.6 (a)). This relationship is expected for the electron transfer in the tunneling regime. When the tip–sample distance decreases, the $\log(R_{gap})$-vs.-z curve exhibits a discontinuity, and R_{gap} drops abruptly to an asymptotic value of 10 kΩ. This suggests a transition from tunneling to contact electron transfer. The interaction force gradient is negative over the range $0 < z < 2.5$ Å (Fig. 8.6 (b)), and this implies that the tip–sample force is attractive. At distances corresponding to $R_{gap} < 100$ kΩ, the force gradient becomes positive and is in the 15–50 N/m range [15b], thereby showing that the tip and sample came into mechanical contact in agreement with the behavior of the $\log(R_{gap})$-vs.-z curve.

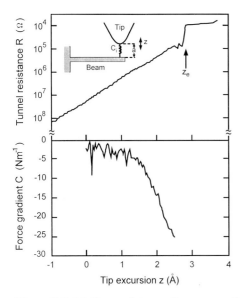

Figure 8.6 (a) Gap resistance R_{gap} versus tip excursion z measured on an iridium surface using an iridium tip. The tip–sample distance decreases with increasing z. The transition to tip–sample contact is indicated by the arrow. (b) Interaction force gradient vs. tip excursion measured simultaneously with gap resistance. The inset in the left-hand upper corner of (a) shows the experimental set-up. (Adapted from Ref. 15c.)

Tip–sample force interactions are present at all distances relevant for STM measurements in UHV and are relatively strong (in the nanonewton range), even in the attractive force regime. These results are important in analyzing the atomic-resolution images of simple metal surfaces recorded with relatively low R_{gap} (from several megaohms to tens of kiloohms). The extents of the atomic corrugations in such images increase exponentially with decreasing R_{gap}. This effect should originate at least in part from the tip force induced lattice relaxation, because such an image variation is also observed in the STM images of HOPG recorded in ambient conditions [10]. Thus, the attempt to enhance the atomic-scale image contrast by adjusting R_{gap} to smaller values might cause an imaging in the tip–sample contact regime. Salmeron and co-workers [12] obtained atomic-resolution images of the $(2\sqrt{3} \times 2\sqrt{3})$ R30° sulfur overlayer on the Re(0001) surface in UHV at low R_{gap} (even at 20 kΩ), for which the tip and sample are in contact according to Dürig et al. [15c] and Gimzewski and Möller [16].

Consideration of the tip–sample force interaction is also helpful in understanding the low values of the local tunneling barriers, ϕ, detected in STM experiments in air and in UHV. For example, the values of ϕ range from 1 to 10^{-4} eV, which are much lower than the typical work function (several electronvolts). The tunneling current I_{tun} is related to the tunneling barrier by

$$I_{tun} \propto \exp(-d\phi)$$

where d is the width of the barrier (i.e., the tip–sample distance). Salmeron et al. [12a] showed that the "normal" values of ϕ (of several electronvolts) can be achieved in UHV experiments on clean rhodium substrate only at R_{gap} values higher than 20 GΩ, and that the dependence of ϕ on R_{gap} (e.g., the collapse of the tunnel barrier) cannot be explained by electronic effects alone. Chen and Hammers [17] related the small values of ϕ found in the UHV experiments on Si(111) 7 × 7 to surface deformation. The real tip–sample separation deviates from the apparent one, measured by the piezodrive displacement, because of the force induced deformation of the sample surface (and the tip). Assuming the validity of the exponential dependence of the tunneling conductance on the true tip–sample separation over the whole STM operation range and taking into account the corrections associated with the surface deformation, they reproduced the experimental relationship between the apparent barrier height and piezodrive displacement.

The available experimental results indicate that tip–sample force interactions are present in STM experiments conducted in air and in UHV. In general, the mechanical contact between the tip and sample surface can be either direct or indirect (e.g., through the liquid contamination layer in air). It is not straightforward to predict what kind of mechanical contact is realized in the STM experiment in UHV, because this depends on the nature of the sample. For the iridium-tip/iridium-sample system under UHV, the mechanical contact (as judged by the discontinuity of the force gradient) and the electronic contact (as judged by the discontinuity in the $\log(R_{gap})$-vs.-z curve) occur nearly at the same tip–sample separation (Fig. 8.6). From STM studies of self-assembled monolayers of mercaptohexadecanol on Au(111) in UHV, Dürig et al. [18] showed that at the tip–sample separations corresponding to R_{gap} in the range 100 MΩ – 1 GΩ, the force gradient is positive (i.e., the tip–sample interaction is repulsive), while the $\log(R_{gap})$-vs.-z curve is still linear (i.e., the tunneling regime applies) (Fig. 8.7). In other words, the tip and sample engage in mechanical interactions in the region where the electron transport is still tunneling-like. (This situation differs from that found for the iridium-tip/iridium-sample system in UHV.) In addition, their results show that the compression of the monolayer occurs prior to the detection of the tunneling current. Estimates for self-assembled monolayers on Au(111) indicate that the lateral dimension of the tip–sample mechanical contact is approximately five times larger than the width of the tunneling current filament [18]. (In the iridium-tip/iridium-sample system, the tip–sample contact area should be much closer to the dimension of the tunneling current filament [15c].) It should be noted that STM imaging of self-assembled monolayers on Au(111) in UHV has features in common with imaging under ambient conditions; the insulating contamination layer on the sample surface in air might play a role similar to that of the self-assembled monolayer on Au(111) in UHV.

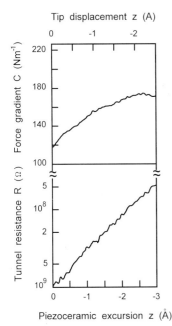

Figure 8.7 Interaction force gradient C_i and gap resistance R_{gap} measured as a function of the tip excursion z for a tungsten tip and a self-assembled monolayer of mercaptohexadecanol on Au(111) in UHV ($I_{set} = 0.1$ nA, $V_{bias} = 100$ mV). (Adapted from Ref. 18.)

In the analysis of the I-vs.-z curve observed for a $(2\sqrt{3} \times 2\sqrt{3})$ R30° sulfur overlayer on an Re(0001) surface, Salmeron et al. [12a] assumed that the 20 Å displacement of the tip actually represents the deformation of the sample surface, which increases the tip–sample contact area. In this case the experimental I-vs.-z curve (Fig. 8.8) is well described by the Sharvin resistance,

$$R = 4\rho l/3\pi a^2$$

where ρ is the specific resistivity of the material, l is the mean free path of an electron, and a is the radius of the circular contact area (the shape of the tip apex is approximated by a sphere). Figure 8.8 reveals that the I-vs.-z curve is quite steep. This explains the detection of atomic-scale images in the quantum-dot contact regime [14].

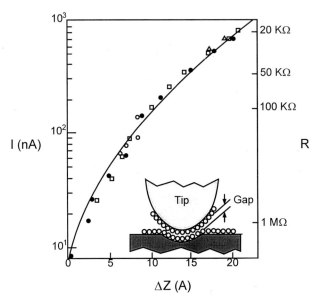

Figure 8.8 Tunneling current and gap resistance as a function of tip displacement for an Re(0001) surface with a $(2\sqrt{3} \times 2\sqrt{3})$R30° sulfur overlayer and a Pt/Rh tip. Tip displacement is measured relative to the initial tunneling conditions of $I_{set} = 8$ nA and $V_{bias} = -16$ mV. The symbols show different runs from the same sample and tip, and the solid line is calculated by assuming that the current increase is due solely to an increase in the area of contact for a conical tip. (Reproduced from Ref. 12a.)

During STM experiments, the imaging may proceed in the tunneling as well as in the contact (either electronic or mechanical) regime. In the latter case as well, the dominant contribution to the current is likely to come from the electronic states of the tip and sample in the vicinity of the Fermi level. Therefore, it is valid to employ the partial electron density plots for the interpretation of the STM images recorded in the contact mode. However, in the contact regime, the tip force may induce macroscopic as well as microscopic deformations of the sample surface. Consequently, a comprehensive analysis of the STM and AFM images ultimately requires a theoretical or experimental estimation of the surface deformation pattern and the subsequent electron density calculations.

8.3 Tip–Sample Interactions in AFM

The analysis of tip–sample force interactions in AFM is more straightforward than in STM, because the normal and lateral forces are measured directly. As discussed in Chapters 2 and 4, the repulsive, long-range, and capillary forces influence the mechanical response of the cantilever, and therefore the image contrast. However, the contrast variations in atomic-scale AFM images obtained by the contact mode are primarily caused by repulsive forces.

8.3.1 Force Interactions on the Atomic Scale

In a number of theoretical studies, the tip–sample force interactions have been examined using ideal tip–sample systems. The tip is commonly approximated by a single-atom tip (or an atomic layer) positioned over the sample, which is represented by one or several atomic slabs. The model systems frequently employed are elementary metals (gold, aluminum, silver, etc.) and graphite. Figure 8.9 shows force-vs.-distance curves (for the parallel and perpendicular force components) calculated for the tip of an aluminum layer and the sample, consisting of five aluminum layers [19]. The strongest attraction occurs when the tip is located over the hollow (H) site of the sur-

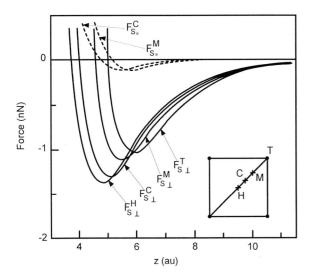

Figure 8.9 Perpendicular ($F_\perp$) and lateral ($F_\parallel$) short-range forces between two thin Al(001) slabs calculated as a function of their separation z. Forces are calculated for the H-, C-, M-, and T-sites shown in the inset. The C- and M-sites lie on the line between the hollow (H) and top (T) sites. (Reproduced from Ref. 19a.)

face. When the tip is positioned over the top (T) of the surface atom, the attractive interaction decreases, and the minimum of the curve is shifted to a larger tip-sample separation. Thus, the force is dominated by the strong ion-ion repulsion for scanning at small separations (less than 3.2 Å), and by the ion-electron attraction for scanning at larger separations. When the tip moves along the surface, the attractive interactions of neighboring surface atoms tend to enhance the normal force component but cancel out the lateral force component. Therefore, the frictional forces are much smaller than the normal ones (Fig. 8.9).

The adsorption energies of a Pd atom over different regions of the graphite lattice were calculated by Zhong and co-workers [20]. As in the case of the aluminum-tip/aluminum-sample system, either the H-sites or the T-sites of the substrate are energetically preferable for the adhesion of a Pd atom. The energy of adhesion is a function of the tip-sample separation so that, during the scanning of a one-atom tip, the apparent surface corrugations depend on the strength of the applied force. The elevated places correspond to the H-sites at low forces (ca. 1 nN), but to the T-sites at higher forces (ca. 10 nN). When the tip was modeled with three atoms, no direct correlation was found between the type of surface sites and the surface corrugation. The variations of the frictional force determined for the scanning of a one-atom Pd tip on HOPG exhibit periodic features which coincide with the lattice parameter of graphite.

The atomic and molecular processes involved in tip-sample contact were studied by molecular dynamics calculations [21]. This method is based on the numerical solutions of Newton's equations for a large assembly of atoms by employing two-body atom-atom potentials as well as three-body potentials [22]. In the molecular dynamics calculations, Landman et al. [23] represented the tip and surfaces by several hundred atoms. Their results reproduced macroscopic phenomena such as surface melting, neck formation, and jump-to-contact, as well as the elastic and plastic regimes of nanoscale indentation, in the gold-nickel system. In the simulation with a four-atom tip, scanning over a silicon surface, it was found that the periodic variations of the calculated normal and lateral forces coincide with the surface unit-cell parameters [23b].

Currently, molecular dynamics calculations are being used to study various aspects of STM and AFM. The limitations of this approach are the short time scale of calculations and the reliability of the atom-atom potentials used.

8.3.2 Surface Deformation

By analogy with the description of the mechanical interaction between two macroscopic solids, Pethica and Oliver [24a] used mechanical parameters such as hardness and surface stiffness (S) to describe the tip-sample force interactions in STM and

AFM. From dynamic force experiments conducted with a diamond tip on copper and sapphire surfaces, they determined the surface stiffness of the examined samples. The stiffness can be used to estimate a contact area and its dependence on the load, because S is related to the contact area a by

$$S = kaE^*$$

where k is a constant between 1.9 and 2.4 [25]. The modulus E^* is expressed as

$$1/E^* = (1 - v_1^2)/E_1 + (1 - v_2^2)/E_2$$

where v_i ($i = 1, 2$) are Poisson's ratios and E_i ($i = 1, 2$) are Young's moduli for the tip and sample [24].

Recently, the use of tip–sample force interactions for the study of nanomechanical surface properties (e.g., hardness and adhesion) has received much attention. For example, an atomic-force microscope can be applied as an advanced nanoindenter for operation with small forces and at small penetration depths. Typically, the conventional nanoindenter operates with applied forces starting from several millinewtons, and minimal penetration depths of ca. 20 nm were considered in practical applications [26]. Burnham and Colton [27a] demonstrated the possibilities of AFM for nanoindentation experiments on materials with different stiffness (e.g., graphite, elastomer, gold). They used applied forces up to 20 µN, and the penetration depths varied up to 100 nm. As expected, elastic behavior was found for graphite and elastomer, and plastic deformation for a gold foil. Tao et al. [27b] showed that for samples with different hardness (e.g., stainless steel, bone, graphite), force-vs.-distance curves in the high-force region exhibit different responses. Applying a pyramidal diamond tip, Bhushan and Koinkar [28] determined the hardness of Si(111) wafers. They produced irreversible indentation marks on the surface by applying forces in the 10–150 µN range and characterized the dimensions of the indentations by performing AFM imaging at smaller forces. In such a way, one can determine the surface hardness at an indentation depth as low as 1 nm, which is well below the minimum depth (20 nm) achieved with the conventional nanoindenter [26].

The mapping of surface mechanical properties can be achieved by collecting the force-vs.-distance curves for a set of mesh points of the examined area [29]. However, the reliability of these methods should be properly checked. In a simpler way, the lateral variations of the surface mechanical properties on a large scale and on an atomic scale can be recorded by the contact-mode AFM imaging with different applied forces. For a number of layered compounds discussed in Chapter 9, AFM imaging with high forces leads to a microscopic deformation of the sample surface. This deformation proceeds through surface relaxation on the atomic scale and is explained in terms of variation in the local hardness.

8.4 Concluding Remarks

The present survey points out several important observations which are essential for understanding STM and AFM imaging. The theoretical study of Ciraci et al. [7] predicts that, as the tip approaches the sample surface, the electron transfer process changes from the conventional tunneling regime changes to the electronic-contact regime before tip-sample mechanical contact occurs. Experimentally, however, a different picture emerges: namely, that tip-sample mechanical interaction occurs prior to electronic contact. Therefore, the tip and sample may undergo geometry relaxation even in the tunneling regime due to their mechanical interactions. This happens typically in ambient-condition measurements (with I_{set} in the nanoampere range) and may also occur in UHV measurements, depending on the nature of samples and the magnitude of R_{gap}.

References

[1] J. Tersoff, D. R. Hamman, *Phys. Rev. B* **1985**, *31*, 85.
[2] A. Baratoff, *Physica* **1984**, *127B*, 143.
[3] S. Ohnishi, M. Tsukada, *Solid State Commun.* **1989**, *71*, 391.
[4] J. Winterlin, J, Wiechers, H. Brune, T. Gritsch, H. Hofer, R. J. Behm, *Phys. Rev. Lett.* **1989**, *62*, 59.
[5] (a) C. J. Chen, *Phys. Rev. Lett.* **1990**, *65*, 448. (b) C. J. Chen, *Phys. Rev. Lett.* **1992**, *69*, 1656.
[6] (a) J. V. Barth, H. Burne, G. Ertl, R J. Behm, *Phys. Rev. B* **1990**, *42*, 937. (b) J. P. Pelz, *Phys. Rev. B* **1991**, *43*, 6746. (c) M. Schmid, H. Stadler, P. Varga, *Phys. Rev. Lett.* **1993**, *70*, 1441.
[7] (a) S. Ciraci, *Ultramicroscopy* **1992**, *42-44*, 16. (b) S. Ciraci, in *Scanning Tunneling Microscopy III* (Eds.: R. Wiesendanger, H.-J. Güntherodt), Springer, Heidelberg, **1993**. (c) S. Ciraci, A. Baratoff, I. P. Batra, *Phys. Rev. B* **1990**, *41*, 2763. (d) S. Ciraci, A. Baratoff, I. P. Batra, *Phys. Rev. B* **1990**, *42*, 7618.
[8] (a) S. Ciraci, I. Batra, *Phys. Rev. B* 1987, *36*, 6194. (b) I. P. Batra, S. Ciraci, *J. Vac. Sci. Technol. A* **1988**, *6*, 313.
[9] (a) J. M. Soler, A. M. Baro, N. Garcia, H. Rohrer, *Phys. Rev. Lett.* **1986**, *57*, 444. (b) H. J. Mamin, E. Ganz, D. W. Abraham, R. E. Thompson, J. Clarke, *Phys. Rev. B* **1986**, *34*, 9015. (c) J. H. Coombs, J. B. Pethica, *IBM J. Res. Develop.* **1986**, *30*, 455.
[10] (a) H. Yamada, T. Fuji, K. Nakayama, *J. Vac. Sci. Technol. A* **1988**, *6*, 293. (b) C. M. Mate, R. Erlandsson, G. M. McClelland, S. Chiang, *Surf. Sci.* **1989**, *208*, 473.
[11] E. Tekman, S. Ciraci, *Phys. Rev. B* **1989**, *40*, 10286.
[12] (a) M. Salmeron, D. F. Ogletree, C. Ocal, H.-C. Wang, G. Neubauer, W. Kolbe, G. Meyers, *J. Vac. Sci. Technol. B* **1991**, *9*, 1347. (b) F. Ogletree, M. Salmeron, *Prog. Solid State Chem.* **1990**, *20*, 235.

[13] R. J. Colton, S. M. Baker, R. J. Driskoll, M. G. Youngquist, J. D. Baldeschwieler, W. J. Kaiser, *J. Vac. Sci. Technol. A* **1988**, *6*, 349.
[14] (a) D. P. E. Smith, G. Binnig, C. F. Quate, *Appl. Phys. Lett.* **1986**, *49*, 1166. (b) R. C. Barrett, J. Nogami, C. F. Quate, *Appl. Phys. Lett.* **1992**, *57*, 992.
[15] (a) U. Dürig, J. Gimzewski, D. W. Pohl, *Phys. Rev. Lett.* **1986**, *57*, 243. (b) U. Dürig, O. Züger, D. W. Pohl, *J. Microsc.* **1988**, *152*, 259. (c) U. Dürig, O. Züger, D. W. Pohl, *Phys. Rev. Lett.* **1990**, *65*, 349.
[16] J. K. Gimzewski, R. Möller, *Phys. Rev. B* **1987**, *36*, 1284.
[17] C. J. Chen, R. J. Hammers, *J. Vac. Sci. Technol. B* 1991, *2*, 503.
[18] U. Dürig, O. Züger, B. Michel, L. Häussling, H. Ringsdorf, *Phys. Rev. B* **1993**, *48*, 1711.
[19] (a) S. Ciraci, E. Tekman, A. Baratoff, I. P. Batra, *Phys. Rev. B* **1992**, *46*, 10411. (b) S. Ciraci, E. Tekman, M. Goksedag, A. Baratoff, I. P. Batra, *Ultramicroscopy* **1992**, *42-44*, 163.
[20] (a) W. Zhong, D. Tomanek, *Phys. Rev. Lett.* **1990**, *64*, 3054. (b) W. Zhong, G. Overney, D. Tomanek, *Europhys. Lett.* **1991**, *15*, 49. (c) G. Overney, W. Zhong, D. Tomanek, *J. Vac. Sci. Technol. B* **1991**, *9*, 479.
[21] Papers in *MRS Bulletin*, **1993** (May), *18* (5).
[22] F. H. Stillinger, T. A. Webber, *Phys. Rev. B* **1985**, *31*, 5262.
[23] (a) U. Landman, W. D. Luedke, N. A. Burnham, R. J. Colton, *Science* **1990**, *248*, 454. (b) U. Landman, W. D. Luedke, A. Nitzan, *Surf. Sci.* **1989**, *10*, L177.
[24] (a) J. B. Pethica, W. C. Oliver, *Phys. Scr.* **1987**, *T19*, 61. (b) J. D. Todd, J. B. Pethica, *J. Phys. Condens. Matter* **1989**, *1*, 9823. (c) T. P. Weihs, Z. Nawaz, S. P. Jarvis, J. B. Pethica, *Appl. Phys. Lett.* 1991, *59*, 353.
[25] K. Kendall, D. Tabor, *Proc. R. Soc., London, Ser. A* **1971**, *A323*, 321.
[26] J. B. Pethica, R. Hutchings, W. C. Oliver, *Philos. Mag. A* **1983**, *48*, 598.
[27] (a) N. A. Burnham, R. Colton, *J. Vac. Sci. Technol. A* **1988**, *7*, 296. (b) N. J. Tao, S. M. Lindsay, S. Lees, *Biophys. J.* **1992**, *63*, 1165.
[28] B. Bhushan, V. N. Koinkar, *Appl. Phys. Lett.* **1994**, *64*, 1653.
[29] (a) D. R. Baselt, J. D. Baldeschwieler, *J. Appl. Phys.* **1994**, *76*, 33. (b) K. O. van der Werf, C. A. J. Putman, B. G. de Grooth, J. Greve, *Appl. Phys. Lett.* **1994**, *65*, 1195. (c) M. Radmacher, J. P. Cleveland, M. Fritz, *Biophys. J.* **1994**, *66*, 2159.

9 Surface Relaxation in STM and AFM Images

Numerous theoretical and experimental results discussed in Chapter 8 indicate that STM and AFM imaging involve strong tip–sample force interactions. It is important to re-analyze the STM and AFM images of HOPG from this viewpoint. To uncover the influence of the tip–sample force interactions on the images, it is necessary to perform measurements at different applied forces. In the force-dependent AFM experiments on layered transition-metal trihalides and tellurides, reversible image variations are found, as will be discussed in this chapter. On a qualitative basis, these variations can be accounted for by considering the local surface hardness, which describes how easily the local region of a given surface atom (i. e., the atom plus its nearest-neighbor atoms) is depressed by the tip. Atoms with different chemical environments have different local hardnesses and hence are depressed to different extents under the tip force. This gives rise to a tip force induced surface corrugation.

9.1 Tip Force Induced Deformation in HOPG

As discussed in Section 5.4.3, atomic-scale STM and AFM images of HOPG exhibit the three-for-hexagon pattern. This pattern of the STM image is explained when the geometry of the graphite surface is assumed to be unaffected by the tip force. In contrast, the three-for-hexagon pattern of the AFM image cannot be understood unless the geometry relaxation of the graphite surface is taken into consideration. In addition, STM images of HOPG occasionally show a giant hexagonal superstructure (moiré) pattern which cannot be explained by electronic effects. In this section, these observations are accounted for by analyzing the way in which the surface graphite layer is relaxed under the tip force.

9.1.1 Three-for-Hexagon Pattern of HOPG

Due to the interlayer interactions occurring through the A-site carbon atoms, the π-electron band levels of HOPG in the vicinity of e_f are more concentrated on the p_π-orbitals of the B-site carbon atoms than on those of the A-site carbon atoms (Chapter 5). Therefore, the $\rho(e_f, r_0)$ plots for HOPG have a higher density on the B-site carbon atoms [1]. The STM and AFM images of HOPG obtained from simultaneous STM/AFM measurements [2] have a three-for-hexagon pattern with an identical peak registry. Consequently, the bright spots of the AFM images, which represent the regions of stronger tip–sample repulsive forces, coincide with those of the STM images (i.e., the B-sites) [1]. This indicates that the tip force depresses the B-site atoms less than the A-site atoms [2–4]. For an unrelaxed graphite surface monolayer, all carbon atoms should be seen with equal brightness in the AFM image.

The tip–sample interactions involving HOPG have been the subject of several first-principle electronic structure studies [5]. For instance, the force-responses of the atomic and hollow sites of the surface graphite monolayer were examined by employing a one-atom tip (e.g., Al [5a] and Pd [5b–d]). These studies show that the repulsive force felt by the tip atom is stronger over the atomic than over the hollow sites. The macroscopic topography change in the HOPG surface under the tip force was studied in terms of empirical potential functions [6]. In all these studies, however, the difference in the force-responses of the carbon atoms at the A- and B-sites of the surface graphite sheet was not examined. In the following, the atomic-scale surface relaxation of HOPG induced by the tip force is analyzed [7].

The local hardness of the A- and B-sites in HOPG can be estimated using the atom–atom potential [8]:

$$E = -Ar^{-6} + B \exp(-Cr)$$

for a nonbonded C$\cdots$C contact of distance r, where $A = 568$ Å^6 kcal/mol (2377 Å^6 kJ/mol), $B = 83.630$ kcal/mol (349.90 kJ/mol), and $C = 3.60$ Å^{-1}. This potential reproduces the interlayer distance of 3.35 Å in bulk HOPG. For simplicity, the local hardness of the A- and B-sites in the surface graphite monolayer is estimated on the basis of a graphite bilayer, because what matters eventually for STM and AFM imaging is the atomic-scale topography change in the surface layer. Here the main concern is not the macroscopic bending of the HOPG sample [6] under the tip force, but the microscopic relaxation of the surface monolayer, i.e., whether or not the B-site carbons are depressed less than the A-site carbons. Therefore, the geometry of the bottom sheet of the graphite bilayer can be assumed to be frozen.

Interesting quantities to calculate are the energies $\Delta E_A^{(0)}$ and $\Delta E_B^{(0)}$ needed to depress the surface carbon atoms at the A- and B-sites, respectively, by Δz_0 toward the underlying layer while keeping all the remaining carbon atom positions fixed. The $\Delta E_A^{(0)}$ and $\Delta E_B^{(0)}$ values calculated as a function of Δz_0 are summarized in Table 9.1, which reveals that, for a given amount of energy supplied, the B-site carbon can be depressed more than the A-site carbon, and the difference in their depressions is smaller than 0.05 Å. To achieve the same extent of depression, the A-site requires more energy, and is therefore harder, than the B-site. Given the fact that the A-site atoms have interlayer C$\cdots$C interactions while the B-site atoms do not, this finding is reasonable but contradicts the conclusion from the simultaneous STM/AFM study. This problem stems from the implicit assumption of the analysis, i.e., the treatment of the tip as a mathematical point tip.

Table 9.1 $\Delta E_A^{(0)}$, $\Delta E_B^{(0)}$, Δr, and ΔE_{str} values calculated as a function of Δz_0 (for $\Delta z_1 = 0$)

Δz_0 (Å)	$\Delta E_A^{(0)}$ (kcal/mol)[a]	$\Delta E_B^{(0)}$ (kcal/mol)	Δr (Å)	ΔE_{str} (kcal/mol)[a]
0.1	0.13	0.09	0.004	0.005
0.2	0.40	0.29	0.014	0.075
0.3	0.87	0.65	0.031	0.37
0.4	1.61	1.23	0.055	1.16
0.5	2.77	2.13	0.086	2.78

[a] 1 kcal/mol = 4.184 kJ/mol.

In a graphite sheet all carbon atoms are linked by the carbon–carbon bond network, and the stretching of the C–C bond raises its energy. Thus, the depression of one carbon atom (directly under the tip atom) by Δz_0 will induce that of its three first-nearest-neighbor (FNN) carbon atoms by Δz_1, that of its six second-nearest-neighbor (SNN) carbon atoms by Δz_2, and so on. Eventually this leads to a circular depression under the tip atom. The C–C bond length (~ 1.42 Å) of the graphite layer is not large compared with the atomic radii of the metal atoms Pt and Ir (1.39 and 1.36 Å, respectively) [9] constituting the Pt/Ir tip. Therefore, even for the most ideal case when the tip apex is given by a single atom, the depression effect of the tip force must be strongly felt by the four adjacent carbon atoms, namely the carbon atom lying directly under the tip atom plus its three FNN carbon atoms (i.e., $\Delta z_0 \sim \Delta z_1 > \Delta z_2 > ...$). On the basis of a graphite bilayer, one can calculate the energy $\Delta E_A^{(1)}$ needed to depress an A-site atom of the surface monolayer by Δz_0 and its three FNN B-site atoms by Δz_1, with the positions of all other atoms frozen, for several values of the ratio $\Delta z_1/\Delta z_0$. Likewise, one can calculate $\Delta E_B^{(1)}$ needed to depress a B-site atom of the surface monolayer by Δz_0 and its three FNN A-site atoms by Δz_1. Figure 9.1 plots the relative energy $\Delta\Delta E^{(1)} = \Delta E_B^{(1)} - \Delta E_A^{(1)}$ calculated as a function of $\Delta z_1/\Delta z_0$ for $\Delta z_0 = 0.3$ Å. Because the C–C bond length and the tip-atom radius are comparable, the ratio $\Delta z_1/\Delta z_0$ should be close to unity. Therefore, in agreement with experiment, the plot of Fig. 9.1 shows that the B-site is more difficult to depress than the A-site (i.e., $\Delta\Delta E^{(1)} > 0$).

For a C_{sp^2}–C_{sp^2} bond, the bond stretch by Δr is estimated to increase the bond energy by ΔE_{str} [10]:

$$\Delta E_{str} = A (\Delta r)^2$$

where $A = 379.8$ kcal/mol Å² (1589 kJ/mol Å²). For $\Delta z_1 = 0$, the Δr value of the graphite C–C bond varies with the Δz_0 value as listed in Table 9.1. Also presented in this table are the ΔE_{str} values calculated (for $\Delta z_1 = 0$) as a function of Δz_0. When $\Delta z_0 < \sim 0.2$ Å, ΔE_{str} is negligible compared with $\Delta E_A^{(1)}$ and $\Delta E_B^{(1)}$, because

Δr is very small. When $\Delta z_0 > \sim 0.3$ Å, ΔE_{str} becomes close to, and eventually larger than, $\Delta E_A^{(1)}$ and $\Delta E_B^{(1)}$. Thus, for $\Delta z_0 > \sim 0.3$ Å, the depression of one carbon should induce that of its three FNN atoms, and that of its six SNN atoms, and so on, even if it is imagined that only the carbon atom right underneath the most protruding tip atom is depressed. The experimental finding that the B-site is harder than the A-site on the surface of HOPG is understandable if the most protruding tip atom exerts force upon the surface mainly through the carbon atom lying directly underneath it plus its three FNN carbon atoms so that $\Delta z_0 \sim \Delta z_1 > \Delta z_2 > \dots$.

Under the tip force, the B-sites become less depressed than the A-sites. This explains the three-for-hexagon pattern of the AFM image and enhances the contrast difference between the A- and B-sites in the STM image. That is, the STM image of HOPG reflects both the electronic and mechanical effects.

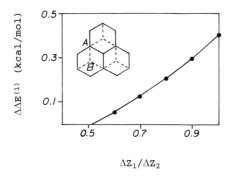

Figure 9.1 Relative energy $\Delta\Delta E^{(1)} = \Delta E_B^{(1)} - \Delta E_A^{(1)}$ calculated as a function of $\Delta z_1/\Delta z_0$ for $\Delta z_0 = 0.3$ Å. Positive $\Delta\Delta E^{(1)}$ values mean that the B-site is harder than the A-site. The inset shows the arrangement of two adjacent graphite layers. 1 kcal/mol = 4.184 kJ/mol.

9.1.2 Hexagonal Moiré Patterns in STM Images

STM images of HOPG occasionally exhibit hexagonal moiré patterns [11, 12]. The large-scale image of Fig. 9.2 (a) shows bright hexagon patterns with a repeat distance of 5.5 nm on both surface terraces. At a high magnification (Fig. 9.2 (b)), the atomic features of the graphite lattice are seen in the bright patches as well as in the regions between them. Other superstructure patterns found for HOPG samples and graphitized polyimide films [11, 12] possess different periods in the 1–12 nm range. The hexagonal moiré patterns of HOPG can be simulated by using two adjacent graphite layers: when one layer is rotated with respect to the other by the angle θ, the period D of the resulting moiré pattern is related to the period d (0.246 nm) of the graphite lattice by [11 c]:

$$2D \sin(\theta/2) = d$$

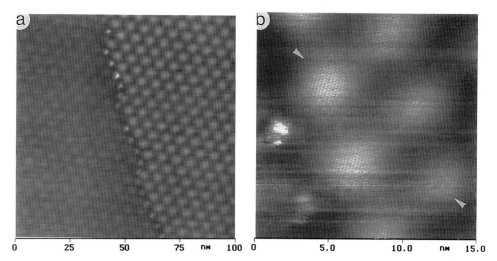

Figure 9.2 (a) Large-scale STM height image of HOPG showing a hexagonal moiré pattern (I_{set} = 0.2 nA, V_{bias} = −1 V). (b) STM height image for a part of the area shown in (a) (I_{set} = 0.2 nA, V_{bias} = −0.4 V). In both images the contrast covers height variations in the 0–1 nm range. The arrows in (b) indicate the direction along which to take a cross-sectional profile (see Fig. 9.3 (c)).

The moiré pattern of Fig. 9.3 (a) was obtained for $\theta = 4.7°$. To understand how such superstructure arrangements give rise to the observed moiré patterns in STM images, it is necessary to consider the tip force induced corrugation in the surface graphite monolayer.

The contribution of a surface atom to the $\rho(r_0, e_f)$ plot, and hence the associated STM image, increases as its distance to the tip decreases and as its electronic contribution to the energy levels around e_f increases. The $\rho(r_0, e_f)$ plot calculations for a variety of layered compounds [13] show that when the subsurface atoms lie more than 1 Å below the surface atoms, the STM patterns are dominated by the topmost surface atoms. Therefore, in terms of electronic contributions directly affecting the tunneling current, the moiré STM patterns of HOPG can only be associated with the surface graphite monolayer. In addition, for the moiré pattern to occur, the surface graphite monolayer cannot remain flat. For layered materials with moiré STM patterns, it has been suggested that the mechanical force exerted by the tip alters the topography of the surface layer according to its local hardness variation, and the imaging of the altered surface layer is responsible for the moiré patterns [13]. It is important to consider this aspect in more detail.

When the surface graphite monolayer is rotated with respect to the underlying graphite lattice, the surface-layer carbon atoms possess a number of different environments. This implies the presence of atomic-scale variations of the local surface hardness in the surface monolayer, which are expected to exhibit a moiré pattern. To examine this feasibility, the hardness map of the surface graphite monolayer was calculated using a graphite bilayer with a moiré pattern for which $\theta = 4.7°$ and $D = 30.0$ Å (Fig. 9.3 (a)) [7]. Each graphite layer was represented by a rhombus shaped sheet made up of 1536 carbon atoms. The map of the repulsive energies $\Delta E^{(1)}$, and also the map of the associated repulsive forces felt by the tip when it compresses the surface (as defined by $\Delta z_0 = 0.3$ Å and $\Delta z_1/\Delta z_0 = 0.8$), were obtained by calculating the $\Delta E^{(1)}$ values at all the carbon atoms of the surface layer. The force map of the surface monolayer is shown in Fig. 9.3 (b), where the large, medium, and small (dot-like) circles represent the atoms with small, medium, and large repulsive forces,

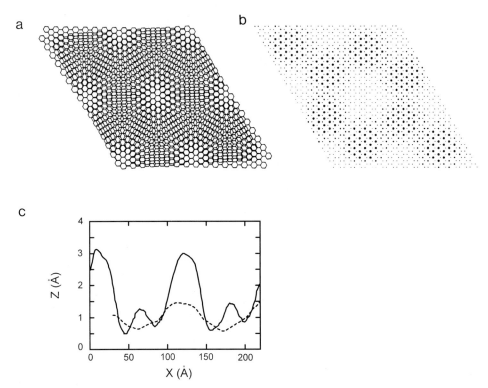

Figure 9.3 (a) Moiré pattern of $\theta = 4.7°$ and $D = 30.0$ Å generated by two adjacent rhombus-shaped graphite layers made up of 1536 carbon atoms. (b) Map of the repulsive forces calculated for the moiré bilayer in (a). The large, medium and small circles represent the contour values of 0.171, 0.182 and 0.209 nN, respectively. These forces are small in absolute value, because they are based only on the interlayer C⋯C interactions. For the discussion of the relative magnitudes of the local forces, the neglect of the C–C bond stretching effect on the forces (which accompanies the surface atom depression) is not significant. (c) Cross-sectional profiles of STM images with moiré pattern (reproduced from Ref. [14]) along a direction similar to that indicated by arrows in Fig. 9.2 (b). The tunneling conditions are $V_{bias} = 72$ mV, $I_{set} = 5.6$ nA for the solid curve, and $V_{bias} = 535$ mV, $I_{set} = 5.6$ nA for the broken curve.

respectively. The atom site with the higher repulsive force resists the compression effect of the tip force more strongly. Thus, the hardest region of the map (i.e., that composed of mostly sites with large repulsive forces) has an interlayer arrangement in which the carbon rings of the adjacent layers are mostly eclipsed (Figs. 9.3(a) and 9.3 (b)). The softest region (i.e., that consisting of mostly sites with small repulsive forces) has an interlayer arrangement resembling that of HOPG, and a region of intermediate hardness is found between every two adjacent softest regions.

Clearly, the variation of the local repulsive force, and hence that of the local hardness, in the surface graphite monolayer exhibit a hexagonal moiré pattern. The harder region is less depressed under the tip force and hence becomes the brighter domain of the moiré STM pattern. Consequently, the map of the local hardness in Fig. 9.3 (b) predicts a moiré STM image containing three regions of different contrast: the brightest domains forming a hexagonal pattern; the darkest domains around each of the brightest ones forming a honeycomb pattern; and the intermediate domains between every two adjacent darkest domains also forming a honeycomb pattern. This prediction is in good agreement with the experimental result of Rong and Kuiper (Fig. 9.3 (c)) [14]. By increasing R_{gap} from 12.9 MΩ to 95.5 MΩ, Rong and Kuiper found not only a substantial decrease in the contrast difference between the brightest and darkest regions but also the loss of a clear distinction between the darkest regions and those of intermediate brightness (Fig. 9.3 (c)) [14]. This result supports the conclusion that the tip force induced topography change of the surface monolayer is essential for the occurrence of moiré patterns in the STM images of layered materials and overlayers on different substrates. It is noted that the observed moiré STM images of HOPG have the three-for-hexagon pattern in the darker as well as in the brighter regions. This can be explained if it is not the surface graphite monolayer but two or more top graphite layers that are rotated with respect to the underlying graphite lattice.

9.2 Wagon-Wheel Patterns of MoSe$_2$ Epilayers on MoS$_2$

In general, superstructure patterns are frequently observed in the STM images of "overlayer" structures whose lattice constants differ from those of the underlying substrate. Examples include organic layers deposited on HOPG and MoS$_2$ substrates (see Chapter 11), single graphite layers grown on Pt(111) [15a], palladium monolayers electrochemically deposited on Ag(111) [15b], FeO adsorbates on Pt(111) [15c], and MoSe$_2$ epilayers grown on MoS$_2$ [16]. To explain these observations, it is necessary to consider how the lattice mismatch at the interface affects the geometry of the topmost layer probed by STM.

As a representative example, one may examine the STM images of MoSe$_2$ epilayers grown by molecular-beam epitaxy onto the isostructural layered compound

MoS$_2$. The lattice parameter of the MoSe$_2$ layer is larger than that of the MoS$_2$ layer by 5%, so the epitaxial growth of MoSe$_2$ on MoS$_2$ leads to a lattice mismatch between the adjacent MoSe$_2$ and MoS$_2$ layers (Fig. 9.4 (a)). The two hexagonal sheets of S and Se atoms at the interface generate an interfacial Se/S arrangement with a "wagon-wheel" pattern, as illustrated in a top view in Fig. 9.4 (b). The STM images obtained for the MoSe$_2$/MoS$_2$ system by Parkinson et al. exhibit a wagon-wheel superstructure pattern (Figs. 9.5 (a)–9.5 (c)) [16]. They found that STM experiments on epilayers up to 10 layers thick clearly show the wagon-wheel pattern. According to the results of Chapter 6, the STM image of the MoSe$_2$/MoS$_2$ system can only show the contributions of the surface Se atoms of the topmost MoSe$_2$ layer. In addition, the VDW interactions at the Se/S interface cannot influence the electronic structure of the topmost MoSe$_2$ layer. Under the tip force, the MoSe$_6$ trigonal prisms of the topmost MoSe$_2$ epilayer are in a number of different environments

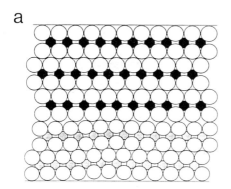

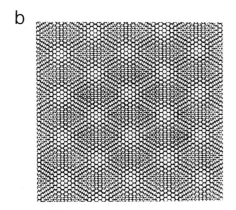

Figure 9.4 (a) Cross-sectional view of MoSe$_2$ epilayers on MoS$_2$. Larger and smaller empty circles represent the Se and S atoms, respectively, while small dark and gray circles represent Mo atoms. (b) Top view of two unrotated hexagonal closest-packed lattices with a 10% lattice mismatch showing a moiré pattern. (Reproduced from Ref. [16].)

due to the wagon-wheel pattern of the Se/S arrangement at the interface. Therefore, the tip force induced surface corrugation of a wagon-wheel type is observed in the STM images of the MoSe$_2$ epilayer (Fig. 9.5 (c)).

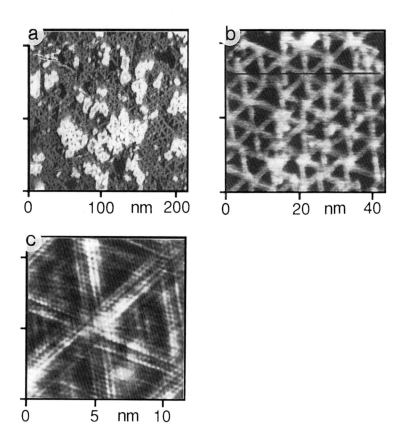

Figure 9.5 (a) STM height image of a four-monolayer coverage of MoSe$_2$ on MoS$_2$ ($V_{bias} = -1.21$ V, $I_{set} = 0.87$ nA). (b) STM height image of a two-monolayer coverage of MoSe$_2$ on MoS$_2$ ($V_{bias} = -0.87$ V, $I_{set} = 0.87$ nA). (c) STM current image of a two-monolayer coverage of MoSe$_2$ on MoS$_2$ ($V_{bias} = -0.75$ V, $I_{set} = 1.6$ nA). The contrast covers height variations in the 0–5 nm range in (a) and (b), and it is proportional to current variations in relevant units in (c). (Reproduced from Ref. [16]; Courtesy of Dr. B. Parkinson.)

9.3 STM and AFM Images of α-RuCl$_3$ and α-MoCl$_3$

In general, the forces applied to the sample surface in AFM are lower than those in STM. To enhance the tip force induced surface relaxation in AFM to the level observed in STM, therefore, it is necessary to carry out high-force AFM imaging. The

AFM images of layered transition-metal trihalides MCl$_3$ (e.g., α-RuCl$_3$, α-MoCl$_3$) undergo reversible changes when the applied force is increased from 30–50 nN to 200–300 nN. The analysis of these images is presented below.

9.3.1 Images of α-RuCl$_3$ at Low Applied Force

The layered transition metal trihalide α-RuCl$_3$ possesses a structure consisting of Cl–Ru–Cl layers. In each layer, the metal-atom sheet is sandwiched between two Cl-atom sheets. The Cl atoms form a hexagonal lattice, and the metal atoms form a honeycomb lattice. Each metal atom is located at an octahedral site (Fig. 9.6). The bulk structure of α-RuCl$_3$ [17] has two crystallographically different Cl atoms, Cl$_H$ and Cl$_L$: the Cl$_H$ atoms lie higher than the Cl$_L$ atoms, and the Cl$_L$ atoms form a honeycomb lattice with the Cl$_H$ atoms occupying the hexagon centers of this lattice. The z-height difference (i.e., Δz) between the Cl$_H$ and Cl$_L$ atoms is negligible ($\Delta z = 0.014$ Å). As will be shown below, it is the interlayer interaction that essentially determines the structural relaxation of the surface Cl atoms of the topmost RuCl$_3$ layer during STM and AFM measurements. Among the trihalides MCl$_3$, only α-RuCl$_3$ exhibits electrical conductivity which is high enough for STM measurements.

Given the oxidation states of Cl$^-$ and Ru^{3+}, the d-electron count for α-RuCl$_3$ is d^5 so that the highest occupied bands (i.e., the t$_{2g}$-block bands) of α-RuCl$_3$ are partially empty. Figure 9.7 shows the total DOS and PDOS plots calculated for a single RuCl$_3$ layer using the EHTB method. This compound is a magnetic semiconductor and becomes antiferromagnetic below 13 K [18]. Therefore, the electronic structure of α-RuCl$_3$ is such that the energy levels lying near the top portion of the t$_{2g}$-block

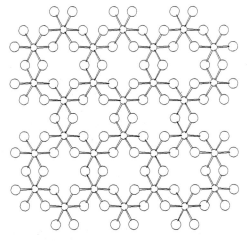

Figure 9.6 Top view of a single RuCl$_3$ layer taken from the crystal structure of α-RuCl$_3$. The metal and Cl atoms are represented by small and large circles, respectively.

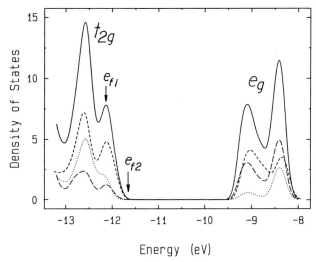

Figure 9.7 Total DOS and PDOS plots calculated for a single RuCl$_3$ layer of α-RuCl$_3$. Only the energy region between the top portion of the t$_{2g}$-block bands and the e$_g$-block bands are presented. ——, total DOS. PDOS values: – – –, for the Ru atom; – – – –, for the in-plane p-orbitals of Cl; · · · ·, for the out-of-plane p-orbital of Cl. e$_{f2}$ refers to the top of the t$_{2g}$-block bands, and e$_{f1}$ is the Fermi level relevant for the normal metallic state of α-RuCl$_3$.

bands are singly filled [19] and hence should be involved in the sample-to-tip as well as tip-to-sample tunneling process. Consequently, the images should not depend on the polarity of the bias voltage; this indeed was found to be the case [20].

The STM and AFM images of α-RuCl$_3$ are shown in Fig. 9.8 (a) and 9.8 (b), respectively. In contrast to the hexagonal arrangement of the surface Cl atoms (Fig. 9.6) and the hexagonal pattern of the AFM image, the STM image has a lower symmetry. Each "distorted centered-hexagon" of Fig. 9.8 (a) is characterized by three sides of 2.91, 3.28, and 3.67 Å [20]. To explain these observations, it is necessary to examine the $\rho(r_0, e_f)$ and $\rho(r_0)$ plots calculated for a single RuCl$_3$ layer [21]. Figure 9.8 (c) shows the $\rho(r_0, e_f)$ plot of a single RuCl$_3$ layer calculated by employing the energy levels lying within 0.27 eV of the top of the t$_{2g}$-block bands (i.e., between e$_{f1}$ and e$_{f2}$ of Fig. 9.7). The essential characteristics of the $\rho(r_0, e_f)$ plot do not depend strongly on the "energy window" used for calculations, as long as the energy window is chosen from the top portion of the t$_{2g}$-block bands. Figure 9.8 (c) shows that (1) only the top surface Cl atoms contribute to the $\rho(r_0, e_f)$ plot, (2) the HED spot associated with each Cl atom is at a location considerably removed from the atomic position, and (3) the intensity pattern of the electron density at each Cl atom is aligned along the a-, b-, or $(-a-b)$-direction to have a local three-fold rotational symmetry at each empty octahedral site.

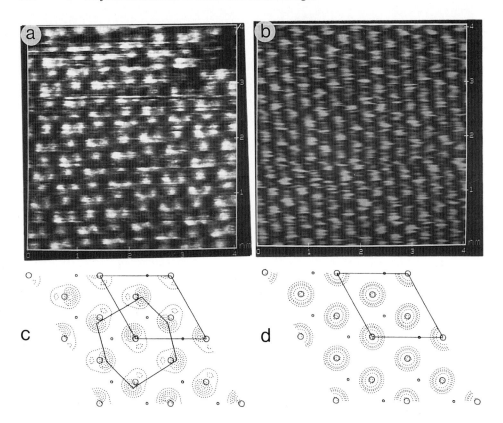

Figure 9.8 (a) STM current image of α-RuCl$_3$ (I_{set} = 5.7 nA, V_{bias} = 300 mV). The contrast variation is proportional to current in relative units. (b) AFM height image of α-RuCl$_3$. The contrast covers height variations in the 0–5 Å range. (c) 2D contour representation of the $\rho(r_0, e_f)$ plot calculated for a single α-RuCl$_3$ layer. The contour values used are 10×10^{-3}, 8×10^{-3}, 6×10^{-3}, and 2×10^{-3} electrons/au^3. (d) 2D contour representation of the $\rho(r_0)$ plot calculated for a single α-RuCl$_3$. The contour values used are 20×10^{-2}, 15×10^{-2}, 10×10^{-2}, 8×10^{-2}, and 5×10^{-2} electrons/au^3. In (c) and (d), the surface Cl and the Ru atoms are represented by large and small circles, respectively, and the rhombus indicates a surface unit cell.

The HED spots of the $\rho(r_0, e_f)$ plot form distorted centered-hexagons with three different sides of 3.08, 3.17, and 4.34 Å, which correspond to the values 2.91, 3.28, and 3.67 Å obtained from the STM images. The $\rho(r_0, e_f)$ plot of α-RuCl$_3$ reproduces the essential characteristics of the low-symmetry STM images. The HED positions (and hence the bright STM spots) are significantly removed from the Cl atom positions because, as shown in Fig. 9.7, the energy levels in the top portion of the t_{2g}-block bands possess the Cl-atom character — largely the in-plane p-orbitals (i.e., p_x and p_y) with some admixture of the out-of-plane p-orbital (i.e., p_z). The resulting hybridized p-orbital of each Cl atom is canted from the local z-axis of the atom, so that the HED spot for each Cl atom occurs at a location removed from the atomic site. In contrast, the $\rho(r_0)$ plot (Fig. 9.8 (d)) shows that the positions of the electron density peaks practically coincide with those of the surface Cl atoms. Thus, the main features of the AFM image agree well with the $\rho(r_0)$ plot.

9.3.2 Images of α-RuCl$_3$ at High Applied Force

The above interpretation of the STM and AFM images of α-RuCl$_3$, based on the electron density plots calculated for a undistorted layer, explains the essential features of the observed images. However, there are several important observations yet to be explained. First, in the STM images at low R_{gap}, the bright spots representing the surface Cl atoms are separated into two groups such that the brighter spots form a honeycomb lattice with the less bright ones occupying its hexagon centers (Fig. 9.9 (a)). Second, AFM images observed at high applied forces (Figs. 9.9 (b) and 9.9 (c)) exhibit a distorted pattern similar to that found for the STM images at low R_{gap}. These findings strongly suggest that the surface of α-RuCl$_3$ undergoes a tip force induced corrugation during STM and high-force AFM imaging.

9.3.3 Tip Force Induced Surface Deformation in α-RuCl$_3$

The above findings cannot be explained in terms of a single RuCl$_3$ layer, because all surface Cl atoms are equivalent in an isolated RuCl$_3$ layer. Therefore, it is necessary to examine the interactions between adjacent RuCl$_3$ layers. The bulk-crystal structure [17] of α-RuCl$_3$ is formed by stacking the RuCl$_3$ layers in such a way that the Cl atoms on the lower surface of one RuCl$_3$ layer are located above the Cl$_3$-triangle centers on the upper surface of the RuCl$_3$ layer lying below. A side projection view of the top two RuCl$_3$ layers under the tip (Fig. 9.10 (a)) demonstrates that this arrangement leads to two kinds of Cl atoms on the bottom surface of the first layer, which are indicated as gray and dark circles in Figs. 9.10 (a) and 9.10 (b). Each "gray" Cl atom lies above the center of a Cl$_3$ triangle possessing a Ru atom underneath, while each "dark" Cl atom lies above the center of a Cl$_3$ triangle with no Ru atom underneath (Fig. 9.10 (a)). Thus, when these Cl atoms are pushed against the top surface of the second layer, the "dark" Cl atoms will be depressed more easily than the "gray" Cl atoms. In other words, the "gray" and "dark" Cl atoms are "hard" and "soft", respectively. As depicted in Fig. 9.10 (b), the hard Cl atoms form a honeycomb lattice with the soft Cl atoms enclosed at the hexagon centers of this lattice.

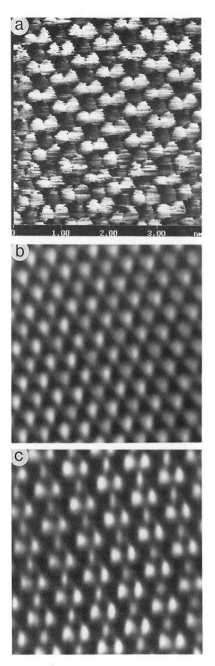

Figure 9.9 (a) STM current image of α-RuCl$_3$ (I_{set} = 27 nA, V_{bias} = 0.79 V). The contrast is proportional to current variations in relative units. AFM height images of α-RuCl$_3$ recorded with applied forces of 70 and 240 nN are shown in (b) and (c), respectively. The contrast covers height variations in the 0–0.4 nm range.

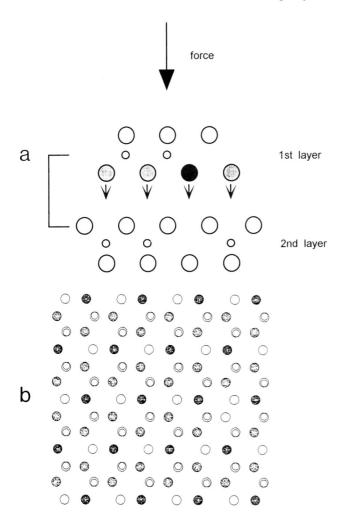

Figure 9.10 Arrangement of the top two RuCl₃ layers in α-RuCl₃: (a) side projection view and (b) top projection view. The small and large circles represent Ru and Cl, respectively. The top projection view includes only the metal atoms and the bottom-surface Cl atoms of the first RuCl₃ layer and the top-surface Cl atoms of the second RuCl₃ layer, as indicated by a bracket in (a). The bottom-surface Cl atoms of the first RuCl₃ layer are either gray (hard) or dark (soft), depending on their chemical environments.

It should be recalled that the bright spots of the AFM images represent the top-surface Cl atoms of the first RuCl₃ layer. Thus, it is necessary to examine how the presence of the hard and soft Cl atoms between adjacent layers affects the depression of the top-surface Cl atoms of the first RuCl₃ layer. A RuCl₃ layer is made up of edge-sharing RuCl₆ octahedra (Fig. 9.11 (a)), so that each top-surface Cl atom is connected to a bottom-surface Cl atom through a Ru₂Cl₂ rhombus (Fig. 9.11 (b)).

Such a rhombus provides the shortest bond linkage between the top- and bottom-surface Cl atoms, so that the force acting on a top-surface Cl atom would be transmitted to the bottom-surface mainly through the Ru_2Cl_2 rhombus containing the Cl atom. To a first approximation, therefore, each Ru_2Cl_2 rhombus may be considered to move as a unit under the tip force. Then, a top-surface Cl atom is hard if the bot-

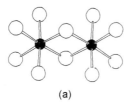

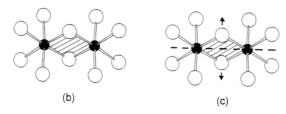

Figure 9.11 (a) Schematic view of two edge-sharing $RuCl_6$ octahedra. (b) Ru_2Cl_2 rhombus of two edge-sharing $RuCl_6$ octahedra (indicated by hatching). (c) Lateral displacements of the Cl atoms by the rotation around the Ru⋯Ru axis of a Ru_2Cl_2 rhombus.

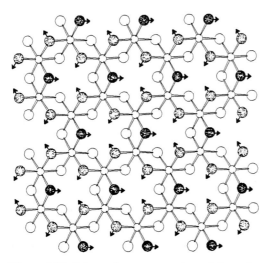

Figure 9.12 Schematic representation of the soft and hard Cl atoms (dark and gray, respectively) on the top surface of the first $RuCl_3$ layer. The small arrows represent the directions of the lateral displacements that the top-surface Cl atoms are expected to undergo.

tom-surface Cl atom of its Ru$_2$Cl$_2$ rhombus is hard, and soft otherwise. Consequently, on the top surface of the first RuCl$_3$ layer, the hard Cl atoms form a honeycomb lattice with the soft Cl atoms occupying its hexagon centers as depicted in Fig. 9.12. This explains the occurrence of hard and soft Cl atoms in the STM and AFM images obtained with strong tip forces.

The plane of each Ru$_2$Cl$_2$ rhombus is not perpendicular but inclined to the plane of the RuCl$_3$ layer (Fig. 9.11 (b)). Consequently, the tip force will not only depress the top-surface Cl atom but it will also rotate the Ru$_2$Cl$_2$ rhombus around the Ru$\cdots$Ru axis (Fig. 9.11 (c)). The latter leads to a lateral displacement of the top-surface Cl atom along the direction perpendicular to the Ru$\cdots$Ru axis. As indicated in Fig. 9.12, the lateral motions of the top-surface Cl atoms lead to smaller Cl$_3$ triangles made up of two hard and one soft atoms. Since the soft atoms are depressed more, the hard atoms appear as pairs along one direction under the strong tip force. This explains why the bright spots pair up in high-force AFM images and why the same effect is enhanced in the STM images obtained at low R_{gap}.

9.3.4 AFM Images of α-MoCl$_3$

The structure of α-MoCl$_3$ is similar to that of α-RuCl$_3$, except that the Mo atoms in the metal-atom sheet of each MoCl$_3$ layer pair up in one direction (Fig. 9.13) [17b]. On the surface of each MoCl$_3$ layer, the z-heights of the Cl$_H$ atoms are greater than those of the Cl$_L$ atoms by 0.285 Å, because the Cl$_H$ atoms lie above the Mo–Mo dimer units. Thus, in the AFM image of α-MoCl$_3$ recorded with a weak

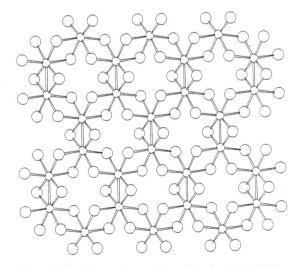

Figure 9.13 Schematic top view of a single MoCl$_3$ layer taken from the crystal structure of α-MoCl$_3$. The metal and Cl atoms are represented by small and large circles, respectively.

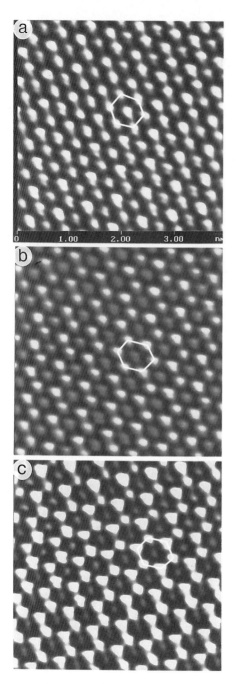

Figure 9.14 AFM height images of α-MoCl$_3$ recorded with (a) 30 nN, (b) 150 nN and (c) 240 nN. The contrast covers height variations in the 0–0.4 nm range. The hexagons have been added to indicate the positions associated with the Cl$_L$ atoms.

applied force, the Cl_H atoms should occur as brighter spots than do the Cl_L atoms. Therefore, the spots forming the honeycomb lattice should be less bright than those present at its hexagon centers. This is indeed the case, as can be seen from the image in Fig. 9.14 (a).

The interlayer arrangement of α-$MoCl_3$ is the same as that of α-$RuCl_3$, i.e., the Cl_H sites are softer than the Cl_L sites. Consequently, under strong applied forces, the Cl_H sites may be depressed more than the Cl_L sites to make the z-height of Cl_H lower than that of Cl_L, and the contrast pattern of α-$MoCl_3$ becomes similar to that of α-$RuCl_3$. This expectation is confirmed by the AFM images of α-$MoCl_3$ recorded at higher forces (Figs. 9.14 (b) and 9.14 (c)).

9.4 Layered Transition-Metal Tellurides MA_xTe_2

Depending on the value of x, layered tellurides MA_xTe_2 (M = Nb, Ta; A = Si, Ge; $\frac{1}{3} < x < \frac{1}{2}$) exhibit commensurate or incommensurate superstructure modulations of a nonelectronic origin. These compounds consist of Te–M/A–Te sandwich layers, which can be described in terms of three zigzag chains (a, b, and c types) made up of face-sharing MTe_6 trigonal prisms (Fig. 9.15). The a- and b-chains each contain one M–M dimer and one A atom (in a "square-planar" site) per unit cell, while the c-chain contains two isolated M atoms per unit cell. Since the unit cell of a commensurate sandwich layer is given by $(ab)_nc$ ($n = 1, 2, 3, \ldots$), the general formulas of commensurate MA_xTe_2 phases are given by $MA_{n/(2n+1)}Te_2$, i.e., $x = n/(2n+1)$. In the commensurate tellurides $NbSi_{1/2}Te_2$ [22], $NbGe_{1/3}Te_2$ [23], $NbGe_{2/5}Te_2$ [24], and $NbGe_{3/7}Te_2$ [25], the Te–Nb/A–Te sandwich layers consist of $(ab)_n$, $(abc)_n$, $(ababc)_n$ and $(abababc)_n$, respectively (Fig. 9.15). Such a periodic order is absent in the incommensurate tellurides $TaSi_{0.355}Te_2$ [26] and $TaSi_{0.414}Te_2$ [27]. It is recalled that a structural modulation is commensurate when the period of modulation is expressed as an integral combination of the lattice translations of the basic unmodulated structure, and incommensurate otherwise. Several incommensurate MA_xTe_2 phases have been studied by single-crystal X-ray diffraction with (3+1)-dimensional superspace group, and the results have been interpreted in terms of a real-space three-dimensional picture in which the Te–M/A–Te sandwich layers have a quasi-periodic distribution of the zigzag chains a, b, and c (e.g., $\ldots(abc)_4(ab)(abc)_3(ab)\ldots$).

9.4.1 Atomic-Scale Deformation in the Commensurate Tellurides

For the commensurate MA_xTe_2 phases, atomic-scale STM and AFM images of their surface Te-atom sheets exhibit the contrast variations corresponding to the crystallographic repeat distances [28]. However, the atomic-scale patterns within the unit cell

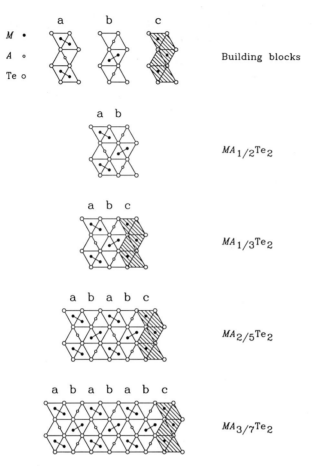

Figure 9.15 Top projection views of the sandwich layers of the commensurate MA_xTe_2 (A = Ge, Si) phases in terms of the building units, the a-, b-, and c-chains. The c-chain contains isolated M atoms, while the a- and b-chains contain M–M pairs and A.

of the observed images differ from those of the density plots calculated using the crystallographic data. The discrepancy is most likely to originate from a slight surface relaxation caused by the tip force. This is confirmed by force-dependent AFM studies.

The alternation of a- and b-chains defines the crystal structure of $NbSi_{1/2}Te_2$. Figure 9.16 shows the AFM images of $NbSi_{1/2}Te_2$ recorded with two different forces. The periodic patterns of the images correspond to the surface unit cell ($b = 0.7888$ nm, $c = 0.6336$ nm), and the individual atomic spots within the unit cell coincide with the geometrical positions of the Te atoms forming a pseudo-hexagonal lattice. The contrast difference between the individual spots increases with increasing applied force. The difference in contrast of the individual atoms within the surface

9.4 Layered Transition-Metal Tellurides Ma_xTe_2

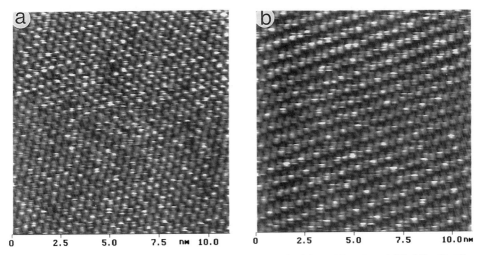

Figure 9.16 AFM height images of $NbSi_{1/2}Te_2$ obtained with (a) 70 nN and (b) 240 nN. The contrast covers height variations in the 0.0–1.0 nm range.

unit cell is clearly seen in the high-force AFM image of Fig. 9.17 (a), in which the brightest spots form the corners of the unit cell (shown by a rectangle) with four dimmer spots located within the cell. As in the case of Fig. 9.16 (b), the rows with alternating brightness are oriented along the b-axis, i.e., perpendicular to the a- and b-chains. The AFM pattern of Fig. 9.17 (a) is similar to that found in the STM image (Fig. 9.17 (b)). The unrelaxed surface structure of $NbSi_{1/2}Te_2$ (Fig. 9.17 (c)) does not exhibit height corrugations larger than 0.014 nm. The observed image variations arise from the tip force induced surface relaxation. The latter reflects the presence of soft and hard environments in the Te⋯Te arrangements between adjacent $NbSi_{1/2}Te_2$ layers [29].

The MA_xTe_2 compounds containing c-chains (e.g., $NbGe_{1/3}Te_2$) exhibit drastic changes in their AFM images under high forces. The repeat unit cell of $NbGe_{1/3}Te_2$ is given by (abc) chains. The images of $NbGe_{1/3}Te_2$ (Figs. 9.18 (a) and 9.18 (b)) show periodically arranged dark rows along the crystallographic a-direction, i.e., along the chain direction. The dark rows are more pronounced at a higher force, and the separation between them corresponds to the distance between the c-chains (1.169 nm). The contrast variations between the rows of spots in the direction perpendicular to the chains is less pronounced than in the case of $NbSi_{1/2}Te_2$. A similar effect is found in the images of $NbGe_{2/5}Te_2$ (Figs. 9.19 (a)–9.19 (d)), each layer of which has the repeat unit cell (ababc). The periodical contrast variation is difficult to recognize at a moderate applied force (Fig. 9.19 (a)) but is strongly enhanced at high forces (Figs. 9.19 (b)–9.19 (d)). Also, the image patterns do not depend on the sample position with respect to the fast-scanning direction. The separation between the dark rows is equal to the distance between the c-chains (1.947 nm). The dark

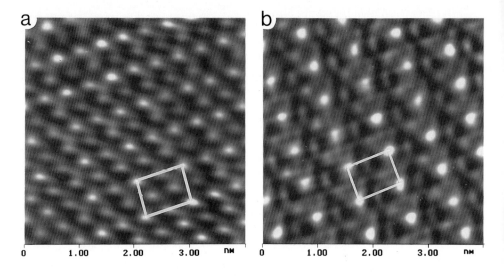

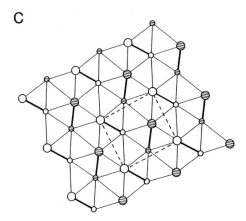

Figure 9.17 (a) AFM image and (b) STM image (I_{set} = 1.8 nA, V_{bias} = 200 mV) recorded on the bc-plane of an NbSi$_{1/2}$Te$_2$ crystal. The contrast covers height variations in the 0.0–1.1 nm range in (a) and is proportional to current variations in relative units in (b). (c) Schematic representation of the two interlocked rectangular sublattices of the surface Te-atom sheet of NbSi$_{1/2}$Te$_2$ (distinguished by shaded and unshaded circles). The higher- and lower-lying Te atoms are represented by large and small circles, respectively.

grooves, assigned to the c-chains, alternate with the wider and bright ribbon patterns representing the (abab)-chains. For completeness, it is necessary to recall the AFM images of the compound NbGe$_{3/7}$Te$_2$ with repeat unit cell (abababc) (Chapter 4, Fig. 4.7). The high-force images also exhibit the pronounced dark rows, the separation between them coinciding with the distance between the c-chains (2.735 nm). Consequently, for the commensurate tellurides NbGe$_{n/(2n+1)}$Te$_2$, the dark rows be-

9.4 *Layered Transition-Metal Tellurides* Ma_xTe_2 173

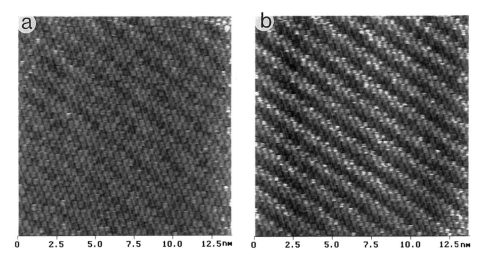

Figure 9.18 AFM height images of $NbGe_{1/3}Te_2$ obtained with (a) 120 nN and (b) 230 nN. The contrast covers height variations in the 0.0–0.5 nm range.

long to the Te atoms of the c-chains. In terms of the local hardness, this means that the Te atoms associated with the c-chains are softer than those associated with the a- and b-chains. This conclusion is supported by the following structural considerations.

The relative local hardness of the surface Te atoms may be discussed on the basis of commensurate $NbGe_{1/3}Te_2$. In the Te–Nb/Ge–Te sandwich layer (Fig. 9.20), the surface Te atom that is engaged in more short Te$\cdots$Nb and Te$\cdots$Ge contacts and in Te–Te dimer formation will be more difficult to depress with the tip force and hence will be harder. There are three sets of equivalent Te atoms in the surface Te-atom sheet [Te(2), Te(4); Te(1), Te(5); Te(3), Te(6)]. Te(2) and Te(4) are isolated, while the rest of the Te atoms form Te–Te dimers. Te(1), Te(2), and Te(3) are similar in terms of Te$\cdots$Nb contacts, and each has three Te$\cdots$Nb contacts smaller than 3 Å (i.e., 2.771, 2.813, and 2.851 Å for Te(2); 2.855, 2.898, and 2.972 Å for Te(1); 2.832, 2.853, and 2.981 Å for Te(3)). The Te$\cdots$Ge contact is long for Te(2) but short for Te(1) and Te(3) (3.824 Å for Te(2), 2.764 Å for Te(1) and 2.809 Å for Te(3)). Consequently, on the surface Te-atom sheet, the isolated Te-atom sites are softer and easier to depress than the Te atoms of the Te–Te dimers. Therefore, the tip forces will induce a corrugation on the surface Te-atom sheet, in which the zigzag chain of the isolated Te atoms is lower in "height" than are the two consecutive zigzag chains of Te atoms associated with the Te–Te dimers. Thus, as depicted in Fig. 9.20, the AFM of $NbGe_{1/3}Te_2$ should have, per unit cell, two bright and one dim zigzag chains of spots representing the surface Te atoms. The observed AFM image is in agreement with this prediction. Note that the presence of the c-chain is signaled by dim Te spots, and that of the a- or b-chain by bright Te spots. These features are expected to be

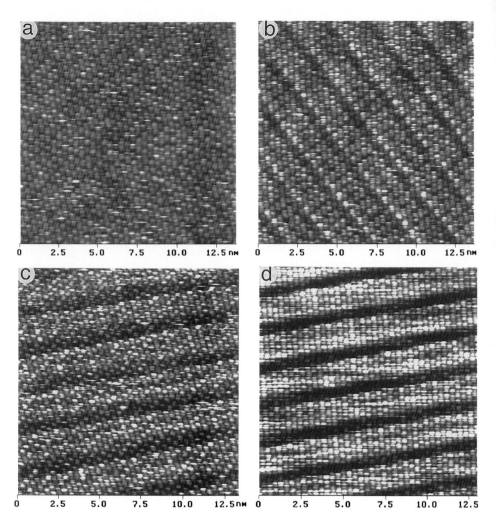

Figure 9.19 AFM height images of NbGe$_{2/5}$Te$_2$ obtained with (a) 70 nN, (b) 220 nN, (c) 140 nN, and (d) 240 nN. In all images the contrast covers height variations in the 0.0–1.0 nm range.

a characteristic pattern of the AFM images of both commensurate and incommensurate tellurides MA$_x$Te$_2$. It should be noted that the tip force induced height difference between the isolated Te atoms and the Te–Te dimers is much greater than that between the Te atoms within each Te–Te dimer (resulting from the interlayer Te $\cdots$ Te arrangement) found for the AFM images of NbSi$_{1/2}$Te$_2$ (Figs. 9.16 and 9.17).

The force-dependent AFM study of the commensurate tellurides MA$_x$Te$_2$ [30] raises an important question concerning the interpretation of their STM images. The STM images of NbGe$_{1/3}$Te$_2$ and NbGe$_{2/5}$Te$_2$ (Fig. 9.21) show that the width of the less-bright rows is considerably larger than that of the corresponding AFM images.

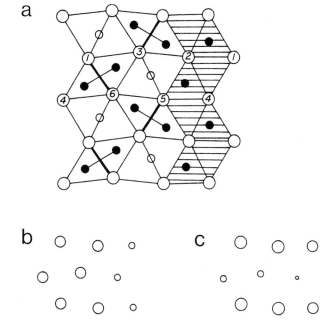

Figure 9.20 (a) Schematic top projection view of the Te–Nb/Ge–Te sandwich layer of NbGe$_{1/3}$Te$_2$. The Te atoms are represented by large empty circles, the Nb atoms by small filled circles, and the Ge atoms by small empty circles. The Nb–Nb dimers are joined by thin lines, and the Te–Te dimers by thick lines. The numbers 1–6 represent the Te(1)–Te(6) atoms, respectively. (b) Schematic AFM pattern expected for the surface Te-atom sheet of commensurate NbGe$_{1/3}$Te$_2$, where the larger circles represent the brighter Te-atom spots. (c) Schematic AFM pattern describing part of the observed image for incommensurate TaGe$_{0.355}$Te$_2$. The larger the circle is, the brighter the Te-atom spot becomes.

Because the tip force is high in STM measurements in air, it is certain that the isolated Te atoms belonging to the c-chains are depressed more than are the Te–Te dimers belonging to the a- and b-chains. In terms of this geometrical factor alone, the isolated Te atoms of the c-chains should appear as less bright than the Te–Te dimers of the a- and b-chains. However, it is not necessary that the more protruding atoms correspond to the brighter spots in STM images. Analysis of the STM images of ReSe$_2$ [31] and NbTe$_2$ (Chapter 6) shows that the bright spots are represented by the lower-lying surface chalcogen atoms. Thus, it is possible that the more depressed isolated Te atoms in MA$_x$Te$_2$ dominate in the STM image. Indeed, recent $\rho(r_0, e_f)$ plot calculations carried out for deformed surface structures confirmed this to be the case [32].

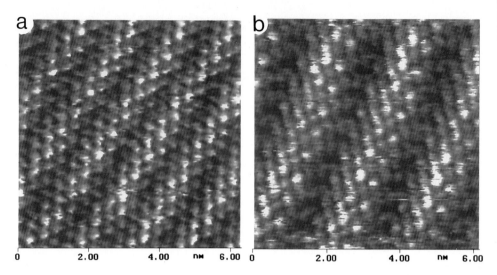

Figure 9.21 (a) STM height image of NbGe$_{1/3}$Te$_2$ (I_{set} = 5 nA, V_{bias} = 50 mV). The contrast covers height variations in the 0.0–1.4 nm range. (b) STM height image of NbGe$_{2/5}$Te$_2$ (I_{set} = 2 nA, V_{bias} = 50 mV). The contrast covers height variations in the 0.0–0.3 nm range.

9.4.2 Structure of Incommensurate Telluride TaGe$_{0.355}$Te$_2$

The fact that the isolated Te atoms belonging to the c-chains are more depressed than other Te atoms can be used for the structural characterization of incommensurate tellurides such as TaGe$_{0.355}$Te$_2$ [33]. A high-resolution AFM image of TaGe$_{0.355}$Te$_2$ is shown in Fig. 9.22 (a), which exhibits a pseudo-hexagonal pattern of bright spots representing the surface Te atoms. Figure 9.22 (b) also shows bright and less-bright rows along the direction perpendicular to the a-, b- and c-chains. This is similar to the pattern observed for NbSi$_{1/2}$Te$_2$ in Fig. 9.17 (a). All the surface Te atoms are seen, and their contrast differences lead to ribbon-like patterns. Therefore, the image of incommensurate TaGe$_{0.355}$Te$_2$ is similar to that of commensurate NbGe$_{1/3}$Te$_2$ obtained at high forces, except that the image of incommensurate TaGe$_{0.355}$Te$_2$ exhibits "ribbons" of different widths. As mentioned above, the presence of a c-chain is signaled by a row of dim Te spots along the chain direction. With this guide, one can identify from Fig. 9.22 (b), from left to right, the chain patterns [(ab)(abc)(abc)(abc)(abc)][(ab)(abc)(abc)(abc)(abc)(abc)][(ab)(abc)(abc)(abc)(abc)][(ab)(abc)...]. Thus, the image exhibits the absence of a c-chain at the intervals of five and six (abc) chains. This finding strongly supports the notion that the origin of the incommensurate MA$_x$Te$_2$ phases is a quasi-periodic distribution of the a-, b- and c-chains in the Te-M/A-Te sandwich layers [27].

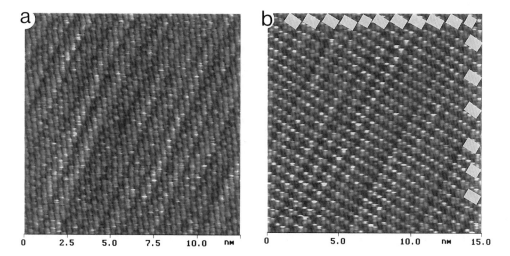

Figure 9.22 AFM height images of TaGe$_{0.355}$Te$_2$. In (a) and (b) the contrast covers height variations in the 0.0–0.7 nm range. The ribbon patterns corresponding to the (ab) and (abc) zigzag chains are marked by white strips of different length in (b).

9.5 Tip Force Induced Changes in AFM Images of NbTe$_2$

As part of the study aimed at understanding the atomic-scale mechanical response of a material under the tip force, force-dependent AFM studies were carried out on layered telluride NbTe$_2$ [34]. As described in Chapter 6, the crystal structure of this compound is characterized by the clustering of the Nb atoms into ribbon chains. Figures 9.23 (a)–9.23 (d) show atomic-scale AFM images of NbTe$_2$, all obtained for a fixed orientation of the sample with respect to the fast-scanning direction [34]. The image of Fig. 9.23 (a) shows a pseudo-hexagonal array of the surface Te atoms (separated by the distance of about 0.36 nm) as well as linear corrugations with a repeat distance of about 1 nm. The latter are caused by the clustering of the underlying Nb atoms. Every three consecutive rows of bright spots in this image correspond to those of the surface Te atoms and signal the presence of an Nb-atom ribbon chain lying directly underneath and running along the rows. For convenience, the direction of the bright rows found in the low-force image (ca. 70 nN, Fig. 9.23 (a)) may be defined as the 0°-direction. When the force is increased to ca. 200 nN, the rows of bright spots rotate their direction by 60° as shown in Fig. 9.23 (b). This rotation is reversible because an abrupt lowering of the applied force in the middle of the high-force scan restores the direction of the low-force image (Fig. 9.23 (c)). An AFM image recorded at an intermediate applied force (ca. 130 nN) is given in Fig. 9.23 (d), which shows bright rows running along the 0° and 60° directions. There occur "3 × 3

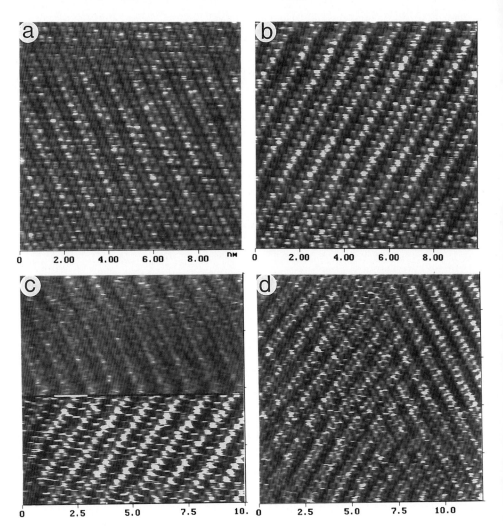

Figure 9.23 AFM height images of NbTe$_2$ obtained with (a) 70 nN, (b) 200 nN, (d) 130 nN. In (c), the lower part was obtained with 200 nN, and the upper part with 70 nN. The image was obtained with the tip scanning from the bottom to the top. The contrast covers height variations in the 0.0–0.3 nm range.

islands" consisting of nine spots in the region where the rows running along the 0° and 60° directions intersect (i.e., the central part of Fig. 9.23 (d)).

As already pointed out, every three consecutive rows of the bright AFM spots mean the presence of an Nb-atom ribbon chain lying directly underneath them and running along the rows. Thus, the rotation of the bright rows from 0° to 60° indicates the switch of the ribbon chains (e.g., from Fig. 9.24 (a) to Fig. 9.24 (c)). The 3 × 3 islands found at the intermediate level of applied force implies the occurrence of

3 × 3 metal clusters (Fig. 9.24 (b)). For the ribbon chains to switch their running direction reversibly under the tip force, the Nb–Nb bond rearrangement must occur reversibly; therefore the strength of the Nb–Nb bonds cannot be high. Indeed, this is evidenced by the long bond length (i.e., 3.329 Å) [35], which is understandable because each Nb atom engages in two linear three-center–two-electron bonds, Nb–Nb–Nb [36].

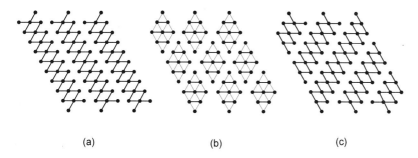

Figure 9.24 Patterns of Nb-atom clustering in the surface NbTe$_2$ layer under (a) low, (b) intermediate, and (c) high applied forces.

It is important to consider why the ribbon chains of the surface NbTe$_2$ layer switch their direction under high force. In the bulk crystal structure of NbTe$_2$ at ambient pressure [35], the ribbon chains in adjacent NbTe$_2$ layers are arranged to be parallel as depicted in Fig. 9.25 (a). Suppose that the sample of NbTe$_2$ under investigation had this arrangement. Then, as the ribbon chains of the surface NbTe$_2$ layer rotate by 60°, the top two adjacent NbTe$_2$ layers adopt a more isotropic arrangement of the ribbon chains (Fig. 9.25 (b)). In this model (hereafter referred to as model *a*), the ribbon chain arrangement of Fig. 9.25 (a) at low force changes to that

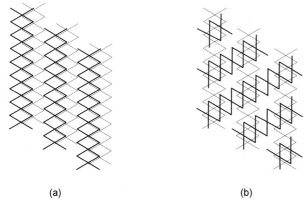

Figure 9.25 Arrangements of the ribbon chains in two adjacent NbTe$_2$ layers: (a) parallel arrangement; (b) 60°-rotated arrangement.

of Fig. 9.25 (b) at high force. It is expected that the tip pressure in the tip–sample contact area is circularly symmetric. According to the Le Chatelier–Braun principle [37], the observed ribbon-chain rotation implies that the sample responds to the mechanical perturbation by adopting a more circularly compact packing. One might propose an alternative model (hereafter referred to as model *b*). NbTe$_2$ being a layered material, it is quite probable that the sample under investigation has several of its top layers already rotated with respect to the lattice lying below. Then, the ribbon-chain arrangement of Fig. 9.25 (b) at low force changes to that of Fig. 9.25 (a) at high force. Therefore the bond switch in the "rotated" surface layers at high force occurs so as to follow the surface corrugations of the "unrotated" lattice lying below.

It is important to see if one can determine which explanation, model *a* or model *b*, is correct. Figure 9.26 shows a hypothetical layer arrangement in NbTe$_2$. Domains A and B have the same ribbon-chain direction but differ in their thickness. Domain C is thick, and its chain direction is rotated from that of domains A and B by 60°. According to model *b*, bond switching under high-force AFM imaging is possible in domain B, but not in domains A and C. According to model *a*, bond switching is expected to occur in all domains. Experimental results indicate that the reversible bond switching phenomenon occurs only in certain domains of the given sample. This finding is consistent with model *b*.

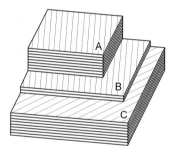

Figure 9.26 Schematic representation of three different domains on the surface of an NbTe$_2$ sample. The thin lines in domains A, B, and C represent the directions of the metal-atom ribbon chains. The chains in domains A and B are parallel to each other but are rotated by 60° with respect to those in domain C. Domains A and C are thick, and domain B is thin.

9.6 Nanoscale Ring Structures of MoS$_2$ and WSe$_2$

Tip–sample force interactions are also important in the analysis of the STM image imperfections of MoS$_2$ and WSe$_2$ caused by point defects. As discussed in Section 7.4.1, the electron-trapped area around a donor dopant on the chalcogen surface is created by slightly lengthening the metal–ligand bonds around the dopant. When the electron-trapped area is compressed by the tip force, the metal–ligand bonds will decrease in length, thereby raising the donor level and making it unfavorable to trap an electron there. The hole-trapped area around an acceptor defect at the chalcogen site will respond differently from the tip force, because the acceptor level is formed as a result of shortening the metal–ligand bond lengths of the hole-trapped area. The compression of this area will further shorten the bonds and enhance the hole-trapping.

It was briefly mentioned in Chapter 4 that the tip force affects the appearance of point defects in STM images, as seen from the nanometer-scale ring structures observed for natural MoS_2 [38a] as well as synthetic WSe_2 [38b, 39]. According to the semiconducting properties of WSe_2, as confirmed by an asymmetrical $I-V$ curve determined experimentally by STM, the sample of WSe_2 is more conductive at positive than at negative V_{bias}. Figures 9.27 (a) and 9.27 (b) show STM images recorded with opposite bias polarity. Two kinds of image imperfections can be readily identified: One is darker, and the other is brighter, than the surrounding at both positive and negative V_{bias}. The brightest spots at the negative V_{bias} become less bright or even vanish at the positive V_{bias}. The image contrast is stronger for the negative than for the positive V_{bias} measurements. The latter observation implies that the tip is closer to the sample surface at the negative V_{bias}, because the sample is less conductive at this bias polarity.

When the sample area is scanned with a lower R_{gap} value, bright spots are converted to rings (Fig. 9.27 (c)). This is caused by stronger tip–sample force interactions, as confirmed by recording STM images at several different R_{gap} values from 980 to 100 MΩ (Figs. 9.28 (a)–9.28 (e)) [38]. These images show a gradual change of two nanometer-scale bright spots at large R_{gap} into rings at small R_{gap} values. When R_{gap} is reduced from 980 to 734 MΩ, the two spots increase their brightness. At R_{gap} of ca. 500 MΩ, they are converted to rings. As R_{gap} is further reduced, the size of each ring increases, as does the central dark area. Therefore, it is tip–sample force interactions that convert bright spots at a large tip–sample distance into rings at a short tip–sample distance.

To explain the formation of ring structures, it is necessary to consider first the origin of the nanometer-scale bright spots observed at large R_{gap}. The analysis of Section 7.4.1 suggests that the nanometer-scale bright spots of WSe_2 originate from donor-type dopant atoms on the surface Se-atom sheet. The dopants are likely to be iodine atoms because I_2 is used as a transport agent [40] in the synthesis of WSe_2. As already mentioned above, the donor level is created by slightly lengthening the W–L (L = Se, I) bonds in the local region around the dopant atom. This bond length relaxation decreases with increasing distance from the center of the electron-trapped area, and the donor level has largely the orbital character of the atoms in the electron-trapped area. At small R_{gap}, the trapped area is more compressed and the W-L bonds will be shortened thereby raising the energy of the donor level. If the compression of the electron-trapped area is strong enough, it will become energetically unfavorable to trap an electron around the dopant, thereby releasing the trapped electron

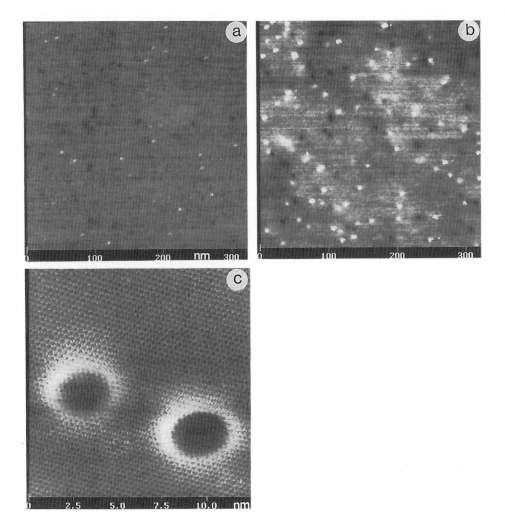

Figure 9.27 STM height images of WSe$_2$: (a) V_{bias} = 800 mV, I_{set} = 1 nA; (b) V_{bias} = −800 mV, I_{set} = 1 nA; (c) V_{bias} = −600 mV, I_{set} = 1 nA. The contrast covers height variations in the 0.0–1.0 nm range. (Reproduced from Ref. [38].)

from the area. The loss of the trapped electron will reduce the brightness of the electron-trapped area to the level of its surroundings. The bond lengthening in the electron-trapped area should decrease with increasing distance from the center of the electron-trapped area (as in the radial distribution of a hydrogen 1s-orbital). Therefore the loss of the trapped electron should occur only when the tip compresses the central part of the electron-trapped area. The compression of a local surface area is a dynamic one, which disappears as the tip moves away from the area. When an elec-

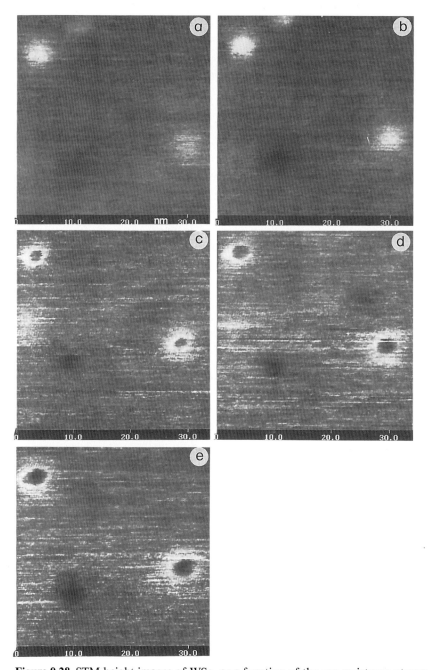

Figure 9.28 STM height images of WSe$_2$ as a function of the gap resistance at negative V_{bias}: (a) $V_{bias} = -980$ mV, $I_{set} = 1$ nA; (b) $V_{bias} = -734$ mV, $I_{set} = 1$ nA; (c) $V_{bias} = -500$ mV, $I_{set} = 1$ nA; (d) $V_{bias} = -980$ mV, $I_{set} = 3$ nA; (e) $V_{bias} = -980$ mV, $I_{set} = 5$ nA. The contrast covers height variations in the 0.0–1.0 nm range. (Reproduced from Ref. [39].)

tron-trapped area is scanned by the tip at small R_{gap}, the compression of the area by the tip force gives rise to the formation of a ring structure.

The above explanation is consistent with the observation that at small R_{gap}, the size of the ring increases (Figs. 9.28 (c)–9.28 (e)). As the tip comes closer to the surface, it can detect the trapped electron at a distance farther away from the center of the electron-trapped area. However, at the shorter tip–sample distance, the tip force will be larger. This will help to free the trapped electron, even at distances farther away from the immediate vicinity of the center of the electron-trapped area. It is noted from Fig. 9.28 that the contrast of the ring center becomes darker than the background at small R_{gap}. Within the proposed tip–sample interaction model, this observation implies that the local hardness of the surface is less in the central area of the ring containing the dopant atom than in the background area, so the ring center is depressed more by the tip force.

9.7 Concluding Remarks

The macroscopic deformation of a sample surface caused by the tip force (Fig. 9.29 (a)) is described by classical theories treating the interacting solids as materials with uniform density distributions. This depression is accompanied by the shrinking of the surface unit cell, which is commonly detected by X-ray diffraction and vibrational spectroscopy measurements. It is practically impossible to recognize this relaxation by STM and AFM because the lattice parameters determined from the images are not accurate enough (Chapter 4). For a surface with several different chemical environments, the extent of the surface atom depression is determined by the local hardness. Thus, as depicted in Fig. 9.29 (b), this introduces an atomic-scale corrugation into the macroscopically depressed surface. Such corrugations can be detected with STM and AFM as shown for a number of layered compounds. The variation of the surface local hardness in layered compounds can originate from the atomic arrangements within each layer and/or from those between adjacent layers. The knowledge of atomic-scale surface relaxation obtained from AFM and STM experiments is essential in developing a theoretical description of mechanical deformation beyond the continuum theory and hence in understanding the mechanical properties of materials on a nanometer scale.

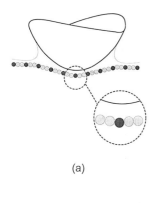

(a)

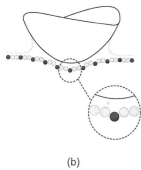

(b)

Figure 9.29 Schematic representation of the surface deformation induced by the tip force. The presence of chemically non-equivalent atoms on the surface is ignored in the macroscopic deformation shown in (a). Under the tip force, the "hard" atomic sites (gray circles) of the surface are depressed less than the "soft" atomic sites (black circles). Consequently, this leads to an atomic-scale surface corrugation into the macroscopically depressed surface under the tip apex.

References

[1] (a) D. Tománek, S. G. Louie, H. J. Mamin, D. W. Abraham, R. E. Thompson, E. Granz, J. Clarke *Phys. Rev. B* **1987**, *35*, 7790. (b) D. Tománek, S. G. Louie, *Phys. Rev. B* **1988**, *37*, 8327.
[2] (a) C. M. Mate, R. Erlandsson, G. M. McClelland, S. Chiang, *Surf. Sci.* **1989**, *208*, 473. (b) M. Salmeron, D. F. Ogletree, C. Ocal, H.-C. Wang, G. Neubauer, W. Kolbe, G. J. Meyers, *J. Vac. Sci. Technol. B* **1991**, *9*, 1347.
[3] D. Marchand, C. Fretigny, N. Lecomte, M. Laguès, J. E. Fischer, *Synth. Met.* **1988**, *23*, 165.
[4] (a) J. M. Soler, A. M. Baro, N. García, H. Rohrer, *Phys. Rev. Lett.* **1986**, *57*, 444. (b) H. J. Mamin, E. Ganz, D. W. Abraham, R. E. Thompson, J. Clarke, *Phys. Rev. B* **1986**, *34*, 9015.

[5] (a) S. Ciraci, A. Baratoff, I. P. Batra, *Phys. Rev. B* **1990**, *41*, 2763. (b) W. Zhong, D. Tománek, *Phys. Rev. Lett.* **1990**, *64*, 3054. (c) W. Zhong, G. Overney, D. Tománek, *Europhys. Lett.* **1991**, *15*, 49. (d) D. Tománek, in *Scanning Tunneling Microscopy III* (Eds.: R. Wiesendanger, H.-J. Güntherodt), Springer, Heidelberg, **1993**, p. 269.

[6] (a) G. Overney in *Scanning Tunneling Microscopy III* (Eds.: R. Wiesendanger, H.-J. Güntherodt), Springer, Heidelberg, **1993**, p. 251, (b) U. Landman, W. D. Luedtke in *Scanning Tunneling Microscopy III* (Eds.: R. Wiesendanger, H.-J. Güntherodt), Springer, Heidelberg, **1993**, p. 207.

[7] M.-H. Whangbo, W. Liang, J. Ren, S. N. Magonov, A. Wawkuschewski, *J. Phys. Chem.* **1994**, *98*, 762.

[8] D. E. Williams, *J. Chem. Phys.* **1967**, *47*, 4680.

[9] U. Müller, *Inorganic Structural Chemistry*, Wiley, New York, **1993**, p. 32.

[10] N. L. Allinger, F. Li, J. C. Tai, *J. Comput. Chem.* **1990**, *11*, 868. The equilibrium distance of a $C_{sp^2}-C_{sp^2}$ bond is taken as 1.434 Å. In our discussion, we use the potential $\Delta E_{str} = 379.8 (\Delta r)^2$ assuming the graphite C–C distance, 1.418 Å, to be the equilibrium distance.

[11] (a) J. W. Lyding, J. S. Hubacek, G. Gammie, S. Skala, R. Brockenbrough, J. R. Shapely, M. P. Keyes, *J. Vac. Sci. Technol. A* **1987**, *6*, 363. (b) M. Kuwabara, D. R. Clarke, D. A. Smith, *Appl. Phys. Lett.* **1990**, *56*, 2396. (c) J. E. Buckley, J. L. Wragg, H. W. White, A. Bruckdorfer, D. L. Worcester, *J. Vac. Sci. Technol. B* **1991**, *9*, 1079.

[12] (a) C.-Y. Liu, H. Chang, A. J. Bard, *Langmuir* **1991**, *7*, 1138. (b) J. Garbarz, E. Lacase, G. Faivre, S. Gauthier, M. Schott, *Philos. Mag. A* **1992**, *65*, 853. (c) B. Nysten, J.-C. Roux, S. Flandrois, C. Daulan, H. Saadaoui, *Phys. Rev. B* **1993**, *48*, 12527.

[13] S. N. Magonov, M.-H. Whangbo, *Adv. Mater.* **1994**, *6*, 355.

[14] Z. Y. Rong, P. Kuiper, *Phys. Rev. B* **1993**, *48*, 17427.

[15] (a) T. A. Land, T. Michely, R. J. Behm, J. C. Hemminger, G. Comsa, *Surf. Sci.* **1992**, *264*, 261. (b) U. Muller, D. Carnal, H. Siegenthaler, E. Schmidt, W. J. Lorentz, W. Obretenov, U. Schmidt, G. Staikov, E. Budevski, *Phys. Rev. B* **1992**, *46*, 12899. (c) H. Galloway, J. J. Benitez, M. Salmeron, *Surf. Sci.* **1993**, *298*, 127.

[16] B. A. Parkinson, F. S. Ohuchi, K. Ueno, A. Koma, *Appl. Phys. Lett.* **1991**, *58*, 472.

[17] (a) E. V. Stroganov, K. V. Ovchinnikov *Vestnik Leningrad. Univ., Ser. Fiz. i Khim.* **1957**, *12*, 152. (b) H. Hillebrecht, Ph.D. Thesis, Albert-Ludwigs University, Freiburg, Germany, **1991**.

[18] J. M. Fletcher, W. E. Gardner, A. C. Fox, G. Topping, *J. Chem. Soc. (A)* **1967**, 1038.

[19] (a) M.-H. Whangbo, *J. Chem. Phys.* **1980**, *73*, 3854. (b) M.-H. Whangbo, *Inorg] Chem.* **1980**, *19*, 1728.

[20] H.-J. Cantow, H. Hillerbrecht, S. N. Magonov, H. W. Rotter, M. Drechsler, G. Thiele, *Angew. Chem., Int. Ed. Engl.* **1990**, *29*, 537.

[21] J. Ren, M.-H. Whangbo, H. Bengel, S. N. Magonov, *J. Phys. Chem.* **1993**, *97*, 4764.

[22] L. Monconduit, M. Evain, R. Brec, J. Rouxel, E. Canadell, *C. R. Acad. Sci. Paris* **1993**, *316*, 25.

[23] (a) L. Monconduit, M. Evain, F. Boucher, R. Brec, J. Rouxel, *Z. Anorg. Allg. Chem.* **1992**, *616*, 177. (b) J. Li, P. J. Caroll, *Mater. Res. Bull.* **1992**, *27*, 1073. (c) E. Canadell, L. Monconduit, M. Evain, R. Brec, J. Rouxel, M.-H. Whangbo, *Inorg. Chem.* **1992**, *32*, 10.

[24] A. van der Lee, M. Evain, M. Mansuetto, L. Monconduit, R. Brec, J. Rouxel, *J. Solid State Chem.* **1994**, *111*, 75.

[25] A. van der Lee, M. Evain, L. Monconduit, R. Brec, S. Smaalen, *J. Phys.: Condens. Mat.* **1994**, *6*, 933.
[26] A. van der Lee, M. Evain, L. Monconduit, R. Brec, J. Rouxel, V. Petricek, *Acta Crystallogr., Sect. B: Struct. Sci.* **1994**, *50*, 119.
[27] M. Evain, A. van der Lee, L. Monconduit, V. Petricek, *Chem. Mater.* **1994**, *6*, 1776.
[28] W. Liang, M.-H. Whangbo, M. Evain, L. Monconduit, R. Brec, H. Bengel, H.-J. Cantow, S. N. Magonov, *Chem. Mater.* **1994**, *6*, 678.
[29] M.-H. Whangbo, M. Evain, H. Bengel, S. N. Magonov, unpublished data.
[30] H. Bengel, H.-J. Cantow, S. N. Magonov, L. Monconduit, M. Evain, M.-H. Whangbo, *Surf. Sci.* **1994**, *321*, L170.
[31] B. A. Parkinson, J. Ren, M.-H. Whangbo, *J. Am. Chem. Soc.*, **1991**, *113*, 7833.
[32] H. Bengel, M. Evain, S. N. Magonov, M.-H. Whangbo, unpublished results.
[33] H. Bengel, H.-J. Cantow, S. N. Magonov, L. Monconduit, M. Evain, W. Liang, M.-H. Whangbo, *Adv. Mater.* **1994**, *6*, 649.
[34] (a) H. Bengel, H.-J. Cantow, S. N. Magonov, M.-H. Whangbo, *Adv. Mater.* **1995**, *6*, 483. (b) H. Bengel, H.-J. Cantow, S. N. Magonov, D. Jung, J. Ren, M.-H. Whangbo, *New J. Chem.*, submitted for publication.
[35] B. E. Brown, *Acta. Crystallogr.* **1966**, *20*, 264.
[36] M.-H. Whangbo, E. Canadell, *J. Am. Chem. Soc.* **1992**, *114*, 9587.
[37] (a) H. Le Chatelier, *Ann. Mines* **1888**, *13*, 200. (b) F. Braun, Z. Physik Chem. **1887**, *1*, 259.
[38] (a) W. M. Heckl, F. Ohnesorge, G. Binnig, M. Specht, M. Hashimi, *J. Vac. Sci. Technol. B* **1991**, *9*, 1072. (b) Th. Schimmel, H. Fuchs, S. Akari, K. Dransfeld, *Appl. Phys. Lett.* **1991**, *58*, 1039.
[39] S. N. Magonov, H.-J. Cantow, M.-H. Whangbo, *Surf. Sci. Lett.* **1994**, *318*, L1175.
[40] F. Levy, *Nuovo Cimento B* **1977**, *38*, 359.

10 Organic Conducting Salts

Organic salts prepared from donor or acceptor molecules can be insulating, semiconducting, or metallic [1, 2]. A number of metallic salts possess electronic and structural instabilities: on lowering the temperature, many of them undergo a metal-to-insulator phase transition, and some become superconductors at temperatures below 14 K [3]. Organic conducting salts are prepared as high-quality and air-stable single crystals, and their crystal samples possess crystallographically well-defined flat surfaces. Thus, ambient-condition STM and AFM measurements can be performed to examine their surface topography and electronic features and also to probe surface processes accompanying the imaging. This chapter describes the STM/AFM characterization of a number of conducting salts derived from the acceptor molecule tetracyanoquinodimethane (TCNQ) and the donor molecule bis(ethylenedithio)tetrathiafulvalene (BEDT-TTF) (Fig. 10.1)

10.1 Crystal and Electronic Structures

The crystal structures of TCNQ salts, TTF-TCNQ, Qn(TCNQ)$_2$, 4EP(TCNQ)$_2$ and TEA(TCNQ)$_2$, are depicted in Fig. 10.2. (Here TTF refers to tetrathiafulvalene, where Qn, 4EP and TEA represent the quinolinium, 4-ethyl-N-methylpyridinium, and triethylammonium cations, respectively; see Fig. 10.1). TTF-TCNQ consists of segregated stacks of TTF and TCNQ aligned along the b-direction. The ab-plane surface contains both the TTF and TCNQ stacks (Fig. 10.2 (a)) [4a]. Qn(TCNQ)$_2$ has segregated stacks of TCNQ and Qn running along the b-direction. The bc-plane surface of Qn(TCNQ)$_2$ can be represented either by TCNQ molecules only, or by Qn cations only (Fig. 10.2 (b)) [4b]. In the ab-plane surface of 4EP(TCNQ)$_2$, there are two kinds of TCNQ molecules at different heights (Fig. 10.2 (c)) [4c]. The ac-plane of TEA(TCNQ)$_2$ has four TCNQ molecules per unit cell, and the surface is strongly corrugated (Fig. 10.2 (d)) [4d].

In BEDT-TTF salts with monovalent anions X$^-$, (BEDT-TTF)$_2$X, the layers of the cations, BEDT-TTF$^{0.5+}$, alternate with the layers of the anions, X$^-$ (Fig. 10.3) [3]. The BEDT-TTF molecules are inclined with respect to the plane of the cation layer, and the molecular arrangements within the cation layer depend on the crystallographic modifications (e.g., α-, β- and κ-phase) [3]. In the α-(BEDT-TTF)$_2$X (X = I$_3^-$, IBr$_2^-$, TlHg(SCN)$_4^-$)-phases, the cation layers contain two different stacks of BEDT-TTF molecules arranged in a herring-bone pattern (Fig. 10.4 (a)). In the β-

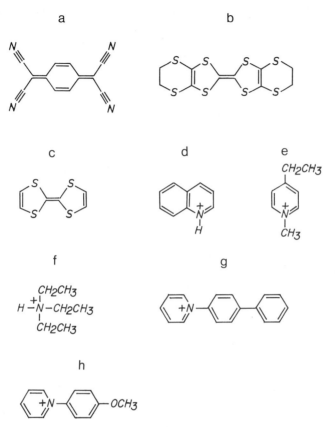

Figure 10.1 Chemical formulas of some of the compounds described in Chapter 10: (a) tetracyanoquinodimethane, TCNQ; (b) bis(ethylenedithio)tetrathiafulvalene, BEDT-TTF; (c) tetrathiafulvalene, TTF; (d) quinolinium cation, Qn^+; (e) 4-ethyl-N-methylpyridinium cation, $4EP^+$; (f) triethylammonium cation, TEA^+; (g) N-biphenylpyridinium cation; and (h) N-(p-methoxyphenyl)pyridinium cation.

(BEDT-TTF)$_2$X (X = I_3^-, IBr_2^-) salts, the cation layers consist of stacks of BEDT-TTF molecules arranged parallel to each other (Fig. 10.4 (b)). In the κ-(BEDT-TTF)$_2$X salts, where X stands for Cu(SCN)$_2$, Cu[N(CN)$_2$]Cl, Cu[N(CN)$_2$]Br, and Cu[N(CN)$_2$]I [5], the cation layers contain BEDT-TTF dimers packed orthogonally to each another (Fig. 10.4 (c)). The anions are either discrete (e.g., I_3^-, IBr_2^-, AuI_2^-, TlHg(SCN)$_4^-$) or polymeric (e.g., Cu(SCN)$_2^-$, Cu[N(CN)$_2$]Br$^-$, etc.). The main crystal faces of these salts are represented by the cation or the anion layer, and the hydrogen atoms of the ethylene groups are the topmost atoms of the cation surfaces.

TTF-TCNQ is described by the oxidation state (TTF$^{+0.59}$)(TCNQ$^{-0.59}$), so the HOMO of TTF and the LUMO of TCNQ are partially filled. The oxidation state of the other TCNQ salts is given by A^+(TCNQ$^{0.5-}$)$_2$ (A = Qn, 4EP, TEA), and the

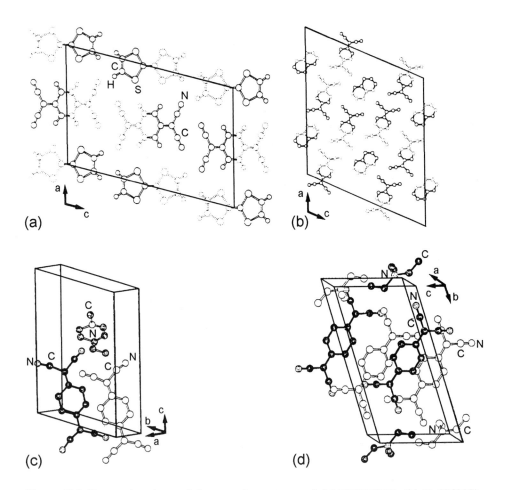

Figure 10.2 Perspective views of the crystal structures of (a) TTF-TCNQ, (b) Qn(TCNQ)$_2$, (c) 4EP(TCNQ)$_2$, and (d) TEA(TCNQ)$_2$.

LUMOs of their TCNQ become partially filled. The BEDT-TTF salts have the oxidation state (BEDT-TTF$^{0.5+}$)$_2$X$^-$, and the HOMO of BEDT-TTF is partially filled. These partially filled levels lead to partially filled bands in the salts, which determine their electrical transport properties. As depicted in Fig. 10.5, the LUMO of TCNQ and the HOMOs of TTF and BEDT-TTF are π-orbitals. Whether the electrical conductivity of a charge-transfer salt is 1D or not depends on how well the HOMOs (or LUMOs) of adjacent donor (or acceptor) molecules overlap in different directions [2a].

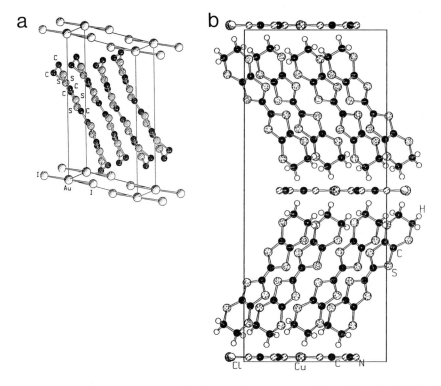

Figure 10.3 Perspective views of the crystal structures of (a) β-(BEDT-TTF)$_2$AuI$_2$ and (b) κ-(BEDT-TTF)$_2$Cu[N(CN)$_2$]Cl.

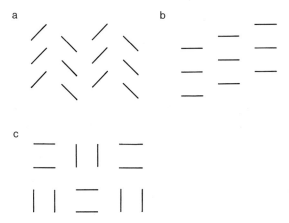

Figure 10.4 Schematic presentation of the molecular packing in the cation layers of (a) the α-phases, (b) the β-phases, and (c) the κ-phases of BEDT-TTF salts.

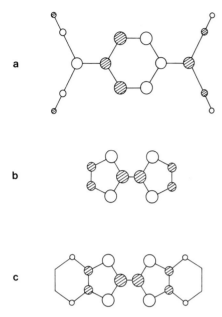

Figure 10.5 Top projection views of (a) the LUMO of TCNQ, (b) the HOMO of TTF, and (c) the HOMO of BEDT-TTF. The open and hatched circles represent the p_π-orbitals of the atoms with positive and negative signs, respectively.

10.2 Early STM Studies of Organic Conductors

In the first STM study of organic conductors, Sleator and Tycko [6] obtained molecular-scale images of TTF-TCNQ exhibiting a periodic pattern corresponding to the *ab*-plane surface. The STM image features within the unit cell are different from the arrangement of the molecular groups on the surface, and this was explained in terms of the HOMO density distribution of TTF and the LUMO density distribution of TCNQ. STM images of κ-(BEDT-TTF)$_2$Cu(SCN)$_2$ [7] exhibit the topography of either the anion or the cation layer (hereafter referred to as anion- and cation-layer STM images, respectively). Yoshimura et al. [7b] interpreted their cation STM image on the basis of the HOMO density distribution of BEDT-TTF and the proximity of the atoms of BEDT-TTF to the tip. Because the BEDT-TTF molecules are inclined with respect to the cation layer (Fig. 10.3), one ethylenedithio-group S atom of BEDT-TTF lies closest to the tip (hereafter, referred to as the highest-lying S atom). Yoshimura et al. suggested that the electron density of the highest-lying S atom is mainly responsible for each individual STM pattern, and the anion layer covering the cation layer provides a perturbation, thereby leading to the observed pattern. This assignment is difficult to reconcile with the fact that the amplitude of an atomic orbital decreases exponentially with increasing distance from the atomic center. The highest-lying S atom lies considerably below the topmost ethylene-group H atoms even when the cation layer constitutes the surface, not to mention when the anion layer does. Consequently, the contributions of the highest-lying S atoms to the $p(r_0, e_f)$ plot should be very small even when the cation layer constitutes the surface. This raises the question "What is the electronic origin of the individual patterns of the cation STM image?" Another fundamental question is "How does the conducting cation layer contribute to an STM image when it is covered with the insulating anion layer?" This question is important also because insulating compounds can be imaged by STM when placed on a conducting substrate [8].

10.3 STM and AFM Imaging of Organic Conductors

Before the STM and AFM images of organic conductors are analyzed, it is important to discuss the surface processes associated with imaging [9a] and the variations of molecular-scale images in different tunneling conditions [7d, 9b].

10.3.1 Surface Processes During Imaging

The tip–sample force interactions in STM and AFM can initiate the removal of the topmost crystal layers during imaging (Chapter 4). This process is enhanced at higher forces in AFM and at smaller R_{gap} in STM. In studies of organic conductors, it is often observed as the degradation of step borders and the growth of surface holes. Figure 10.6 presents large-scale STM images of β-(BEDT-TTF)$_2$AuI$_2$ recorded in successive scans [9a]. These images show surface layers with a step height of ca. 1.8 nm. The surface layers are gradually restored after each scan at positive bias voltages (Figs. 10.6 (a)–10.6 (f)) but are degraded at negative bias voltages (Figs. 10.6 (f)–10.6 (i)). The reconstruction starts again after switching the polarity to positive values (Figs. 10.6 (j)–10.6 (n)). The crystal growth in the tip–sample local environment may be similar to that in the electrochemical crystallization used to prepare this salt. The organic material for crystal growth comes from the debris on the sample surface. In general the surface growth proceeds much more (three to ten times) slowly than the degradation, and in many cases scanning at positive bias only retards the degradation.

The removal of the topmost layers by the tip is also observed in the atomic-scale images. The surface step is seen as a horizontal line in the center of the image of κ-(BEDT-TTF)$_2$I$_3$ (Fig. 10.7) [9a]. Nondestructive imaging of this area is achieved only when the scanning proceeds from the higher terrace (in the upper part of the image) to the lower one. Scanning in the opposite direction causes removal of the higher terrace. The similarity of the molecular-scale image patterns on both terraces indicates that the cation/anion double layer is removed during the degradation. (The image pattern in Fig. 10.6 most likely corresponds to the cation layer; see Section 10.5.)

10.3 STM and AFM Imaging of Organic Conductors 195

Figure 10.6 (a)–(o) Successive STM height images of the *ab*-plane of the β-(BEDT-TTF)$_2$AuI$_2$ crystal. (a)–(f), (j)–(o) I_{set} = 1 nA, V_{bias} = +100 mV ; (g)–(i) I_{set} = 1 nA, V_{bias} = −100 mV. The contrast covers height variations in the 0–10 nm range.

Figure 10.7 STM height image of the *ab*-plane of κ-(BEDT-TTF)$_2$I$_3$ (I_{set} = 1 nA, V_{bias} = 100 mV). The contrast covers height variations in the 0–1 nm range.

Dynamic surface processes are also observed in molecular-scale imaging of BEDT-TTF salts. The motion of a defect associated with a lattice vacancy is observed in the atomic-scale images of α-(BEDT-TTF)$_2$I$_3$ (Fig. 10.8) [9a]. The two defects indicated by arrows at the top and bottom of the images behave differently in the successive scans. Relative to the "top" defect, the "bottom" defect changes its position from scan to scan. Initially, both are located in the same row of the image, and then the "bottom" defect gradually moves up and to the right, and finally escapes from the imaging area. Because the cation layers are typically seen in the STM images of the α-phases, the observed defects are related to cation vacancies. Then, migration of this type of defect involves a translation of the large donor molecules, which is probably facilitated by the tip force applied to the sample during scanning.

10.3.2 Molecular-Scale Images

In molecular-scale STM and AFM imaging of organic crystals, variations of image patterns are often observed. In STM these effects can be caused by changes of R_{gap}, as shown by the images of α-(BEDT-TTF)$_2$TlHg(SCN)$_4$ (Fig. 10.9) [9b]. These images were obtained at R_{gap} = 100, 15, 7.5 and 5 MΩ. In the images observed at R_{gap} = 100 and 15 MΩ, the bright and dim rows with poorly resolved details alternate. This contrast difference eventually vanishes when R_{gap} is reduced below 15 MΩ, most likely because of an increase in the tip–surface force interaction, although the details of a unit cell pattern become well resolved. On further decreasing R_{gap} to 5 MΩ, the molecular-scale features exhibit even higher contrast (Fig. 10.9 (d)). In general, the R_{gap}-dependent image changes are reversible. At low R_{gap}, the images are less stable, and the contrast and size of the atomic-scale features vary frequently in successive scans.

10.3 STM and AFM Imaging of Organic Conductors 197

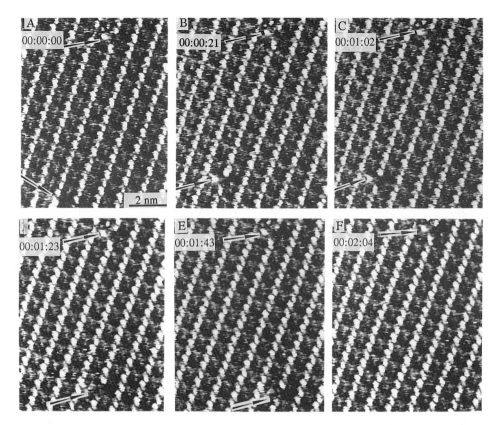

Figure 10.8 (a)–(f) Successive STM current images of the *ab*-plane of an α-(BEDT-TTF)$_2$I$_3$ crystal. The relative time of registration is shown in the upper left-hand corner. Defects in the inclined rows are marked by arrows. (a), (b) $I_{set} = 1$ nA, $V_{bias} = -100$ mV; (c)–(f) $I_{set} = 1$ nA, $V_{bias} = +100$ mV. The contrast is proportional to current variations in relative units.

In such tunneling conditions the current-vs.-voltage curves become linear (see Fig. 8.5), so that the imaging is probably performed in the contact regime.

The image changes described above indicate a complexity in the interpretation of STM images of organic conductors. It is difficult to determine what an "ideal" STM image (i.e., the one least disturbed by the tip–surface force interaction and hence best mimicking the partial electron density distribution of the surface) should look like. In general, the tip–surface interaction is reduced by increasing R_{gap}, but this lowers the image resolution. Consequently, an STM image close to the "ideal" one should retain the contrast difference of the images at large R_{gap}, and also exhibit well-resolved features of a unit cell pattern as found at small R_{gap}. This criterion is best satisfied by the image shown in Fig. 10.9 (b). The R_{gap} value for an "ideal" image might vary from place to place on the sample surface, but the dependence on R_{gap} of the image variation is typical of organic conducting salts. To understand a

Figure 10.9 STM current images of the ac-plane surface of $(BEDT\text{-}TTF)_2TlHg(SCN)_4$. (a) $I_{set} = 1$ nA, $V_{bias} = 100$ mV; (b) $I_{set} = 2$ nA, $V_{bias} = 30$ mV; (c) $I_{set} = 2$ nA, $V_{bias} = 15$ mV; (d) $I_{set} = 3$ nA, $V_{bias} = 15$ mV. In all images the contrast is proportional to current variations in relative units in the same range.

probable cause for the image variation as a function of R_{gap}, it is necessary to know with what atoms the bright STM spots are associated. As will be shown in Section 10.5, the bright spots of the STM images are assigned to the surface ethylene-group H atoms. It is not energetically difficult to change the ethylene-group conformations in the six-membered rings of BEDT-TTF [10]. Therefore, the tip–surface force interaction alters the ethylene-group conformations of the cations on the surface; hence it changes the heights of the ethylene-group H atoms, and consequently causes a variation in the unit cell patterns of the STM image.

10.4 Analysis of the Images of TCNQ Salts

10.4.1 TTF-TCNQ

STM and AFM images of the *ab*-plane surface (Figs. 10.10 (a) and 10.10 (b), respectively) exhibit parallel rows of atomic-scale patterns aligned along the *b*-axis, which are consistent with the stacks of TTF and TCNQ molecules lying in this plane. According to the crystal structure of TTF-TCNQ, the H atoms of TTF and the N atoms of TCNQ protrude most on the *ab*-plane surface (Fig. 10.11). AFM and STM images of the *ab*-plane surface of TTF-TCNQ are compared with the total and partial electron density plots in Fig. 10.12. Although the H atoms of TTF are protrude more than the N atoms of TCNQ (in terms of the atomic positions), the N atoms have a higher electron density than the H atoms in the $\rho(r_0)$ plot (Fig. 10.12 (b)) because the N atom is larger than the H atom. Of the two H atoms in TTF, the one protruding more shows a much stronger electron density. The bright spots of the AFM image (Fig. 10.12 (a)) show separations of ca. 0.48 nm and ca. 0.38 nm along the *a*-direction. The larger separation corresponds to the N···N distance in the terminal C(CN)$_2$ group of TCNQ, and the smaller one to the distance from the most protruding H atom in TTF to its nearest-neighbor N atom in TCNQ along the *a*-direction.

In the STM image of the *ab*-plane surface (Fig. 10.12 (c)), there are well-pronounced triplets plus one or two weaker spots within a unit cell. This pattern was first explained by calculating the electron densities associated with the HOMO of

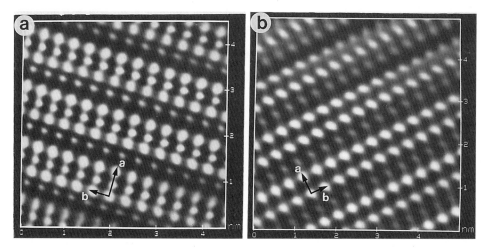

Figure 10.10 (a) STM current image of the *ab*-face of TTF-TCNQ (I_{set} = 3 nA, V_{bias} = 200 mV). The contrast is proportional to current variations in relative units. (b) AFM height image of the *ab*-face of TTF-TCNQ. The contrast covers height variations in the 0–1 nm range.

TTF and the LUMO of TCNQ [6]. This approach led to the result similar to the $p(r_0, e_f)$ plot (Fig. 10.12 (d)), which shows triplet spots (in a bent shape) at TCNQ and a single spot at TTF [11]. However, it should be noted that STM images of TTF-TCNQ can be significantly different from that just described. This is exemplified in Fig. 10.13, which reveals that (a) two small spots with unequal contrast appear at the position of TTF, and (b) of the three spots belonging to TCNQ, the middle one becomes the brightest. These observations are difficult to explain unless it is assumed that the TTF and TCNQ molecules on the surface undergo a certain geometric defor-

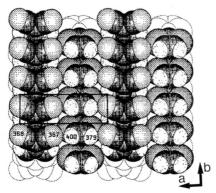

Figure 10.11 Molecular arrangement of the *ab*-plane of TTF-TCNQ taken from the bulk crystallographic structure. The numbers in the circles representing surface atoms indicate their relative heights (pm).

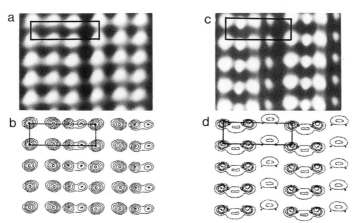

Figure 10.12 (a) Part of the AFM image of Fig. 10.10 (b). (b) $p(r_0)$ surface plot on the *ab*-face of TTF-TCNQ. Contour values used are 0.35×10^{-1}, 0.20×10^{-1}, 0.10×10^{-1}, and 0.05×10^{-1} electrons/au^3. The small circles represent the uppermost surface atoms. (c) Part of the STM image of Fig. 10.10 (a). (d) $p(r_0, e_f)$ surface plot on the *ab*-face of TTF-TCNQ. Contour values used are 0.60×10^{-3}, 0.20×10^{-3}, 0.10×10^{-3}, 0.05×10^{-3}, and 0.01×10^{-3} electrons/au^3.

mation under the tip force. Observation (a) can be associated with the most protruding H atoms in TTF if these atoms contribute to the HOMO of TTF. Since the HOMO is a π-orbital, it can happen if the C–H bonds are bent out of the π-plane of the TTF molecule. This possibility is consistent with the observation that the spots belonging to the H atoms of TTF in the STM image (Fig. 10.13) are located on a line that is not perpendicular to the stack direction. Observation (b) can be explained if the terminal $C(CN)_2$ unit of TCNQ is bent out of the π-framework of TCNQ so that the carbon center of the $C(CN)_2$ unit becomes slightly pyramidal. The partial density plots calculated as a function of this bending angle show that this is indeed the case [12a]. Similarly, for the explanation of the AFM images of the mixed-valence chain compound $[Pt(en)_2][Pt(en)_2X_2](ClO_4)_4$ [12b] and the STM images of the fluoranthenyl cation salt $(FA)_2PF_6$ [12c], it has been found necessary to invoke surface relaxations induced by tip force.

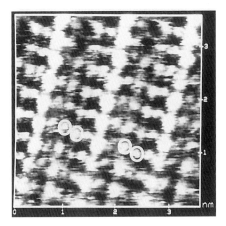

Figure 10.13 STM current image of the *ab*-plane surface of a TTF-TCNQ crystal ($I_{tun} = 1$ nA, $V_{bias} = 34$ mV). The contrast is proportional to current variations in relative units. The spots assigned to the H atoms of TTF are circled.

10.4.2 Qn(TCNQ)$_2$

The *bc*-plane surface of Qn(TCNQ)$_2$ shown in Fig. 10.14 (a) consists of TCNQ molecules in Fig. 10.14 (a). Alternatively, the surface can be made up of Qn cations, but such a surface would be less mechanically stable because of its loose packing. The STM image and the $\rho(r_0, e_f)$ plot calculated for the layer TCNQ molecules are shown in Figs. 10.14 (b) and 10.14 (c), respectively. The STM image exhibits periodic rows [11]. The repeat distances between the rows and within each row correspond to the crystallographic constants $c/2$ and b, respectively [4b]. In addition, the neighboring rows consist of triplet patterns. The $\rho(r_0, e_f)$ plot shows a complex arrangement of patterns (Fig. 10.14 (c)), in which the intensity distribution within each triplet is not uniform because the TCNQ molecules are inclined with respect to the bc-plane. The crystal structure of Qn(TCNQ)$_2$ shows that the two cyano-group N atoms of TCNQ on the surface differ in height by 0.05 nm. In essence, the features of the STM image correspond to the $\rho(r_0, e_f)$ plot calculated for the TCNQ layer.

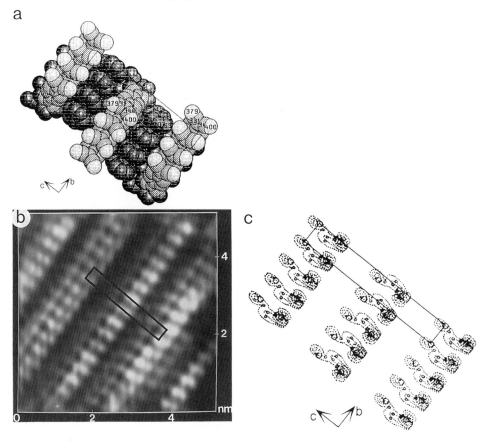

Figure 10.14 (a) Molecular arrangement of the bc-face of Qn(TCNQ)$_2$ taken from the bulk crystal structure. The numbers within the circles representing surface atoms indicate their relative heights (pm). (b) FFT-filtered STM current image of the bc-face of a Qn(TCNQ)$_2$ crystal (I_{set} = 1 nA, V_{bias} = 100 mV). The contrast is proportional to current variations in relative units. (c) $\rho(r_0, e_f)$ surface plot calculated for the TCNQ layer on the bc-face of Qn(TCNQ)$_2$. Contour values used are 0.30×10^{-2}, 0.20×10^{-2}, 0.10×10^{-2}, 0.05×10^{-2}, 0.01×10^{-2}, 0.005×10^{-2}, and 0.001×10^{-2} electrons/au^3.

10.4.3 4EP(TCNQ)$_2$

The arrangement of the TCNQ molecules in the ab-face layer of 4EP(TCNQ)$_2$ is shown in Fig. 10.15 (a). There are two TCNQ molecules per unit cell, which have different heights. In each TCNQ, the most protruding atom is one cyano-group N atom. The STM image of the ab-face layer (Fig. 10.15 (b)) shows a pattern of one bright and one dim spot per unit cell. This feature is in good agreement with the $\rho(r_0, e_f)$ plot calculated for the ab-plane surface of TCNQ molecules (Fig. 10.15 (c)). The electron density plot shows patterns located on the surface N atoms, their intensity being stronger on the more protruding ones. The patterns of the STM image are consistent with those of the partial electron density plot.

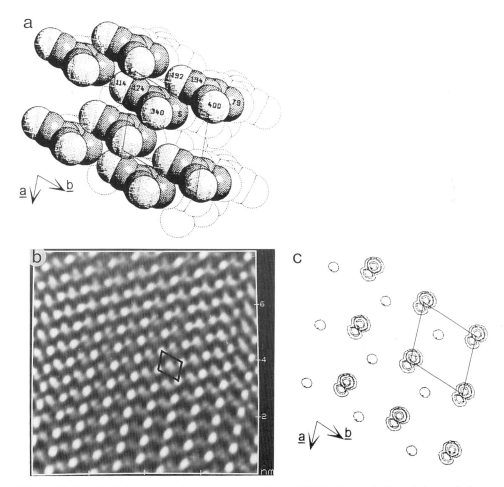

Figure 10.15 (a) Molecular surface structure of the TCNQ layer of the ab-face of the 4EP(TCNQ)$_2$ crystal taken from the bulk crystal structure. The numbers within the circles representing surface atoms indicate their relative heights (pm). (b) FFT-filtered STM current image of the ab-face of the 4EP(TCNQ)$_2$ crystal ($I_{set} = 0.75$ nA, $V_{bias} = -375$ mV). The contrast is proportional to current variations in relative units. (c) $\rho(r_0, e_f)$ surface plot on the TCNQ layer on the ab-face of 4EP(TCNQ)$_2$. Contour values used are 0.30×10^{-2}, 0.10×10^{-2}, 0.05×10^{-2}, 0.01×10^{-2}, 0.005×10^{-2}, and 0.001×10^{-2} electrons/au^3.

204   10 Organic Conducting Salts

10.4.4 TEA(TCNQ)$_2$

Figure 10.16 (a) shows the arrangement of the *ac*-plane surface of TCNQ molecules in TEA(TCNQ)$_2$. There are four TCNQ molecules in a unit cell, and the most protruded atom of each TCNQ molecule is one cyano-group N atom. These protruding N atoms in the TCNQ molecules have a strong height corrugation (up to 20 nm). The STM image of the *ac*-plane surface is given in Fig. 10.16 (b). As shown in Fig. 10.16 (c), the $\rho(r_0, e_f)$ plot calculated for the layer of TCNQ molecules shows

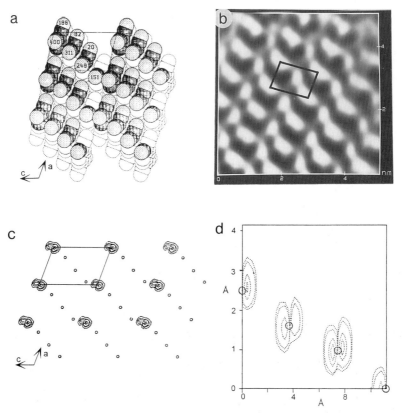

Figure 10.16 (a) Molecular surface structure of the TCNQ layer of the *ac*-face of the TEA(TCNQ)$_2$ crystal taken from the bulk crystal structure. The numbers within the circles representing surface atoms indicate their relative heights (pm). (b) FFT-filtered STM current image of the *ac*-face of the TEA(TCNQ)$_2$ crystal (I_{set} = 1.3 nA, V_{bias} = 100 mV). The contrast is proportional to current variations in relative units. (c) $\rho(r_0, e_f)$ surface plot on the TCNQ layer on the *ab*-face of TEA(TCNQ)$_2$. Contour values used are 0.10×10^{-1}, 0.05×10^{-1}, 0.01×10^{-1}, 0.005×10^{-1}, and 0.001×10^{-1} electrons/au^3. (d) Cross-sectional view of the plot on the plane that is perpendicular to the *ac*-plane, aligned along the (a+c)-direction and passing through the most protruding N atoms of a unit cell. Contour values used are 0.10×10^{-1}, 0.05×10^{-1}, 0.01×10^{-1}, and 0.005×10^{-1} electrons/au^3.

10.4 Analysis of the Images of TCNQ Salts

that, within a unit cell, only the N atom of the most protruding TCNQ molecule has a substantial electron density. This predicts one pattern per unit cell and is therefore not consistent with the STM image of Fig. 10.16 (b), which shows bright strips running along the $(a+c)$-direction, i.e., all four TCNQ molecules of a unit cell are imaged. Figure 10.16 (d) shows the partial electron density of the layer of TCNQ molecules on the cross-section that is perpendicular to the ac-plane and contains the most protruding N atoms of the TCNQ molecules. This plot suggests that the STM image may have three or four patterns along the diagonal direction of a unit cell when the tip follows the surface corrugation instead of keeping a constant height above the surface. This suggestion is consistent with the observed STM image shown in Fig. 10.16 (b). Consequently, the STM features are assigned to the electron densities of the protruding N atoms of the TCNQ molecules.

10.4.5 TCNQ Salts with Substituted Phenylpyridines

So far, the analysis of STM and AFM images has been conducted for TCNQ salts with known crystallographic structure. For some TCNQ salts the quality of the samples does not allow a single-crystal X-ray structure determination. Such crystals can be characterized by STM. Figure 10.17 shows images of TCNQ salts with N-(p-methoxyphenyl)pyridinium and N-biphenylpyridinium cations, which have been chosen from an STM study of several TCNQ salts with substituted phenylpyridines [13]. Most such images possess periodic patterns with rows of bright spots resembling molecular stacks. The repeat distances along the stack direction are in the 0.40–0.45 nm range, whereas the separations between stacks are ca. 1.41 nm and ca.

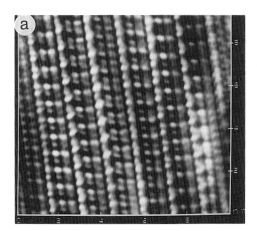 

Figure 10.17 (a), (b) STM current images of TCNQ salts with N-(p-methoxyphenyl)pyridinium ($I_{set} = 0.8$ nA, $V_{bias} = 66$ mV) and N-biphenylpyridinium ($I_{set} = 1$ nA, $V_{bias} = 100$ mV), respectively. The contrast is proportional to current variations in relative units.

3.83 nm in the images of Figs. 10.17 (a) and 10.17 (b), respectively. Since the rows of the most pronounced spots typically show the LUMO density of TCNQ, the presence of different stack periods indicates various inclinations of the TCNQ molecular plane with respect to the stacking direction. It is of interest to note the dimerization pattern along the stack in the image in Fig. 10.17 (a). A similar effect was found in the STM image of the $(FA)_2PF_6$ salt [14]. In general, STM as well as AFM images can be used for the preliminary structural characterization of crystals.

10.5 Analysis of the Images of BEDT-TTF Salts

STM and AFM studies have been performed on a large number of BEDT-TTF salts. Representative molecular-scale images of the α-, β- and κ-phases of these salts are discussed and interpreted in the following.

10.5.1 Cation-Layer Images of α-Phases

The cation layers of α-phase BEDT-TTF salts are formed of stacks of crystallographically nonequivalent A, A', B and C cations. The cation-layer surfaces are presented by the ethylene-group H atoms of BEDT-TTF. The height of these atoms belonging to adjacent stacks is similar in α-$(BEDT-TTF)_2I_3$ (Fig. 10.18 (a)) but differs by 0.3 Å in α-$(BEDT-TTF)_2TlHg(SCN)_4$ (Fig. 10.19 (a)). The $\rho(r_0)$ plots of the cation layers of both salts (Figs. 10.18 (b) and 10.19 (c)) mimic the surface topography and are dominated by the contributions of the most protruding hydrogen atoms [9b, 15]. Crystallographically different cations provide $\rho(r_0)$ patterns of different shape and intensity, which is stronger for the more protruding hydrogen atoms. Therefore, B and C cations have stronger contributions to the $\rho(r_0)$ plot of α-$(BEDT-TTF)_2TlHg(SCN)_4$.

The main features of the $\rho(r_0)$ plots can be found in the AFM images. The image of α-$(BEDT-TTF)_2I_3$ consists of two different rows of patterns aligned along the a-axis (Fig. 10.18 (c)). The bright spots have a similar height, and there are four patterns per unit cell. Clearly, the rows of bright spots, which can be assigned to the stacks of B and C cations, alternate with rows of less bright spots (A and A' cations) in the AFM image of α-$(BEDT-TTF)_2TlHg(SCN)_4$ (Fig. 10.19 (b)). In general, the AFM images resolve the presence of different cations in the surface layers and their height differences.

Figures 10.20 (a) and 10.21 (a) respectively show highly resolved STM images of α-$(BEDT-TTF)_2I_3$ and α-$(BEDT-TTF)_2TlHg(SCN)_4$ salts obtained at moderate R_{gap} values. Both images exhibit more details than the AFM images, and they are characterized by alternating rows with different width and with bright spots separated by

10.5 Analysis of the Images of BEDT-TTF Salts

1.4–2.5 Å. Four sets of spots within the unit cell of these images can be assigned to the four cations of the surface unit cell. The calculations of the $\rho(r_0, e_f)$ plots provide a rational interpretation of these images and an answer to the question "Which atoms of BEDT-TTF contribute to the images?"

The $\rho(r_0, e_f)$ plots of the cation layers of both salts exhibit a narrow pattern along the stacks of the B and C cations, and a wide pattern along the stacks of the

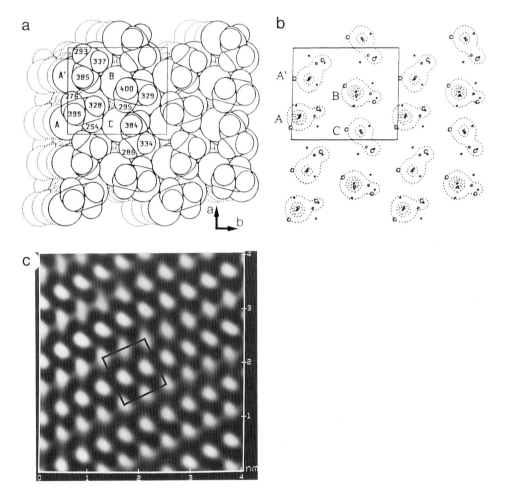

Figure 10.18 (a) Molecular arrangement of BEDT-TTF cations in the *ab*-plane of α-(BEDT-TTF)$_2$I$_3$. The H, C, and S atoms of the dithioethylene group are represented by the small, medium, and large circles, respectively. The numbers within the circles representing the surface atoms indicate their relative heights (pm). (b) $\rho(r_0)$ plot of the cation layer of α-(BEDT-TTF)$_2$I^3. The contour values are 0.4×10^{-1}, 0.3×10^{-1}, 0.2×10^{-1}, 0.1×10^{-1}, and 0.01×10^{-1} electrons/au^3. (c) AFM height images of the *ab*-face of α-(BEDT-TTF)$_2$I$_3$. The contrast covers height variations in the 0–1 nm range.

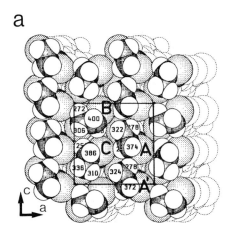
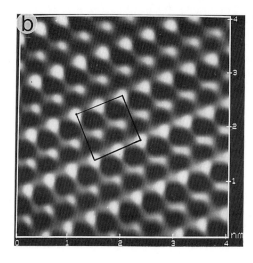
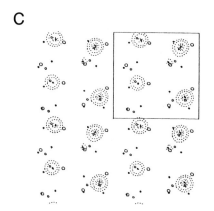

Figure 10.19 Molecular arrangement of BEDT-TTF cations in the *ac*-plane of α-(BEDT-TTF)$_2$TlHg(SCN)$_4$. The H, C, and S atoms of the dithioethylene group are represented by the small, medium, and large circles, respectively. The numbers within the circles representing the surface atoms indicate their relative heights (pm). (b) $\rho(r_0)$ plot of the cation layer of α-(BEDT-TTF)$_2$TlHg(SCN)$_4$. The contour values are 0.5×10^{-1}, 0.35×10^{-1}, 0.25×10^{-1}, 0.1×10^{-1}, and 0.05×10^{-1} electrons/au^3. (c) AFM force image of the *ac*-face of α-(BEDT-TTF)$_2$-TlHg(SCN)$_4$. The contrast is proportional to force variations in relative units.

A and A′ cations (Figs. 10.20 (b) and 10.21 (b)) [9b, 15]. Likewise, rows of patterns with different thickness alternate in the STM images. According to the partial density plots, the STM patterns of a unit cell can be assigned as follows. The four bright spots are associated with the four H atoms of the A and A′ cations. The three bright spots arise from three of the four H atoms of the B and C cations, and the fourth H atom is not seen because it lies lower than the other three.

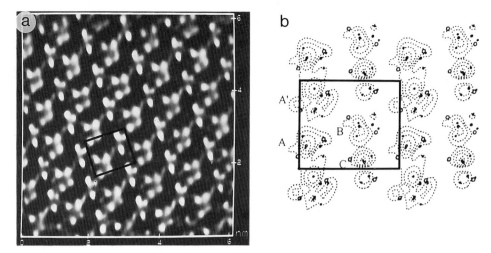

Figure 10.20 (a) STM current image of the *ab*-face of α-(BEDT-TTF)$_2$I$_3$ (I_{set} = 4.6 nA, V_{bias} = 89 mV). The contrast is proportional to current variations in relative units. (b) $\rho(r_0, e_f)$ plot of the cation layer of α-(BEDT-TTF)$_2$I$_3$. The contour values are 0.35 × 10^{-6}, 0.135 × 10^{-6}, 0.05 × 10^{-6}, and 0.02 × 10^{-6} electrons/au^3. The H, C, and S atoms of the dithioethylene group are represented by the small, medium, and large circles, respectively.

As shown in Fig. 10.5 (c), the HOMO of BEDT-TTF is dominated by the p$_\pi$-orbitals of the TTF portion. Through hyperconjugative interactions, the σ- and σ*-orbitals of the C–H and S–C bonds of the dithioethylene groups contribute weakly to the HOMO of BEDT-TTF. To demonstrate that such small orbital components of the dithioethylene groups determine the essential patterns of the $\rho(r_0, e_f)$ plot, a hypothetical cation layer made up of the truncated donor molecules was constructed. The latter is obtained by replacing the two ethylene groups of BEDT-TTF with hydrogen atoms (with S–H = 1.34 Å) (Fig. 10.21 (c)). The $\rho(r_0, e_f)$ plot calculated for the hypothetical cation layer shows that the p$_\pi$-orbitals of the HOMO contribute very little to the $\rho(r_0, e_f)$ plot (Fig. 10.21 (d)). This is because the p$_\pi$-orbitals of the HOMO are far removed from the tip, and the amplitudes of the atomic orbitals decrease exponentially with increasing distance from their centers. The $\rho(r_0, e_f)$ plot is dominated by the ethylene-group atoms (in particular, the H atoms) of the BEDT-TTFs. The orbital contributions of the ethylene groups to the HOMO of BEDT-TTF are small, but the H and C atoms of one ethylene unit determine the essential patterns of the $\rho(r_0, e_f)$ plot because of their proximity to the tip.

10.5.2 HOMO Density of β-(BEDT-TTF)$_2$I$_3$

A perspective view of the structure of β-(BEDT-TTF)$_2$I$_3$ is shown in Fig. 10.22 (a). The cation and anion layers of β-(BEDT-TTF)$_2$I$_3$ are parallel to the (001) surface,

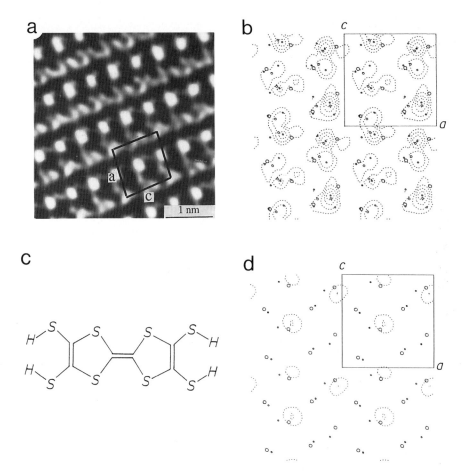

Figure 10.21 (a) STM current image of the *ac*-face of α-(BEDT-TTF)$_2$TlHg(SCN)$_4$ (I_{set} = 2 nA, V_{bias} = 30 mV). The contrast is proportional to current variations in relative units. (b) $\rho(r_0, e_f)$ plot of the cation layer of α-(BEDT-TTF)$_2$TlHg(SCN)$_4$. The contour values are 350×10^{-8}, 135×10^{-8}, 50×10^{-8}, 18×10^{-8}, and 6.7×10^{-8} electrons/au^3. The H, C, and S atoms of the dithioethylene group are represented by the small, medium, and large circles, respectively. (c) Hypothetical donor molecule derived from BEDT-TTF after replacing the ethylene groups with H atoms. (d) $\rho(r_0, e_f)$ plot calculated for the layer made up of the hypothetical donor molecules as arranged in α-(BEDT-TTF)$_2$TlHg(SCN)$_4$. The contour values are 18×10^{-8} and 6.7×10^{-8} electrons/au^3.

and there are two crystallographically equivalent BEDT-TTF molecules and one I$_3^-$ anion per unit cell. Figure 10.22 (b) shows the arrangement of the BEDT-TTF molecules on the (011) surface. This surface is described by the unit cell parameters a = 6.62 Å, d = 18.37 Å and φ = 85.0°, where d is the length of the vector $\mathbf{d} = \mathbf{c} - \mathbf{b}$. The long molecular axis of BEDT-TTF makes an angle of approximately 21° with the d-axis. The π-plane of each BEDT-TTF is not completely parallel to the (011) surface due to a very small tilting around the short and the long molecular axes, so the atoms of each BEDT-TTF molecule are at unequal heights on the (011) surface. For each BEDT-TTF, the most protruding atom is one H atom of the higher-lying ethylene group. This atom lies about 1.3 Å higher than the most protruding S atom of the ethylenedithio group, and about 1.9 Å higher than the two most protruding S atoms of the TTF moiety.

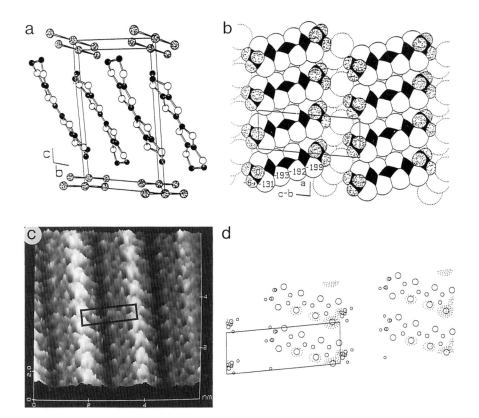

Figure 10.22 (a) Perspective view of the structure of β-(BEDT-TTF)$_2$I$_3$. The S, C, and I atoms are represented by open, filled, and gray circles, respectively. For simplicity, the H atoms are not shown. (b) Arrangement of the BEDT-TTF molecules on the (011) surface. The S, C, and H atoms are represented by open, filled and gray continuous circles, respectively, and the I atoms by open dotted circles. The numbers within the circles representing the atoms indicate their relative heights (pm) with respect to the most protruding H atom. (c) STM current image of the (011) surface of β-(BEDT-TTF)$_2$I$_3$. The contrast is proportional to current variations in relative units. (d) $\rho(r_0, e_f)$ plot calculated for the (011) surface of β-(BEDT-TTF)$_2$I$_3$. The contour values used are 25×10^{-6}, 20×10^{-6}, 15×10^{-6}, 10×10^{-6}, and 5×10^{-6} electrons/au^3. The H, C, and S atoms of the BEDT-TTF molecules are represented by small, medium, and large circles, respectively.

The STM current image obtained on the (011) surface of β-(BEDT-TTF)$_2$I$_3$ is shown in Fig. 10.22 (c). This image is characterized by rows of one large bright and two small bright spots per unit cell running along the crystallographic a-direction. The $\rho(r_0, e_f)$ plot of the (011) plane surface calculated on the basis of the bulk crystallographic structure is shown in Fig. 10.22 (d) [16]. For each BEDT-TTF, the HED spots are located on the most protruding H atom and the most protruding S atom of the ethylenedithio group. The H atoms contribute slightly less than the S atom although they lie 1.3 Å higher, because the sulfur orbital has a larger contribution to the HOMO of BEDT-TTF than does the hydrogen orbital. For each BEDT-TTF, the next HED spots are located on the two most protruding S atoms of the TTF moiety. The general patterns of the HED spots found for the $\rho(r_0, e_f)$ plot are in agreement with those of the STM image.

10.5.3 Cation-Layer Images of κ-Phases

Figure 10.3 (b) shows a perspective view of the crystal structure of κ-(BEDT-TTF)$_2$Cu[N(CN)$_2$]X (X = Cl, Br) [5b, c], which has two cation layers and two anion layers (parallel to the ac-plane) per unit cell. Projection views of the cation and anion layers of κ-(BEDT-TTF)$_2$Cu[N(CN)$_2$]Br are shown in Figs. 10.23 (a) and 10.23 (b), respectively. The cation and anion layers of κ-(BEDT-TTF)$_2$Cu(SCN)$_2$ are parallel to the bc-plane (Fig. 10.23 (c)) [5a]. The STM images observed for the cation layers of κ-(BEDT-TTF)$_2$Cu(SCN)$_2$ and its isostructural salts are nearly the same, regardless of the bias polarity. The atomic-scale resolved patterns of κ-(BEDT-TTF)$_2$Cu(SCN)$_2$ observed at R_{gap} = 50 MΩ (Fig. 10.24 (a)) are converted into only four large spots within a unit cell when recorded with lower R_{gap} (e.g., 12 MΩ in Fig. 10.24 (b)). Such an R_{gap}-dependence of the STM image is found in other κ-phase BEDT-TTF salts also. Comparison of these images with the $\rho(r_0, e_f)$ plot (Fig. 10.24 (c)) shows that the essential image features are associated with the electronic contributions of the topmost H atoms of the cations [17]. A typical AFM image recorded on the cation surfaces of the κ-phases is shown in Fig. 10.25 (a). This image possesses four bright patterns per unit cell and is consistent with the $\rho(r_0)$ plot (Fig. 10.25 b), so that each bright spot of the image is associated with one BEDT-TTF molecule.

10.5.4 Anion-Layer Images of κ-Phases

The loosely packed, nonconducting anion layer is more likely to be destroyed by the scanning tip than the tightly packed, conducting cation layer. The anions of κ-phases form polymer-like structures and thus are mechanically more stable than those of α- and β- phases made up of discrete anions. Therefore, STM and AFM images of the

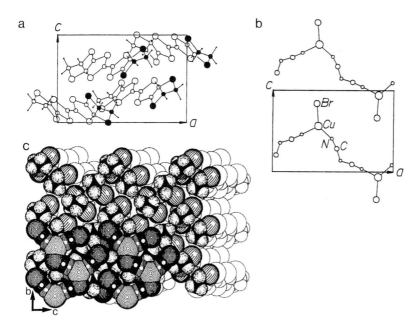

Figure 10.23 (a) Projection view of the cation layer (*ac*-plane) of κ-(BEDT-TTF)$_2$Cu[N(CN)$_2$]Br. For clarity, the ethylenedithio groups of the BEDT-TTF cations closer to the top surface are shown by shading the C and S atoms. (b) Projection view of the anion layer (*ac*-plane) of κ-(BEDT-TTF)$_2$Cu[N(CN)$_2$]Br. (c) Schematic view on the *bc*-plane surface of κ-(BEDT-TTF)$_2$Cu(SCN)$_2$. The lower part of surface consists of anions, Cu(SCN)$_2^-$, and the upper part of BEDT-TTF cations. All atoms of the anions and the topmost atoms (H, C, and S) of the cations are shaded.

anion layers of κ-phases are occasionally found. Figures 10.26 (a) and 10.27 (a) respectively show the AFM image of κ-(BEDT-TTF)$_2$Cu[N(CN)$_2$]Cl and the STM image of κ-(BEDT-TTF)$_2$Cu(SCN)$_2$. Both images resemble the molecular arrangement of the anion layers. The $\rho(r_0)$ plot calculated for the anion layer of κ-(BEDT-TTF)$_2$Cu[N(CN)$_2$]Cl (Fig. 10.26 (b)) is similar to the image in Fig. 10.26 (a) [17]. The bright spots of the AFM image are assigned to the chain atoms with density peaks, and the less bright spots to the Cl atoms.

Observation of the anion layer by STM (Fig. 10.27 (a)) leads to a fundamental question: "Why it is possible to image insulating compounds by STM when they are placed on a metallic substrate?" To probe this question, one can construct a cation/anion double layer on the basis of the crystal structure of κ-(BEDT-TTF)$_2$Cu(SCN)$_2$ and then calculate the $\rho(r_0, e_f)$ plot of its anion-layer surface [17]. Figure 10.27 (b) shows the resulting $\rho(r_0, e_f)$ plot, which is totally dominated by the contributions of the anion atoms. The high electron density is found on every second Cu(SCN)$_2^-$ unit, in excellent agreement with the anion-layer STM image (Fig. 10.27 (a)). The anion

214   10 Organic Conducting Salts

layer has two Cu(SCN)$_2^-$ units per unit cell, with slightly different heights and tilting. The highest-lying atom of the higher-lying Cu(SCN)$_2^-$ unit is sulfur and lies about 0.32 Å higher than that of the lower-lying Cu(SCN)$_2^-$ unit, which is copper. It is the higher-lying Cu(SCN)$_2^-$ units that have the greater density in the $\rho(r_0, e_f)$ plot. Two Cu(SCN)$_2^-$ anions in a unit cell of the anion layer are equivalent in the bulk crystal, but this is not the case when they are part of the anion layer forming the crystal surface.

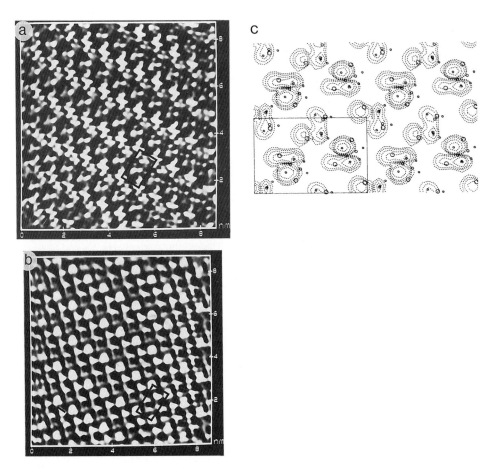

Figure 10.24 STM current images of the bc-plane surface of κ-(BEDT-TTF)$_2$Cu(SCN)$_2$. (a) $I_{set} = 1$ nA, $V_{bias} = 50$ mV; (b) $I_{set} = 1$ nA, $V_{bias} = 12$ mV. The contrast is proportional to current variations in relative units. (c) $\rho(r_0, e_f)$ plot calculated for the cation layer of κ-(BEDT-TTF)$_2$Cu(SCN)$_2$. The contour values are 40×10^{-8}, 30×10^{-8}, 20×10^{-8}, 10×10^{-8}, and 5×10^{-8} electrons/au^3. The H, C, and S atoms of the ethylenedithio group are represented by the small, medium, and large circles, respectively.

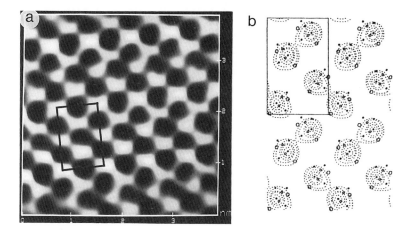

Figure 10.25 (a) AFM force image of the *ac*-plane surface of κ-(BEDT-TTF)$_2$Cu[N(CN)$_2$]Br. The contrast is proportional to force variations in relative units. (b) $\rho(r_0)$ plot calculated for the cation layer of κ-(BEDT-TTF)$_2$Cu[N(CN)$_2$]Br. The contour values are 2.0×10^{-2}, 0.8×10^{-2}, 0.4×10^{-2}, 0.1×10^{-2}, and 0.04×10^{-2} electrons/au^3. The H, C, and S atoms of the ethylenedithio group are represented by the small, medium, and large circles, respectively.

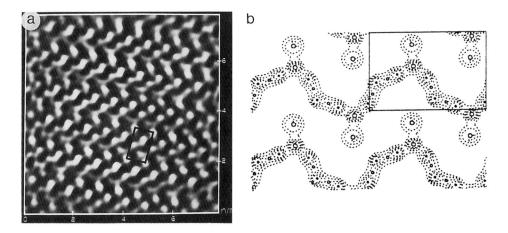

Figure 10.26 (a) AFM force image of the *ac*-plane surface of κ-(BEDT-TTF)$_2$Cu[N(CN)$_2$]Cl. The contrast is proportional to force variations in relative units. (b) $\rho(r_0)$ plot calculated for the anion layer of κ-(BEDT-TTF)$_2$Cu[N(CN)$_2$]Cl. The contour values are 30×10^{-2}, 25×10^{-2}, 20×10^{-2}, 15×10^{-2}, 10×10^{-2}, and 5×10^{-2} electrons/au^3. The Cu atoms of κ-(BEDT-TTF)$_2$Cu[N(CN)$_2$]Cl are represented by the largest circles, and the Cl atoms by the second-largest circles.

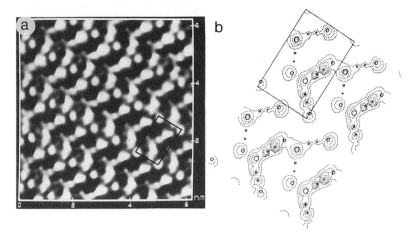

Figure 10.27 (a) STM current image of the *bc*-plane surface of κ-(BEDT-TTF)$_2$Cu(SCN)$_2$ ($I_{tun} = 1$ nA, $V_{bias} = 12$ mV). The contrast is proportional to current variations in relative units. (b) $\rho(r_0, e_f)$ plot calculated for the cation/anion double layer of κ-(BEDT-TTF)$_2$-Cu(SCN)$_2$ with the tip placed above the anion layer. The contour values are 200×10^{-8}, 100×10^{-8}, 50×10^{-8}, 20×10^{-8}, and 5×10^{-8} electrons/au^3. The Cu atoms of the Cu(SCN)$_2^-$ units are represented by the largest circles, and the S atoms by the second-largest circles.

10.6 Concluding Remarks

The results of this chapter show that the use of the partial and total electron density plots is invaluable in understanding the STM and AFM images of organic conducting salts. The $\rho(r_0, e_f)$ plots calculated on the basis of the bulk crystal structures reveal that the fine features of the STM images arise primarily from the contributions of the outermost atoms of the surface molecules.

For the BEDT-TTF salts, the cation layers are more frequently observed in the STM and AFM images than are the anion layers. The electron density plots indicate that the cation layer images are dominated by contributions from the H atoms of the ethylene groups. STM imaging of the anion layers of the κ-phase BEDT-TTF salts is surprising, because of their nonconducting nature. This observation is explained in terms of the $\rho(r_0, e_f)$ plot calculated for a cation/anion double layer. This plot shows that the mixing of the anion-layer orbitals into the valence bands (largely made up of the HOMOs of BEDT-TTF) of the cation layer is not strong, but is sufficient for the anion layer to be imaged by STM. The mixing of the anion-layer orbitals into the valence bands of the cation layer is comparable in magnitude with that of the ethylene-group σ-orbitals into the HOMO of BEDT-TTF. In general, this orbital-mixing mechanism should be applicable to the STM imaging of insulating compounds on metallic substrates (for details, see Chapter 11).

The main features of the images of organic conducting salts have been explained without considering tip force induced surface relaxation. However, several observations indicate the presence of tip–sample force interactions. They are the removal of surface layers, the migration of molecular defects, and the changes in the positions of the topmost molecular groups.

References

[1] (a) I. F. Shchegolev, *Phys. Stat. Sol.* **1972**, *12*, 9. (b) L. B. Coleman, M. J. Cohen, D. J. Sandman, F. G. Yamagishi, A. F. Garito, A. J. Heeger, *Solid State Commun.* **1973**, *12*, 1125. (c) J. Ferraris, D. Cowan, V. Walatka, Jr., J. H. Perlstein, *J. Am. Chem. Soc.* **1973**, *95*, 948.

[2] (a) J. M. Williams, H. H. Wang, T. J. Emge, U. Geiser, M. A. Beno, P. C. Leung, K. D. Carlson, R. J. Thorn, A. J. Schultz, M.-H. Whangbo, *Prog. Inorg. Chem.* **1987**, *35*, 51. (b) R. Moret, J. P. Pouget, *Crystal Chemistry and Properties of Materials with Quasi-One-Dimensional Structures*, (Ed.: J. Rouxel), Reidel, Dordrecht, The Netherlands, **1986**, p. 87. (c) H. Friend, D. Jérome, *J. Phys. C: Solid State Phys.* **1979**, *12*, 1441.

[3] (a) V. Z. Kresin and W. A. Little (Eds.), *Organic Superconductivity*, Plenum, New York, **1990**. (b) J. M. Williams, A. J. Schultz, U. Geiser, K. D. Carlson, A. M. Kini, H. H. Wang, W.-K. Kwok, M.-H. Whangbo, J. E. Schirber, *Science*, **1991**, *252*, 151. (c) J. M. Williams, J. R. Ferraro, R. J. Thorn, K. D. Carlson, U. Geiser, H. H. Wang, A. M. Kini, M.-H. Whangbo, *Organic Superconductors*, Prentice Hall, New York, **1992**.

[4] (a) T. J. Kistenmacher, T. E. Philips, D. O. Cowan, *Acta Crystallogr. B* **1974**, *30*, 763. (b) H. Kobayashi, F. Marumo, Y. Saito, *Acta Crystallogr. B* **1971**, *27*, 373. (c) S. N. Magonov, S. Kempf, H. Rotter, H.-J. Cantow, *Synth. Metals* **1991**, *40*, 73. (d) H. Kobayashi, Y. Ohashi, F. Marumo, Y. Saito, *Acta Crystallogr. B* **1970**, *26*, 459.

[5] (a) H. Urayama, H. Yamochi, G. Saito, K. Nozawa, T. Sugamo, M. Kinoshita, S. Saito, K. Oshima, A. Kawamoto, J. Tanaka, *Chem. Lett.* **1988**, 55. (b) A. M. Kini, U. Geiser, H. H. Wang, K. D. Carlson, J. M. Williams, W. K. Kwok, K. G. Vandervoort, J. E. Thompson, D. Stupka, D. Jung, M.-H. Whangbo, *Inorg. Chem.* **1990**, *29*, 2555. (c) J. M. Williams, A. M. Kini, H. H. Wang, K. D. Carlson, U. Geiser, L. K. Montgomery, G. J. Pyrka, D. M. Watkins, J. M. Kommers, S. J. Boryschuck, A. V. S. Crouch, W. K. Kwok, J. E. Schirber, D. L. Overmyer, D. Jung, M.-H. Whangbo, *Inorg. Chem.* **1990**, *29*, 3272.

[6] T. Sleator, R. Tycko, *Phys. Rev. Lett.* **1988**, *60*, 1418.

[7] (a) C. Bai, C. Dai, C. Zhu, Z. Chen, G. Huang, X. Wu, D. Zhu, J. D. Baldeschwieler, *J. Vac. Sci. Technol.* **1990**, *A8*, 484. (b) M. Yoshimura, H. Shigekawa, H. Nejoh, G. Saito, Y. Saito, A. Kawazu, *Phys. Rev. B,* **1991**, *43*, 13590; M. Yoshimura, H. Shigekawa, H. Yamochi, G. Saito, A. Kawazu, *Phys. Rev. B,* **1991**, *44*, 1970; (c) R. Fainchtein, S. D'Arcangelis, S. S. Yang, D. O. Cowan, *Science* **1992**, *256*, 1012. (d) S. N. Magonov, G. Bar, E. Keller, E. B. Yagubskii, E. E. Laukhina, H.-J. Cantow, *Ultramicroscopy* **1992**, *42–44*, 1009.

[8] (a) J. S. Foster, J. Frommer, *Nature (London)* **1988**, *333*, 542. (b) G. McGonigal, R. Bernhardt, D. Thomson, *Appl. Phys. Lett.* **1990**, *57*, 28.

[9] (a) S. N. Magonov, G. Bar, A. Ya. Gorenberg, H.-J. Cantow, *Adv. Mater.* **1993**, *5*, 453. (b) S. N. Magonov, G. Bar, H.-J. Cantow, J. Paradis, J., Ren, M.-H. Whangbo, E. B. Yagubskii, *J. Phys. Chem.* **1993**, *97*, 9170.

[10] M.-H. Whangbo, D. Jung, J. Ren, M. Evain, J. J. Novoa, F. Mota, S. Alvarez, J. M. Williams, M. A. Beno, A. Kini, H. H. Wang, J. R. Ferraro, *The Physics and Chemistry of Organic Superconductors*, (Eds.: G. Saito, S. Kagoshima), Springer, **1990**, p 262.

[11] S. N. Magonov, G. Bar, H.-J. Cantow, J. Ren, M.-H. Whangbo, *Synth. Met.* **1994**, *62*, 159.

[12] (a) J. Ren, M.-H. Whangbo, G. Bar, S. N. Magonov, unpublished work. (b) G. Bar, B. Scott, S. R. Johnson, B. I. Swanson, J. Ren, M.-H. Whangbo, *Chem. Mater.*, **1995**, *7*, 391. (c) K. S. Lee, D.-K. Seo, M.-H. Whangbo, G. Bar, S. N. Magonov, to be published.

[13] S. N. Magonov, G. Bar, H.-J. Cantow, D. Kabbeck-Kupijai, G. Kossmehl, *Ber. Bunsenges. Phys. Chem.* **1993**, *97*, 1478.

[14] G. Bar, S. N. Magonov, H.-J. Cantow, J. Gmeiner, M. Schwörer, *Ultramicroscopy* **1992**, *42-44*, 644.

[15] S. N. Magonov, G. Bar, H.-J. Cantow, J. Paradis, J. Ren, M.-H. Whangbo, *Synth. Met.* **1994**, *62*, 83.

[16. G. Bar, S. N. Magonov, W. Liang, M.-H. Whangbo, *Synth. Met.* **1995**, *72*, 189.

[17] G. Bar, S. N. Magonov, J.-H. Cantow, N. D. Kushch, E. B. Yagubskii, W. Liang, J. Ren, M.-H. Whangbo *New J. Chem.* **1993**, *17*, 439.

11 Organic Adsorbates at Liquid/Solid Interfaces

The molecular-scale imaging of insulating organic compounds adsorbed on conducting substrates was an exciting event in the development of STM applications. STM images exhibiting molecular-structural details are now known for a large number of adsorbed organic molecules. However, structural characterization of the organic adsorbates by STM has met only limited success because it is not simple to deduce the structural information from the observed images alone. This chapter describes the imaging of organic molecules adsorbed on substrates. The image analysis provides valuable information about the molecular organization in the layers at the solid/liquid interface.

11.1 STM of Organic Adsorbates

Organic molecules placed in the gap between two electrodes modify the characteristics of the electron tunneling process. The inelastic interaction of the tunneling current with the organic molecules leads to peaks in the d^2I/dV^2-versus-voltage curves. These peaks appear at the voltages corresponding to the energies of the vibrational modes of the organic molecules. The study of this phenomenon is the subject of inelastic tunneling spectroscopy [1] and can be carried out in the STM set-up [2]. The detection of insulating molecules by STM has been demonstrated in studies of isolated atomic and molecular adsorbates on semiconductor and metal surfaces [3] as well as liquid crystals and normal alkanes on graphite [4]. The structural features of organic molecules such as aromatic rings and alkyl chains are recognizable in the molecular-scale images. This is useful for probing the molecular order in adsorbates. When an isotropic or partially ordered liquid is in contact with a solid surface, thermodynamically stable adsorbed layers are formed at the liquid/solid interface [5]. The molecular order in such systems have been studied by calorimetry, ellipsometry, and second-harmonic generation [6], and STM complements these techniques.

11.1.1 Organic Compounds and Substrates

The structural characteristics of the organic compounds used in STM experiments described in this chapter are presented in Fig. 11.1. Normal alkanes C_nH_{2n+2} are most stable when their carbon frameworks adopt the all-*trans* conformation

(Fig. 11.1 (a)). Cycloalkanes $(CH_2)_n$ for $n > 20$ possess a structure in which two parallel stems [i. e., $(CH_2)_{(n-8)/2}$ units] in all-*trans* conformation are joined by two molecular folds [i. e., $(CH_2)_4$ units] in (*ggtgg*) conformation to form a rectangular ring [7] (Figs. 11.1 (b) and (c)). The planes of the all-*trans* carbon frameworks of the two stems are perpendicular to the molecular plane. In 4-*n*-hexyl- and 4-*n*-octyl-4'-cyanobiphenyls (hereafter referred to as 6CB and 8CB, respectively), the alkyl chains in all-*trans* conformation cannot be lined up along the long axis of the biphenyl group (Fig. 11.1 (d)). The overall shapes of 6CB and 8CB become linear when the alkyl chain adopts a helix-like conformation by incorporating several *gauche* arrangements in the carbon skeleton.

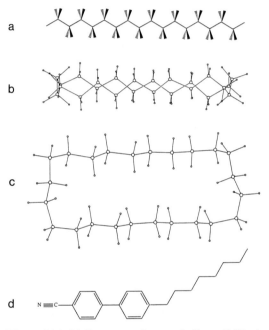

Figure 11.1 (a) Structure of normal alkane C_8H_{18} in all-*trans* conformation. (b), (c) Two projection views of cycloalkane $(CH_2)_{24}$. (d) Structure of 4-octyl-4'-cyanobiphenyl.

For the substrate, the atomically flat surface of HOPG and corrugated surface A of β-Nb$_3$I$_8$ (Chapter 6) were used. The surface hexagonal unit cell of β-Nb$_3$I$_8$ ($a = 0.760$ nm) is substantially larger than that of graphite ($a = 0.246$ nm) because one iodine atom within the unit cell of β-Nb$_3$I$_8$ protrudes farther than the remaining three iodine atoms by 0.060 nm. Therefore this corrugated surface is characterized by "grooves" running in three directions (related by 60° rotation), and every nonprotruding iodine atom is common to two grooves (Fig. 11.2 (a)). In large-scale STM and AFM images of β-Nb$_3$I$_8$, there occur triangular defects with their sides oriented along the main crystallographic directions of the surface (Fig. 11.2 (b)).

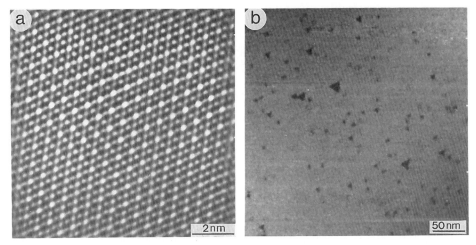

Figure 11.2 (a) AFM height image of β-Nb$_3$I$_8$. (b) STM height image of β-Nb$_3$I$_8$ (I_{set} = 1.3 nA, V_{bias} = 120 mV). The contrast covers height variations in the 0.0–0.6 nm range in (a) and in the 0–2 nm range in (b).

11.1.2 STM Imaging at Liquid/Solid Interfaces

For STM experiments a drop of neat material (6CB and 8CB) and of saturated alkane [C$_{36}$H$_{74}$, (CH$_2$)$_{48}$ and (CH$_2$)$_{72}$] solution in dodecane is deposited on the substrate, and the tip is immersed in the drop during the scanning. A stable imaging of adsorbates at the liquid/solid interface requires optimum tunneling conditions (i.e., optimum I_{set} and V_{bias}) and, probably, an optimum tip shape. However, the images are reproducible, once obtained. The organic layers on the conducting substrates were imaged with R_{gap} in the 0.3–6.0 MΩ range, which is larger than the typical values for the imaging of the substrates. When R_{gap} is gradually reduced, the image of the adsorbed molecules is eventually replaced with that of the substrate because, at small enough tip–sample separation, the tip pushes away the adsorbed molecules.

The tip force induced surface modifications are illustrated in Fig. 11.3, which shows three different scale STM images of cycloalkanes (CH$_2$)$_{48}$ on graphite obtained by zooming out stepwise. Within the hole made by etching the adsorbed layer (Fig. 11.3 (b)), one sees linear striations associated with the lamellar structure of the (CH$_2$)$_{48}$ alkane layer (Fig. 11.3 (a)). Thus, the scanning tip destroys a major part of the adsorbed layers before reaching a well-ordered lamellar sheet of alkanes [8]. The shape of the etched area in Fig. 11.3 (b) differs from the rectangular "windows" observed in a small-scale scanning of transition-metal chalcogenides and organic conducting salts [9]. This difference may be due to a partial re-adsorption of the cycloalkanes from the solution during the scanning. The thickness of the etched layer in Fig. 11.3 (b) is ca. 0.6 nm. A larger-scale scan (3 μm × 3 μm^2) shows "islands" belonging to the topmost, incomplete, layer of 0.6 nm in thickness (Fig. 11.3 (c)). Thus, two less stable adsorbed layers covering the lamellar layer are removed from the surface by the tip when the scanning area is reduced to a smaller size (Fig. 11.3 (d)). The scanning rate is usually increased when the scanning area is reduced. Nevertheless, under a small-scale scan, a particular area exposed long enough for tip–sample interactions to occur is finally destroyed.

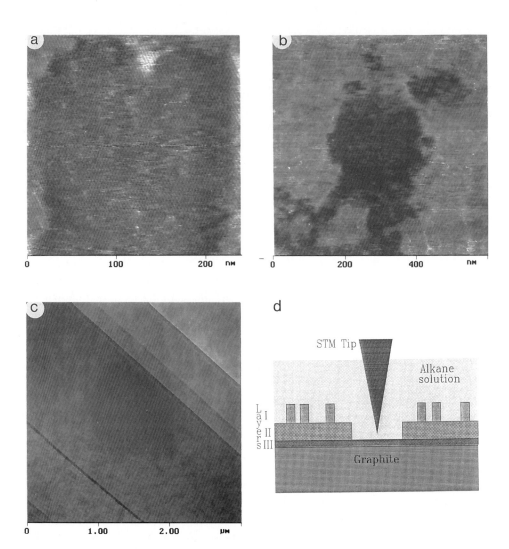

Figure 11.3 (a)–(c) STM images of cycloalkane $(CH_2)_{48}$ on graphite registered by zooming out stepwise ($I_{set} = 0.3$ nA, $V_{bias} = 1.8$ V) [8]. The contrast covers height variations in the 0–1 nm range in (a), in the 0–3 nm range in (b), and in the 0–5 nm range in (c). (d) Sketch of a multilayered alkane adsorbate.

The etching of the alkane adsorbates reveals its multilayered structure. The higher-lying layers are expected to have a poor mechanical stability due to a decrease in the order and density of the adsorbate with an increase in the distance from the substrate surface. When cycloalkanes $(CH_2)_{48}$ are added to the solution of normal alkanes $C_{36}H_{74}$, the cycloalkanes replace the normal alkanes in the first layer on graphite [8]. The image of Fig. 11.4 shows that the topmost lamellar structure with a period of ca. 4.9 nm corresponds to the $C_{36}H_{74}$ layer. Within the holes of this layer, one can observe the stripes of shorter period (ca. 3.9 nm) belonging to the $(CH_2)_{48}$ layer. This suggests that adhesion to graphite is stronger for $(CH_2)_{48}$ than for $C_{36}H_{74}$.

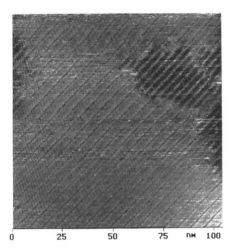

Figure 11.4 STM height image of an ordered adsorbate consisting of the underlying $(CH_2)_{48}$ sheet and $C_{36}H_{74}$ overlayer (I_{set} = 0.3 nA, V_{bias} = −1.8 V). The contrast covers height variations in the 0.0–0.5 nm range.

11.2 STM of Normal and Cyclic Alkane Layers

11.2.1 Images of Normal Alkanes on HOPG

The adsorption of long-chain alkanes C_nH_{2n+2} with $n > 22$ onto graphite leads to stable layers, thereby allowing molecular-scale STM imaging [4b, 10]. Domains with a lamellar order in which individual alkane chains are oriented along the main crystallographic directions of graphite are frequently found (Fig. 11.5). The width of the lamella is the same as the length of an alkane chain in all-*trans* conformation, which supports the earlier conclusion drawn from calorimetric measurements that alkane chains within the adsorbed layers are oriented parallel to the substrate [6a]. Important questions concerning the molecular arrangements in the adsorbed layers are the following. What is the orientation of the alkane all-*trans* carbon skeleton with respect to the graphite plane? Why can the contrast of the lamella borders be darker or brighter than that of the lamella inside [10c]? And what is the origin of the superstructure patterns occasionally observed?

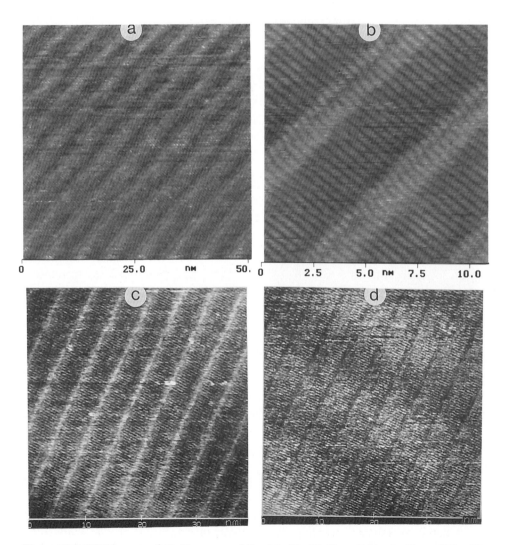

Figure 11.5 STM images of $C_{36}H_{74}$ on graphite: (a), (b), (d) (I_{set} = 0.54 nA, V_{bias} = 265 mV). (c) (I_{set} = 0.54 nA, V_{bias} = 265 mV). The contrast covers height variations in the 0.0–0.2 nm range.

Early attempts to explain the image patterns of organic adsorbates relied on the HOMO and LUMO electron densities of the adsorbed molecules [11a, b]. This approach is not fully satisfactory because the substrate is not taken into account. A better description is achieved by calculating the electron density of a molecule adsorbed on a small cluster of atoms representing the substrate. This approximation led to a reasonable correspondence between the calculated electron density patterns and STM images for naphthalene, azulene, and related molecules adsorbed on Pt(111) [11c]. In the STM images of the organic conducting salt κ-(BEDT-TTF)$_2$Cu(NCS)$_2$, insulating anion layers have been observed (Chapter 10). Such an image is well described by the $\rho(r_0, e_f)$ plot calculated for a cation/anion double layer with the tip located above the anion layer [12]. The anion atomic orbitals mix slightly into the conducting band and dominate the patterns of the $\rho(r_0, e_f)$ plot due to the proximity of the anion atoms to the tip. The relationship between the insulating anion layer and the conducting cation layer in κ-(BEDT-TTF)$_2$Cu(SCN)$_2$ is analogous to that between the insulating layer of alkanes and a conducting graphite surface. Indeed, the essential features of the STM images of alkanes adsorbed on graphite are explained in terms of the $\rho(r_0, e_f)$ plots calculated for a model double layer made up of a sheet of propane molecules C_3H_8 (i.e., the shortest analog with a bent C–C–C framework) adsorbed on a graphite sheet [13]. From the viewpoint of localized molecular orbitals, two-electron–two-center bonds such as C–C and C–H are transferable from one homolog to another in the C_nH_{2n+2} series. Thus, the C–C and C–H bonds of propane are similar to those of long-chain hydrocarbons, and the two C–H bonds of a methyl group in C_3H_8 not in the plane of the carbon skeleton are very similar in electronic structure to any methylene hydrogen atoms of long-chain alkanes. The use of the propane molecule is thus justified for the simulation of long-chain hydrocarbons.

For convenience, propane molecules were placed in the center of every "seven-hexagon" cluster of graphite and the $\rho(r_0, e_f)$ plot was calculated for the tip over the propane monolayer. For the monolayer of propane molecules, two different orientations of propane molecules are of interest, i.e., with the carbon-skeleton plane either parallel or perpendicular to the graphite sheet (Fig. 11.6 (a) and 11.7 (a)). The $\rho(r_0, e_f)$ plots calculated for the propane/graphite double layers with the two extreme orientations of propane show that neither the graphite carbon atoms nor the carbon atoms of the adsorbed molecules contribute to the $\rho(r_0, e_f)$ plots, and the contributions of the most protruding H atoms dominate in all the $\rho(r_0, e_f)$ plots (Fig. 11.6 (b) and 11.7 (b)). The electron density above the adsorbed insulating molecules is small compared with that over the surface layer atoms of layered compounds but is detected by STM when a low I_{set} is applied.

From the $\rho(r_0, e_f)$ plots calculated for the propane/graphite double layer, it is straightforward to guess the STM patterns expected for long-chain normal alkanes adsorbed on graphite. Provided that a long-chain alkane molecule adsorbed on graphite has the plane of its all-*trans* framework perpendicular to the graphite sur-

226 11 *Organic Adsorbates at Liquid/Solid Interfaces*

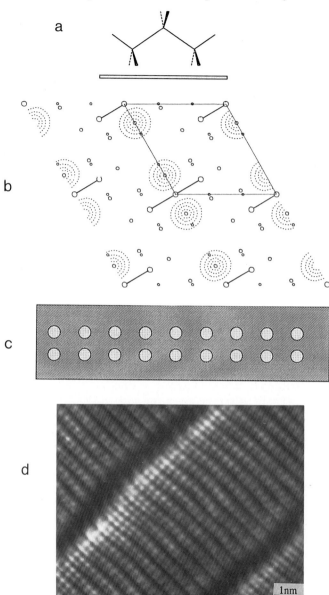

Figure 11.6 (a) Schematic view of a propane molecule on a graphite sheet with its carbon skeleton perpendicular to the graphite sheet. (b) 2D contour representation of the $\rho(r_0, e_f)$ plot calculated for a propane/graphite double layer with the carbon skeleton of propane perpendicular to the graphite sheet, where the contour values used are 4×10^{-7}, 3×10^{-7}, 2×10^{-7}, and 1×10^{-7} electrons/au^3. The large and medium circles represent the C and H atoms of propane, respectively, and the small circles represent the graphite C atoms. (c) STM pattern expected for a normal alkane adsorbed on graphite when its all-*trans* carbon skeleton is perpendicular to the graphite surface. (d) STM height image of $C_{36}H_{74}$ on graphite ($I_{set} = 0.23$ nA, $V_{bias} = -1.7$ V). The contrast covers height variations in the 0–5 Å range.

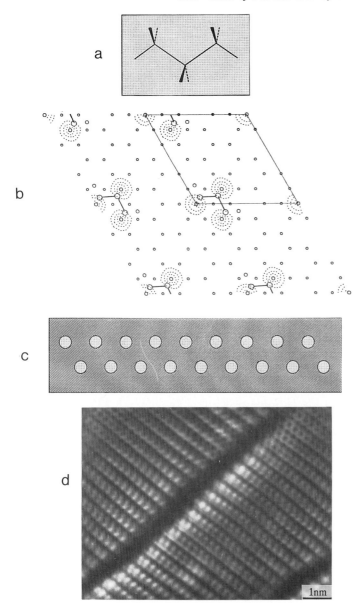

Figure 11.7 (a) Schematic view of a propane molecule on a graphite sheet with its carbon skeleton parallel to the graphite sheet. (b) 2D contour representation of the $\rho(r_0, e_f)$ plot calculated for a propane/graphite double layer with the carbon skeleton of propane parallel to the graphite sheet, where the contour values used are 6×10^{-7}, 3×10^{-7}, and 1×10^{-7} electrons/au^3. The large and medium circles represent the C and H atoms of propane, respectively, and the small circles represent the graphite C atoms. (c) STM pattern expected for a normal alkane adsorbed on graphite when its all-*trans* carbon skeleton is parallel to the graphite surface. (d) STM height image of $C_{36}H_{74}$ on graphite ($I_{set} = 0.23$ nA, $V_{bias} = -1.7$ V). The contrast covers height variations in the 0–5 Å range.

face, the hydrogen atoms of every second methylene should appear as two bright spots (separated by 1.8 Å) perpendicular to the chain direction, thereby forming a doublet-chain pattern (with adjacent doublets separated by ~2.5 Å) (Fig. 11.6 (c)). Depending on the tip quality or tunneling conditions, each doublet may merge into one elongated shape. If the all-*trans* carbon framework is parallel to the graphite surface, then the top hydrogen atom of every methylene group should appear as a bright spot, thereby giving rise to a zigzag pattern for a long-chain alkane (Fig. 11.7 (c)). Within this zigzag pattern, the bright spots are separated by ~2.5 Å along each row and by ~3.2 Å between the two rows.

Both the doublet-chain and the zigzag patterns expected for long-chain alkanes are actually found in the STM images of normal alkanes (Figs. 11.6 (d) and 11.7 (d)). The images can change from a doublet-chain pattern (Fig. 11.6 (d)) to a zigzag-chain pattern (Fig. 11.7 (d)) in successive scans, which can be explained in terms of the tumbling motions of the alkanes on the graphite surface. Because the cohesive interactions between the alkanes within the adsorbed layer and between the alkanes and the graphite surface are weak, the tumbling of the alkanes from one orientation to another should be easy, and the two orientations should be very similar in energy.

The variations in the contrast of the lamellar borders of long-chain alkanes [10c] can be assigned to a slight conformational change at the ends of normal alkane chains within the adsorbed layer. A brighter lamellar border is expected if the ends protrude slightly more than the rest of the chain, which will occur when the adjacent lamellae are pushed toward each other. The probability of observing a brighter lamellar border is higher if the all-*trans* skeleton is parallel to graphite, because a small deviation of the terminal carbon atom position from the all-*trans* conformation will lift the terminal methyl groups to a higher position. When the adjacent lamellae are not pushed toward to each other, the lamellar boundaries will appear darker than the alkane images. The probability of observing darker lamellar boundaries is higher if the carbon skeleton is perpendicular to graphite, because one terminal CH_3 group lies lower than the other CH_3 group (for even-numbered alkanes such as $C_{36}H_{74}$). Thus, the contrast variations within and between the lamellae are all explained in terms of the mobility of the alkanes in the adsorbed layers.

11.2.2 Molecular Order of Cycloalkane Adsorbates on HOPG

Large-scale STM images in Figs. 11.8 (a) and 11.8 (b) show cycloalkane layers adsorbed on the graphite terrace. These layers consist of lamellar domains oriented in different directions. As will be discussed below, the equidistant linear patterns of bright STM spots correspond to the borders of the cycloalkane lamellae. The rows of dimmer patterns perpendicular to the lamellar borders correspond to the all-*trans* stems of cycloalkanes. In contrast to the case of normal alkanes, no contrast variation of the lamellar borders has been observed with cycloalkanes. In addition, there

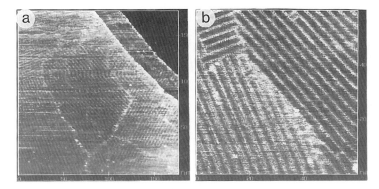

Figure 11.8 STM images of $(CH_2)_{48}$ on graphite ($I_{set} = 0.3$ nA, $V_{bias} = -1.3$ V). The contrast covers height variations in the 0.0–1.5 nm range in (a) and in the 0.0–1.0 nm range in (b).

is no correlation between the orientations of the cycloalkane lamellae in different domains and the main crystallographic directions of the graphite surface (Fig. 11.8 (b)).

The molecular-scale images of cycloalkanes in Figs. 11.9 (a) and (b) show details of the lamellar order. Within each lamella, the dim parallel rows are separated by 0.45 nm, which corresponds to the distance between the neighboring all-*trans* stems of cycloalkanes. The spots within each dim row are separated by 0.25 nm, which agrees with the distance between the hydrogen atoms of every second methylene group of the stems. Thus, the image of an adsorbed alkane is dominated by the contributions of the topmost hydrogen atoms, and the molecular-scale features of the $(CH_2)_{48}$ image are in good agreement with the arrangement in which the cycloalkane lies flat on graphite. The lamellar feature of the STM image has a molecular arrangement of cycloalkanes analogous to that found in the (101) crystallographic plane of the cycloalkane crystal (Fig. 11.9 (c)). Since the most protruding atoms of the cycloalkane constitute two corners of the rectangular ring positioned diagonally with respect to the center of the molecule (Figs. 11.9 (b) and (c)), they give rise to the brightest spots in the STM image. The individual dim spots within the lamella represent the unresolved doublets of the hydrogen atoms of every second methylene group in the all-*trans* stems. The explanation of the brightest STM spots in terms of the topography of the cyclic alkane layer is supported by the AFM image of $(CH_2)_{48}$ adsorbate on graphite, which also exhibits rows of elevated patterns at the lamellar borders [14].

Cycloalkanes are more rigid in structure than normal alkanes, so that the adhesion to graphite is more favorable for a cycloalkane than for a normal alkane. This agrees with the observations mentioned above (Fig. 11.4). Unlike the case of a normal alkane, there has been no evidence for the tumbling of cycloalkane $(CH_2)_{48}$ molecules in the adsorbed layer. Imperfections in molecular order are found in the images of $(CH_2)_{48}$. Figure 11.10 (a) shows orientation defects, which result when short arrays of cycloalkane molecules are incorporated in the lamella at 90° rotation with re-

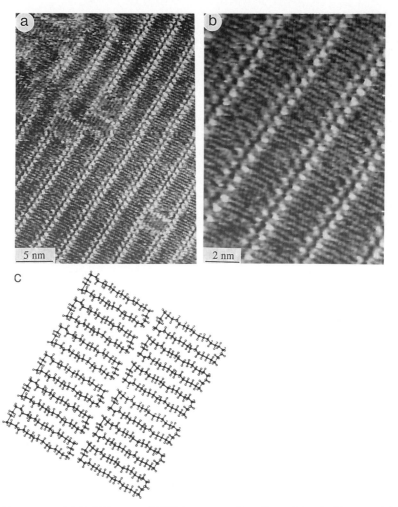

Figure 11.9 (a) STM image of $(CH_2)_{48}$ on graphite (I_{set} = 0.3 nA, V_{bias} = −1.1 V). The contrast covers height variations in the 0.0–1.0 nm range. (b) Zoomed-in part of the image shown in (a). (c) Schematic representation of the lamellar arrangement of $(CH_2)_{36}$ molecules in the (10l) crystallographic plane.

spect to the other molecules. Also, a linear dislocation is found at the lower left-hand corner. In the adsorbate from mixed $(CH_2)_{48}$ and $(CH_2)_{72}$ solution, lamellae of different widths are seen within the same layer (Fig. 11.10 (b)). Two lamellae of $(CH_2)_{72}$ are similar in width to three lamellae of $(CH_2)_{48}$. In some places, arrays of $(CH_2)_{48}$ molecules are incorporated within a $(CH_2)_{72}$ domain.

Though the arrangement of $(CH_2)_{48}$ molecules within the adsorbed layer resembles that in the (10l) plane of a $(CH_2)_{48}$ crystal, the lamellae of a given layer tend

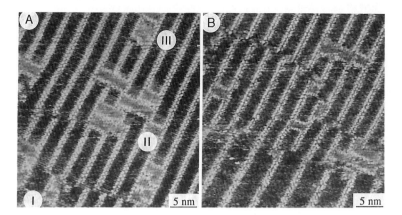

Figure 11.10 (a) STM image of $(CH_2)_{48}$ on graphite ($I_{set} = 0.3$ nA, $V_{bias} = -1.1$ V) showing adsorbate defects. (b) STM image of the adsorbate from a mixed solution of $(CH_2)_{48}$ and $(CH_2)_{72}$ on graphite ($I_{set} = 0.3$ nA, $V_{bias} = -1.1$ V). The contrast covers height variations in the 0.0–0.5 nm range.

to orient orthogonally to those of the underlying layer, as seen in Figs. 11.11 (a)–11.11 (c). Such a layered arrangement differs from the packing motif found in bulk crystals [15]. Several islands of the homoepitaxial layer are seen as bright patches in Figs. 11.11 (a) and 11.11 (b), and as a bigger island in the center of Fig. 11.11 (c). That these islands are homoepitaxial adsorbates is supported by the finding that they move and are destroyed in successive scans. Figure 11.12 (a) shows an $(CH_2)_{48}$ adsorbate, where a top domain exhibits stripes oriented nearly perpendicular to those of the underlying layer. This island has been removed by a tip from the field of view, as seen in the successive image in Fig. 11.12 (b). Such observations lead to an important question: how many adsorbate overlayers can be seen by STM? This question was probed by constructing n hypothetical overlayers ($n = 1$–3) of methane molecules (CH_4) on a graphite sheet [7] and calculating the corresponding $\rho(r_0, e_f)$ plots with the tip above the layers of the molecules. These calculations show that only the most protruding hydrogen atoms of the topmost overlayer appear in the density plots [7]. Consequently, for multilayered alkane adsorbates on graphite, only the most protruding hydrogen atoms of the topmost overlayer should be seen in the STM image. This finding helps one to gain insight into the origin of the superstructure patterns observed in certain images. For example, in Fig. 11.13, the superstructure pattern looks as if two adjacent layers of cycloalkane molecules overlap with the directions of their lamellae rotated by 60°. This image suggests that the topography of the topmost layer is distorted by the topography of the layer lying immediately below, and this distorted topmost overlayer is seen by STM. Superstructure images with overlapping layers at rotation angles of 30° and 90° have also been found.

Figure 11.11 (a) STM height image of $(CH_2)_{48}$ on graphite ($I_{set} = 0.3$ nA, $V_{bias} = -1.5$ V). (b) Zoomed-in part of (a). (c) STM height image of $(CH_2)_{48}$ on graphite ($I_{set} = 0.27$ nA, $V_{bias} = -1.1$ V). The contrast covers height variations in the 0.0–1.0 nm range in (a) and (c).

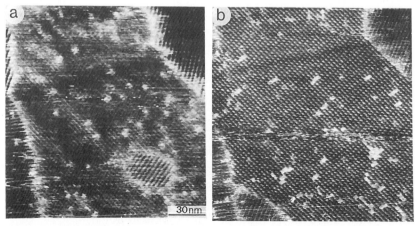

Figure 11.12 Two successive STM images of $(CH_2)_{48}$ adsorbate on graphite ($I_{set} = 0.60$ nA, $V_{bias} = 760$ mV). The contrast covers height variations in the 0–1 nm range.

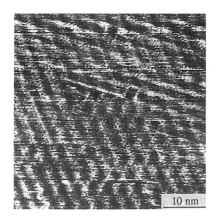

Figure 11.13 STM height image of $(CH_2)_{48}$ on graphite (I_{set} = 0.3 nA, V_{bias} = −1.4 V). The contrast covers height variations in the 0.0–0.5 nm range.

11.3 Influence of Substrate on Adsorbate Structure

Comparative studies of molecular order of normal and cyclic alkanes and *n*-alkylcyanobiphenyls on graphite and β-Nb_3I_8 [16] allow one to gain insight into how a substrate structure affects molecular organization at a liquid/solid interface. These substrates differ not only in their surface atom composition (carbon vs. iodine) but also in the structures (flat vs. corrugated) of their surfaces.

11.3.1 Molecular-Scale Images of Normal Alkanes on β-Nb_3I_8

The repeat distance of a long alkane chain (excluding the two end methyl groups) is quite similar to the lattice parameter of graphite (about 0.25 nm). A lamellar ordering of normal alkanes is also found on the surface of MoS_2 [17], although MoS_2 has a different lattice parameter (a = 0.316 nm). Thus, an atomically flat substrate seems to provide a geometrical constraint for the adsorbed molecules and to enhance intermolecular interactions within the adsorbate layer to form a two-dimensionally ordered arrangement. The calorimetric analysis of the submonolayer region of the adsorption isotherms suggests that there are substantial lateral interactions between adsorbate alkane molecules [18]. Weakly attractive interactions between alkane molecules via the intermolecular H⋯H contacts [19] should help to stabilize the lamellar order of the adsorbed alkanes.

STM measurements of alkanes on β-Nb_3I_8 provide insight into how a substrate structure might influence the molecular order in adsorbed layers. With cycloalkanes $(CH_2)_{48}$ and $(CH_2)_{72}$, adsorbed layers have not been observed on this substrate. The image of normal alkane $C_{36}H_{74}$ on β-Nb_3I_8 (Fig. 11.14 (a)) is quite different from that found on graphite. It exhibits a superstructure pattern with periodicity of 3.8 nm, i.e., five times the repeat distance of β-Nb_3I_8. In a zoomed-in part of this image (Fig. 11.14 (b)), individual spots with a separation of ca. 0.7 nm are distinguished. Four bright spots form the center of each superstructure "hill." The occurrence of this particular alkane adsorption can be related to the corrugated nature of surface A of β-Nb_3I_8. The grooves of this surface can trap individual chains of normal alkanes. Since the grooves run in three directions related by 60° rotation, the formation of a lamella-type layer such as that found on graphite will be prevented. The 3.80 nm repeat distance of the superstructure pattern is shorter than the length of the $C_{36}H_{74}$ chain in the all-*trans* conformation (ca. 4.79 nm), so that each groove in a unit cell of the superstructure pattern can accommodate at most 30 consecutive C-C bonds of $C_{36}H_{74}$. Thus, a chain trapped in each groove cannot be in the all-*trans* conformation, and its end portions (probably of the size of four consecutive C-C bonds) must "bend back" by adopting *gauche* conformations in the bending regions. Consequently, the end portions of each trapped chain will protrude more and hence will appear as the brightest spots in the superstructure image. This rationalization implies that the surface of the first overlayer is more corrugated than that of β-Nb_3I_8, which would make it difficult to grow a second overlayer.

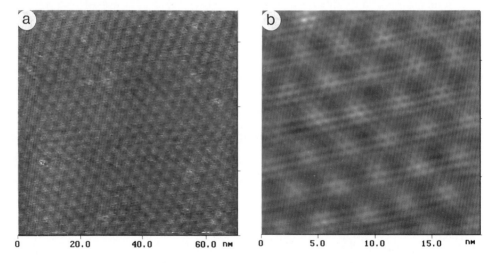

Figure 11.14 (a) STM height image of $C_{36}H_{74}$ on β-Nb_3I_8 (I_{set} = 0.68 nA, V_{bias} = 485 mV). The contrast covers height variations in the 0.0–0.3 nm range. (b) Zoomed-in part of (a) after FFT filtering.

11.3.2 4-Alkyl-4'-cyanobiphenyls on HOPG

Molecular-scale STM images have been reported for 4-alkyl-4'-cyanobiphenyls deposited on graphite [4a, 20]. Large-scale images of 6CB on graphite (Fig. 11.15 (a)) reveal the domain structures of the adsorbates. The bright striations seen at a higher magnification (Fig. 11.15 (b)) are related to the rows of 6CBs (see below). Each row

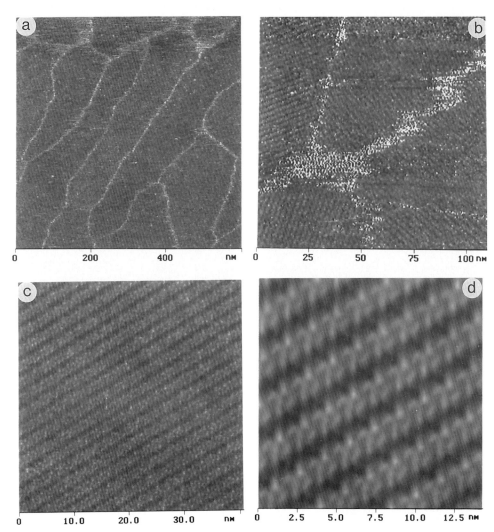

Figure 11.15 (a)–(c) STM height images of 6CB adsorbate on graphite: (a) $I_{set} = 0.25$ nA, $V_{bias} = 190$ mV), (b) $I_{set} = 0.25$ nA, $V_{bias} = 1.84$ V; and (c) $I_{set} = 0.23$ nA, $V_{bias} = -1.77$ V. (d) Zoomed-in part of (c) after FFT filtering. The contrast covers height variations in the 0–2 nm range in (a), in the 0.0–0.5 nm range in (b), in the 0.0–0.7 nm range in (c), and in the 0.0–0.2 nm range in (d).

is aligned along the main crystallographic directions of graphite, and the rows of the neighboring domains are related by 60° rotation. In the molecular-scale image (Fig. 11.15 (c)), linear patterns with separations of 2.5 nm are the dominating features. Every row has a complex repeat pattern (Fig. 11.15 (d)), which has been discussed in terms of strong interdigitation of the head and tail groups of molecules forming the lamellar structures [21].

Molecular-scale images of 8CB adsorbed on graphite are given in Fig. 11.16. The bright rows are separated by ca. 3.8 nm, which is about twice the length of 8CB. This

Figure 11.16 (a) STM height image of 8CB on graphite ($I_{set} = 0.45$ nA, $V_{bias} = -833$ mV). (b) Zoomed-in part of (a) after FFT filtering. (c) STM height image of 8CB on graphite ($I_{set} = 0.30$ nA, $V_{bias} = -780$ mV). The contrast covers height variations in the 0.0–2.0 nm range in (a) and (b), and in the 0.0–1.0 nm range in (c). The image in (c) was provided by Digital Instruments, Inc.

implies that the molecules of the neighboring rows have a minimal interdigitation. The π-orbitals of the cyanobiphenyl groups possess energies similar to those of the graphite π-orbitals. For a given 4-alkyl-4'-cyanobiphenyl molecule adsorbed on graphite, the π-orbitals of the cyanobiphenyl group mix more strongly into the graphite π-orbitals than do the orbitals of the alkyl group. Therefore, the cyanobiphenyl groups are more likely to exhibit brighter contrast in the STM image than are the alkyl groups. This is confirmed by the high-resolution STM image of 8CB in Figure 11.16 (c), where the bright ring-like structures resemble the biphenyl moieties. The images of Figure 11.16 indicate a "head-to-head" arrangement of 8CBs within the adsorbed layers. It should be noted that the relative contributions of the biphenyl and alkyl groups to the image also depend upon the topography of the adsorbate layer. Thus, the alkyl chains might appear as brighter spots, as suggested in the analysis of the 6CB image [21].

11.3.3 4-Alkyl-4'-cyanobiphenyls on β-Nb$_3$I$_8$

Large-scale STM images of 8CB on β-Nb$_3$I$_8$ (Fig. 11.17 (a) and Fig. 11.17 (b)) show a slow growth of adsorbate layers in a drop of liquid 8CB deposited on β-Nb$_3$I$_8$. The image in Fig. 11.17 (b) reveals that the grown overlayer (ca. 0.8 nm in height) has two types of regions; one is featureless (F-domain), and the other has a "hexagonal" pattern (H-domain). The period of the H-domain is about 13–14 nm. The border region between the F- and H-domains is shown at higher magnification in

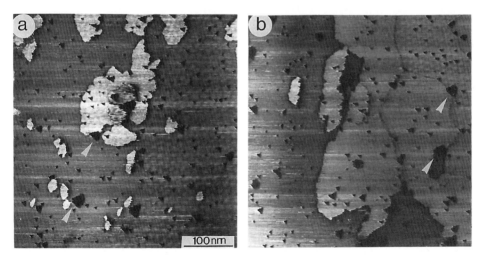

Figure 11.17 STM height images of 8CB on β-Nb$_3$I$_8$ (I_{set} = 0.62 nA, V_{bias} = 390 mV). The image in (b) was recorded 1 h later than the image in (a). The arrowheads indicate identical substrate defects. The contrast covers height variations in the 0.0–4.0 nm range.

Fig. 11.18 (a), which reveals that both domains have numerous periodic features. Although the resolution is poor, it is clear that the F-domain has striped patterns (the lower part of Fig. 11.18 (a)). A better-resolved molecular-scale image of the F-domain (Fig. 11.18 (b) shows that there are three chain patterns of bright spots per unit cell, and each chain pattern consists of three bright spots per unit cell. The repeat distance of the pseudo-hexagonal pattern is ca. 2.2 nm, i.e., three times the cell parameter of β-Nb$_3$I$_8$. The length of 8CB is about 2.0 nm when the 8CB molecule is made linear. This happens when the alkyl chain adopts a helix-like conformation, which can make a portion of the alkyl chain protrude, as does each benzene ring. This would make each 8CB appear as three bright spots if it is trapped in a groove of the β-Nb$_3$I$_8$ surface. Therefore, the linear patterns of Fig. 11.18 (b) suggest that the surface uses parallel grooves to trap 8CB molecules, and each unit cell of the pseudo-hexagonal pattern contains three 8CBs.

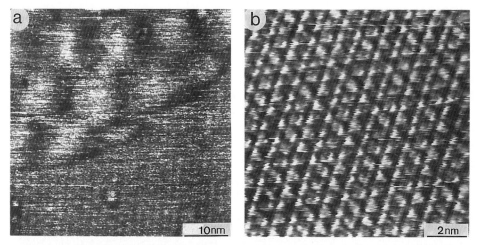

Figure 11.18 STM height images of 8CB on β-Nb$_3$I$_8$: (a) I_{set} = 0.51 nA, V_{bias} = 169 mV; (b) I_{set} = 0.53 nA, V_{bias} = − 352 mV. The contrast covers height variations in the 0.0–0.4 nm range in (a) and in the 0–1 nm range in (b).

Although the large-scale images of the F- and H-domains are strikingly different, the two domains may not differ strongly in structure because, in successive scanning, an H-domain is found to transform into a F-domain. Figures 11.19 (a) and 11.19 (b) show that F-domains grow with time, and an H-domain surrounded by F-domains is eventually transformed to become an F-domain [16a]. This finding suggests that an F-domain is slightly more stable than an H-domain. The possible nature of the difference between the F- and H-domains can be discussed in terms of the STM images of 6CBs adsorbed on β-Nb$_3$I$_8$ (Figs. 11.20 (a)–11.20 (c)). The F- and H-domains possess triangular shapes, and the edges of the F-domain are oriented along

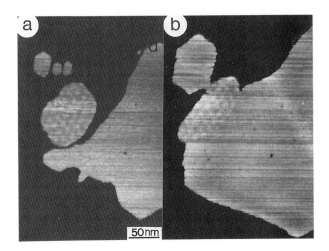

Figure 11.19 Successive STM height images of 8CB on β-Nb$_3$I$_8$ (I_{set} = 0.5 nA, V_{bias} = 163 mV). The contrast covers height variations in the 0.0–2.0 nm range.

the main crystallographic directions of β-Nb$_3$I$_8$ because they coincide with the sides of triangular defects (see the insets in Figs. 11.20 (a) and 11.20 (b)). The edges of the H-domain are rotated by ca. 30° with respect to those of the F-domain. As discussed above, the F-domain can be considered to occur when the β-Nb$_3$I$_8$ surface uses parallel grooves to accommodate 4-alkyl-4′-cyanobiphenyl molecules.

A zoomed-in view of several bright patches of the H-domain (Fig. 11.20 (c)) shows striped features of width ca. 0.7 nm running along the crystallographic (or the groove) directions of the substrate surface. This finding is similar in nature to the superstructure patterns found for graphite (see Chapter 9), which are considered to be a consequence of a layer of graphite that is misoriented with respect to the underlying graphite lattice. Likewise, one might consider the superstructure of the H-domain to result from a misalignment of an adsorbate layer with respect to either the underlying adsorbate layer or the substrate. It is important to probe the molecular order within the overlayer of the H-domain. Temperature-dependent STM measurements show [16a] that the H-domain consists of "nanoparticles" that appear as bright patches of 7–8 nm in diameter (Fig. 11.21). This suggests that the bright patches of the H-domain are certain clusters of 8CBs. Since the substrate surface has grooves running in three directions related by 60° rotation, it is possible for the surface to trap the molecules by employing the grooves to create the superstructure patterns as observed in the H-domain.

The adsorbed layers of 6CBs and 8CBs on β-Nb$_3$I$_8$ are stable at temperatures up to 50 °C [16a], which is higher than the temperatures at which these liquid-crystalline compounds become isotropic liquids (27.0 and 40.5 °C for 6CB and 8CB, respec-

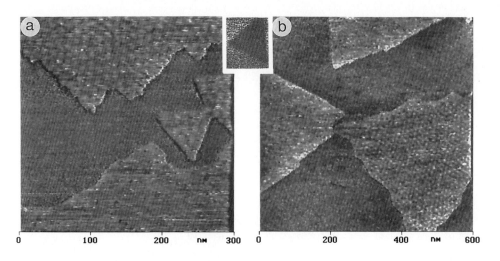
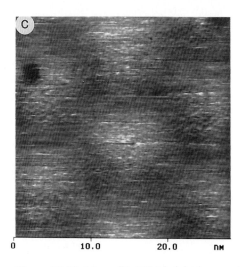

Figure 11.20 (a) − (b) STM height images in different regions of the 6CB adsorbate on β-Nb$_3$I$_8$ (I_{set} = 0.3 nA, V_{bias} = 176 mV). The orientation of a substrate defect is shown in the inset. (c) Zoomed-in STM image of an H-domain (I_{set} = 1 nA, V_{bias} = 30 mV). The contrast covers height variations in the 0.0–2.0 nm range in (a) and (b), and in the 0.0–1.6 nm range in (c).

tively). Thus, 4-alkyl-4′-cyanobiphenyls are anchored on the β-Nb$_3$I$_8$ surface with stronger interactions than those responsible for the liquid-crystal behavior of these compounds. This is easily accounted for because 4-alkyl-4′-cyanobiphenyls trapped in the grooves of the β-Nb$_3$I$_8$ surface should have a reduced mobility.

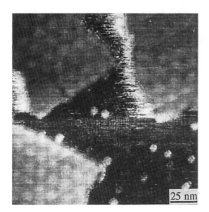

Figure 11.21 STM height image of 8CB on β-Nb$_3$I$_8$ registered at room temperature after the sample was heated to 50 °C and then cooled (I_{set} = 0.5 nA, V_{bias} = 180 mV). The contrast covers height variations in the 0.0–1.0 nm range.

11.4 Concluding Remarks

STM studies of organic adsorbates at liquid/solid interfaces allow one to probe complex multilayered structures of molecular adsorbates. In most cases, the molecular order at the liquid/solid interface is found in the immediate vicinity of the substrate surface. The topmost layers are usually etched away by the tip, because they are less compact (i. e., less stable mechanically) than the ones lying directly above the substrate and also because they are farther away from the conducting substrate. This makes it difficult to characterize their structures properly. This difficulty may be overcome by operating at low currents (in the picoampere range) with the development of more-sensitive instrumentation.

The substrate topography has a strong influence on the molecular order of adsorbed layers. On the atomically flat surface of graphite, organic molecules tend to form a lamellar structure. On the grooved surface of β-Nb$_3$I$_8$, however, long chain-like molecules do not form a lamellar structure, probably because of the trapping of the molecules in the grooves. The difficulty of imaging cycloalkanes (CH$_2$)$_{48}$ and (CH$_2$)$_{72}$ on β-Nb$_3$I$_8$ suggests that they cannot form an ordered overlayer structure because of a mismatch between the molecular shape and surface topography.

References

[1] P. K. Hansma, in *Vibrational Spectroscopy of Molecules on Surfaces* (Eds.: J. T. Yates, T. E. Madey), Plenum, **1987**, p. 135.
[2] D. P. E. Smith, M. D. Kirk, C. F. Quate, *J. Chem. Phys.* **1987**, *86*, 6034.

[3] (a) J. Wintterlin, R. J. Behm, in *Scanning Tunneling Microscopy I* (Eds.: R. Wiesendanger, H.-J. Guntherodt), Springer Verlag, Heidelberg, **1993**, p. 39. (b) S. Chiang, in *Scanning Tunneling Microscopy I* (Eds.: R. Wiesendanger, H.-J. Güntherodt), Springer, Heidelberg, **1993**, p. 181.

[4] (a) J. Foster, J. Frommer, *Nature (London)* **1988**, *60*, 1418. (b) G. C. McGonigal, R. H. Bernhardt, D. J. Thomson, *Appl. Phys. Lett.* **1990**, *57*, 28. (c) J. Frommer, *Angew. Chem. Int. Ed. Engl.* **1992**, *104*, 1298. (d) M. Allen, G. Büldt, W. J. Seikhaus, R. J. Tench, S. Subbiah, *Ultramicroscopy*, **1992**, *42*, 1049.

[5] (a) J. J. Kipling, *Adsorption from Solutions of Non-electrolytes*, Academic Press, London, **1965**. (b) R. H. Ottewill, C. H. Rochester, A. L. Smith (Eds.), *Adsorption from Solution*, Academic Press, London, **1983**. (c) B. Jerome, *Rep. Prog. Phys.* **1991**, *54*, 391.

[6] (a) A. Groszek, *Proc. R. Soc. London, Ser. A* **1970**, *314*, 473. (b) G. Findenegg, M. Liphard, *Carbon* **1987**, *25*, 119. (c) E. G. Bortchagovsky, L. N. Tarakhan, *Phys. Rev. B* **1993**, *47*, 2431. (d) Y. R. Shen, *Nature (London)* **1989**, *337*, 519.

[7] A. Wawkushewski, H.-J. Cantow, S. N. Magonov, M. Möller, W. Liang, M.-H. Whangbo, *Adv. Mater.* **1993**, *5*, 821.

[8] A. Wawkushewski, H.-J. Cantow, S. N. Magonov, *Langmuir* **1993**, *9*, 2778.

[9] (a) B. Parkinson, *J. Am. Chem. Soc.* **1991**, *112*, 7498. (b) S. N. Magonov, G. Bar, A. Ya. Gorenberg, E. B. Yagubskii, H.-J. Cantow, *Adv. Mater.* **1993**, *5*, 453.

[10] (a) J. P. Rabe, S. Buchholz, *Science* **1991**, *253*, 424. (b) J. P. Rabe, S. Buchholz, *Phys. Rev. Lett.* **1991**, *66*, 2096. (c) G. Watel, F. Thibaudau, J. Cousty, *Surf. Sci. Lett.* **1993**, *281*, L297. (d) F. Thibaudau, G. Watel, J. Cousty, *Surf. Sci. Lett.* **1993**, *281*, L303.

[11] (a) P.H. Lippel, R. J. Wilson, M. D. Miller, Ch. Wöll, S. Chiang, *Phys. Rev. Lett.* **1989**, *62*, 171. (b) V. M. Hallmark, S. Chiang, J. K. Brown, Ch. Wöll, *Phys. Rev. Lett.* **1991**, *66*, 48. (c) V. M. Hallmark, S. Chiang, K.-P. Meinhardt, K. Hafner, *Phys. Rev. Lett.* **1993**, *70*, 3740.

[12] (a) G. Bar, S. N. Magonov, H.-J. Cantow, N. D. Kushch, E. B. Yagubskii, W. Liang, J. Ren, M.-H. Whangbo, *New J. Chem.* **1993**, *17*, 439.

[13] W. Liang, M.-H. Whangbo, A. Wawkuschewski, H.-J. Cantow, S. N. Magonov, *Adv. Mater.* **1993**, *5*, 817.

[14] A. Wawkuschewski, S. N. Magonov, unpublished data.

[15] (a) P. Groth, *Acta Chem. Scand.* **1979**, *A33*, 199. (b) T. Trzebiatowski, M. Drager, G. R. Strobl, *Makromol. Chem.* **1982**, *183*, 731.

[16] (a) S. N. Magonov, A. Wawkuschewski, G. Bar, P. Zönnchen, H.-J. Cantow, *Thin Solid Films* **1994**, *243*, 419. (b) S. N. Magonov, A. Wawkuschewski, H.-J. Cantow, W. Liang, M.-H. Whangbo, *Appl. Phys. A* **1994**, *59*, 119.

[17] S. Cincotti, J. P. Rabe, *Appl. Phys. Lett.* **1993**, *62*, 3531.

[18] G. H. Findinegg, M. Liphard, *Carbon* **1987**, *25*, 119.

[19] J. J. Novoa, M. -H. Whangbo, J. M. Williams, *J. Chem. Phys.* **1991**, *94*, 4835.

[20] (a) D. P. E. Smith, J. Horber, Ch. Gerber, G. Binnig, *Science* **1989**, *245*, 43. (b) D. P. E. Smith, J. P. Horber, G. Binnig, H. Nejoh, *Nature (London)* **1990**, *344*, 641. (c) M. Hara, Y. Iwakabe, K. Tochigi, H. Sasabe, A. F. Carito, A. Yamado, *Nature (London)* **1990**, *344*, 228.

[21] S. L. Brandow, D. P. DiLella, R. J. Colton, *J. Vac. Sci. Technol. B* **1991**, *9*, 1115.

12 Self-Assembled Structures

This chapter describes the use of scanning probe techniques in characterizing the self-organized structures of organic amphiphiles, i.e., compounds with hydrophilic groups (e.g., OH) at one end and hydrophobic groups (e.g., alkyl) at the other. The self-organization of amphiphiles in aqueous solution is driven by weak intermolecular interactions and gives rise to layered structures (typically bilayers) as well as supramolecular structures such as micelles, fiber-like assemblies, etc. Scanning probe techniques can be used to characterize the structures of both types. By using STM and AFM, one can determine the morphology and molecular order of organic layers and thus complement studies with other microscopical and diffraction methods. The supramolecular structures of amphiphiles have largely been investigated with transmission electron microscopy (TEM). However, TEM micrographs do not provide accurate information about the heights of the structures, because they represent 2D projections of the electron density distribution. Scanning probe techniques have a high lateral resolution and are sensitive to the height variations on the sample surfaces. Therefore, they can provide a vital information about the self-assembled structures of amphiphiles deposited on substrates.

12.1 Scanning Probe Microscopy Studies of Thin Organic Films

Before the layered and supramolecular structures of amphiphiles are described, it is worthwhile to examine how the structures of thin organic layers have been characterized by STM and AFM. For this purpose, the STM and AFM studies of alkanethiols on Au(111) and Langmuir–Blodgett (LB) films are briefly surveyed in this section.

12.1.1 Morphology and Molecular Order

Organic films with thickness in the submicron range are important in many fields of nanotechnology [1]. A number of different approaches are employed to prepare such films. Monolayers of thiols RSH on Au surfaces are prepared via the formation of S–Au bonds. This chemical bonding insures the mechanical stability of the resulting thin film, and a chemical modification of the group R can yield surfaces with specific functional properties. Thus monolayers of thiols on Au(111) surfaces have become the subject of intensive studies [2].

Molecular-scale STM and AFM images of alkanethiol layers reveal ($\sqrt{3} \times \sqrt{3}$) R30° patterns with a repeat distance of ca. 0.5 nm, whereas the hexagonal lattice of the Au(111) surface has a period of ca. 0.29 nm. It was suggested that the STM images originate from the sulfur atoms bound to Au atoms [3]. Indeed, during molecular-scale STM imaging of alkanethiols in air, and even in vacuum, the tip is found to indent the organic layer (Chapter 8). The tip–sample force interactions are typically smaller in AFM than in STM, so that the periodic lattice detected in the AFM images is related to that part of the SR groups which is farther removed from the substrate than the S atoms. It might be possible to detect the STM images of the anchored head-groups or the topmost tail-groups, depending upon tunneling conditions [4c]. Recently, STM imaging at high R_{gap} (hundreds of gigaohms) has been applied to avoid mechanical surface damage and to detect the morphology and nanoscale order of the topmost tail-groups. Such measurements show that the frequently observed holes in the STM images of alkanethiol layers correspond to depressed spots on the gold surface, which are produced by solution-etching during the preparation of the self-assembled layers [5].

The STM imaging of thin organic layers on conducting substrates is a difficult problem from the viewpoints of both detection and interpretation (see Chapter 11). Therefore, AFM is more often used for the examination of such layers. Thin organic films prepared on flat substrates (e.g., mica, graphite, silicon wafer) by the LB technique [1b] are suitable for AFM examination. Images of large surface areas up to 150 µm × 150 µm are useful for checking the homogeneity of LB films, and preliminary inspection of the sample with an optical microscope is needed when film surfaces possess heterogeneous features. AFM can be used to examine the morphology and molecular order of LB films (which may be influenced by the nature of the substrate surface) as well as the structural reorganizations within the film caused by various treatments [6]. Molecular-scale AFM images exhibit the periodic patterns of molecular arrangement, from which one can estimate the packing density within the thin films. In addition, molecular-scale surface imperfections can be detected by AFM. For example, linear dislocations and domain boundaries were found in LB films of cadmium and barium arachidates [7], and vacancies of several nanometers in diameter were observed in alkanethiol layers on Au(111) [8].

12.1.2 Nanomechanical Properties

The nanomechanical properties of surfaces inherently affect AFM measurements and contribute to the resulting images. To learn about these properties, it is necessary to perform AFM experiments at different forces and analyze the force-vs.-distance curves (Chapter 4). These curves can be used to estimate the elastic and viscoelastic moduli of thin films in the vertical direction and the influence of the substrate on these properties. They can also provide information about the tip–sample adhesion.

Surface domains with different friction properties can be examined by analyzing the lateral force images. The applicability of AFM to the study of adhesion, friction, and wear of thin organic films has already been demonstrated [9]. In studies of two-component organic films, the domains of different components are found to give different responses to lateral force measurements [10].

When AFM experiments are performed at low forces, the surface forces may make the dominant contribution to the image contrast. In model experiments, tips functionalized with hydrophobic or hydrophilic overlayers were used to examine surfaces with specially prepared hydrophobic/hydrophilic patterns [10b, c]. The tip–sample force interactions were indeed found to be different in regions of different chemical nature. Similar observations were made for the force interaction between biologically functionalized tips and surfaces [11]. This suggests a selective AFM imaging. However, when unknown surfaces are examined, it is difficult to interpret the results because several factors influence the image contrast. Even for a single-component system, surface domains of different morphologies can give different vertical and lateral force responses due to the variations in their nanomechanical properties (e.g., hardness, indentation depth, etc.). Therefore, systematic experimental and theoretical efforts are necessary to correlate the surface nanomechanics with the intrinsic properties of the sample surface (e.g., chemical constitution, molecular structure, and packing).

12.2 Self-Organization of Amphiphiles

In aqueous solutions amphiphiles undergo spontaneous self-association to form molecular aggregates and complex supramolecular assemblies. Lipids and biological cells are typical examples of naturally occurring self-organized structures [12]. Recently, synthetic amphiphiles have attracted much attention because of their potential application as advanced nanomaterials [13].

12.2.1 Basic Principles

The typical molecular aggregates found in amphiphile systems are micelles and flat bilayers (Fig. 12.1). The thermodynamic principles of self-organization are described in the monograph by Israelachvili [12]. The self-organization of individual molecules (monomers) into aggregates proceeds when the cohesive interactions between the monomers increase with an increase in the aggregate size, eventually leading to a thermodynamic equilibrium between various supramolecular assemblies. For the simple structures such as "rods", "sheets" and "spheres", the chemical potential per molecule decreases progressively as the number of molecules increases. However, linear aggregates are energetically more favorable than 2D disk-like assemblies, which are in turn more favorable than 3D structures [12]. The self-assembling process is initiated at a critical concentration of amphiphile, and the latter depends upon the specifics of the intermolecular interactions.

The geometry of amphiphile molecules has a strong influence on the shape of their supramolecular structures. Taking into consideration the packing of amphiphile molecules, the formation of micelles (Fig. 12.1 (a)) is favored for amphiphiles with large head-groups, and that of bilayers (Fig. 12.1 (b)) for amphiphiles with small head-groups. The thermodynamic stability of a micelle, which exhibits a strongly curved surface, may derive from the incorporation of water molecules and their hydrogen bonding with the neighboring groups. Micelles and bilayers can further self-assemble to form more complex supramolecular structures. For example, the fusion of micelles might lead to micellar rods, and micellar rods may form multiple helices, and tubular structures (see below). Figure 12.2 shows TEM micrographs of fibrous supramolecular structures of two amphiphiles, N-(n-octyl)-D-gluconamide and N-(n-decyl)-N'-D-maltosylsemicarbazone.

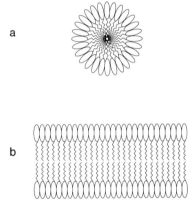

Figure 12.1 Schematic representation of the common assemblies of amphiphile molecules: (a) micelle and (b) bilayer. The hydrophilic and hydrophobic groups of an amphiphile molecule are represented by an ellipse and a wiggly chain, respectively.

12.2.2 Sample Preparation and AFM Imaging

The amphiphile series comprising N-(n-alkyl)-D-gluconamides (AGAs) [14] and N-(n-alkyl)-N'-D-maltosylsemicarbazones (AMSs) [15], consist of saccharide heads and n-alkyl chain tails (Fig. 12.3). (For simplicity, the AGAs with alkyl = heptyl, octyl, nonyl, and decyl will be referred to as HGA, OGA, NGA, and DGA, respectively. Likewise, AMSs with alkyl = decyl and hexadecyl will be called 10MS and 16MS, respectively.) In aqueous solutions, these molecules form supramolecular assemblies leading to gelation at concentrations below 1 % [14, 15]. To observe these assemblies

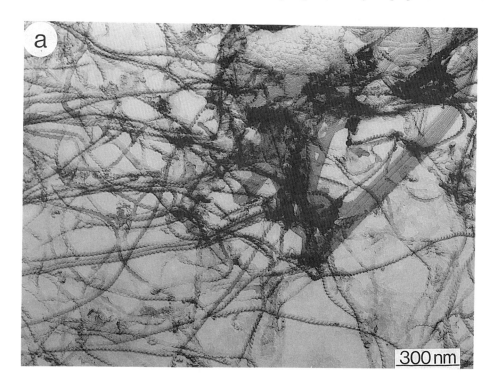

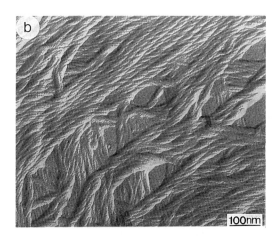

Figure 12.2 Transmission electron microscopy micrographs of (a) N-(n-octyl)-D-gluconamide and (b) N-(n-decyl)-N'-D-maltosylsemicarbazone. The micrographs were obtained from replicas of freeze-quenched specimens of the amphiphile gels. (Courtesy of Dr. B. Pfannemüller and Dr. S. Demharter)

(a) 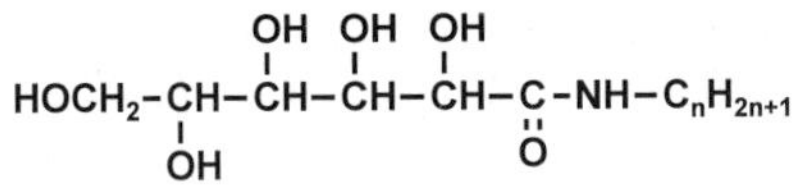 $HOCH_2-CH-CH-CH-CH-CH-C-NH-C_nH_{2n+1}$ with OH groups on carbons and C=O

(b) 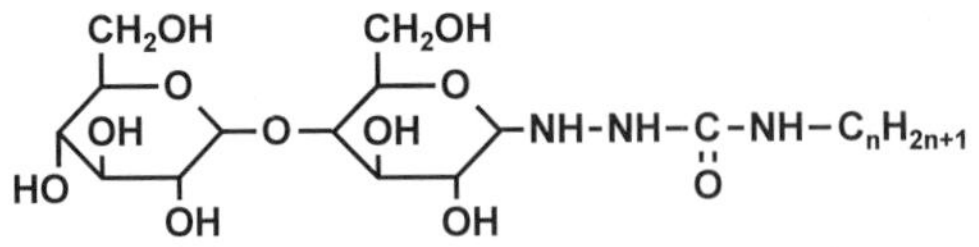 N-(n-alkyl)-N'-D-maltosylsemicarbazone structure: two sugar rings (each with CH₂OH, OH groups) linked by O, connected to $-NH-NH-C(=O)-NH-C_nH_{2n+1}$

Figure 12.3 Chemical formulas of (a) N-(n-alkyl)-D-gluconamide and (b) N-(n-alkyl)-N'-D-maltosylsemicarbazone.

by AFM, drops of gel solutions are deposited on substrates (mica, graphite, silicon wafer), the deposited drops are soaked away to obtain thin adsorbates on the substrates, and measurements are carried out on these partially dried adsorbates in air. The AGAs and 16MS have a poor solubility at room temperature, so their adsorbates are prepared using the solutions heated to 80 °C; 10MS has a better solubility, so its adsorbate can be prepared without heating the solution. The self-assembled structures are affected by the nature of the substrate surface. One important factor is the drying speed, which is slower on the hydrophilic mica than on the hydrophobic graphite.

Inspection of the amphiphile adsorbates with an optical microscope reveals aggregates with different morphologies, and microcrystals. The optical micrographs in Figs. 12.4 (a)–12.4 (f) show the dendrites, ring-like patterns, and microcrystals of the AGA adsorbates as well as the less ordered agglomerates and microcrystals of the 10MS adsorbate. The AGA adsorbates exhibit microcrystals with different habits (e.g., "rectangular," "rhombus-like" with an acute angle in the 50–70° range). The single-crystal X-ray studies of AGAs are based on the rectangular crystals (Fig. 12.4 (c)) [16]. Surface structures with corrugations of several microns cannot be examined by AFM (Chapter 4). Therefore AFM measurements should be carried out on relatively flat surface regions, which are free from optically distinguishable adsorbates (e. g., the place marked with an arrow in Fig. 12.4 (e), as well as on the surfaces of microcrystals.

For ambient-condition AFM experiments on amphiphiles, the tip–sample force interactions and the drying of the sample are two important factors to consider. The minimization of the applied force is necessary because the self-assembled structures are relatively soft. The force-vs.-distance curves recorded on the AGA adsorbates using a silicon tip in air show a small hysteresis in the approach–contact–disconnection cycle. This allows one to perform ambient-condition measurements in the contact mode with moderate forces in the 20–40 nN range. It is desirable to conduct measurements of the amphiphiles in water, where these compounds are in equilibrium with the aqueous medium and where low-force imaging might be realized. Un-

12.2 *Self-Organization of Amphiphiles* 249

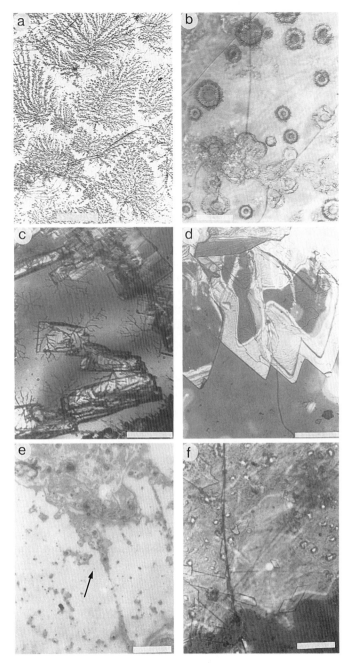

Figure 12.4 Optical micrographs of (a) dendrite-like structures of OGA adsorbate on graphite; (b) circular aggregates found in NGA adsorbate on mica; (c) rectangular microcrystals found in OGA adsorbates on mica; (d) rhombus-like microcrystals found in DGA adsorbate on mica; (e) 10MS adsorbates on mica; (f) 10MS adsorbates with microcrystals. In all micrographs the scale bars correspond to 100 μm.

fortunately, amphiphile adsorbates do not stick well to the substrate under water, so such experiments cannot be performed. At present, it is not well understood how the drying process and aging affect the self-assembled structures at ambient conditions. Nevertheless, the supramolecular structures of the AGAs found by TEM with freeze-quenched samples and by AFM on the adsorbates possess similar features. This suggests that the water molecules bound to the supramolecular structures are mostly retained in the freshly prepared adsorbates at ambient conditions. It is desirable to probe the structural rearrangements in the supramolecular assemblies caused by dehydration at room and elevated temperatures. So far, this attempt has not been successful because the dehydrated samples are damaged by the AFM tip, even at minimal available forces.

12.3 AFM Study of *N*-(n-alkyl)-D-gluconamides

The structures of many amphiphiles in their single crystals have been determined by X-ray diffraction studies [16]. This structural information may not be sufficient for describing the molecular conformation and packing in self-assembled structures in aqueous solution, because such structures most likely incorporate some water molecules around them whereas the single crystals are anhydrous. IR, NMR and powder X-ray diffraction studies [17] provide information about the molecular conformation in self-assemblies, but interpretation of the data is complicated because several different self-organized aggregates coexist within the sample under examination. So far, supramolecular amphiphile structures have been investigated mainly by TEM [18, 19].

12.3.1 Crystal Structures

Figure 12.5 (a) shows the conformation of OGA found in the single crystal. The carbon frameworks of both the hydrophilic and the hydrophobic parts of the OGA molecule adopt the all-*trans* conformation, and the molecule is bent. The single-crystal X-ray analyses of the AGAs show that these crystals have the triclinic $P1$ space group when the number of alkyl groups is odd; when there is an even number of alkyl groups the crystals are in the monoclinic $P2_1$ space group. The crystal possesses a layered structure, each layer is composed of bent molecules, and these molecules are inclined with respect to the plane of the layer (Fig. 12.5 (b)) [16a]. The terminal OH group of the OGA head is "turned away" from the surface so that the hydrophobic $-CH_2-$ groups form the surface of the layer (Fig. 12.5 (c)). (This arrangement is probably responsible for the unusual head-to-tail packing of the adjacent layers in AGA crystals.) Consequently, the thickness (ca. 1.6 nm) of a single molecular layer is less than the length of the OGA molecule as found in the crystal structure (ca. 1.9 nm). The molecular packing of the head-groups on the surface is described by an almost rectangular unit cell ($a = 0.525$ nm, $c = 0.4805$ nm, $\beta = 94.96\,°C$). The molecular conformation and packing found for OGA are also observed for other AGA homologues [16a, b].

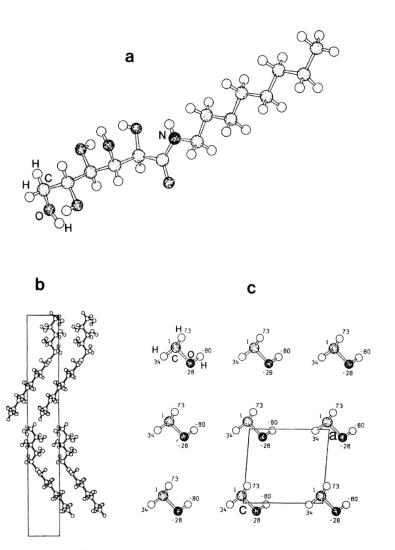

Figure 12.5 (a) Perspective view of the structure of the OGA molecule found in the bulk crystal. (b) Projection view of the molecular packing in the OGA crystal on the *bc*-plane. (c) Projection view of the arrangement of the topmost OGA groups on the *ac*-plane, where the numbers represent the relative heights of the atoms in pm.

252   12 Self-Assembled Structures

The large-scale AFM image of the OGA adsorbate on graphite (Fig. 12.6 (a)) reveals multilayered structures and microcrystals with flat terraces separated by steps. A close-up image of the microcrystal lying on the pedestal of rectangular platelets is shown in Fig. 12.6 (b). The cross-sectional profile along the line A–A of this image

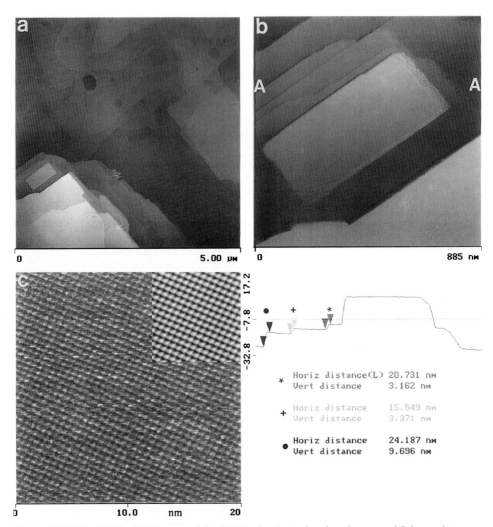

Figure 12.6 (a) AFM height image of the OGA adsorbate showing the area with layered structures and rectangular crystals. (b) AFM height image of the OGA crystal lying on the rectangular platelets in (a). The cross-sectional profile presented below the image was measured along the line A–A, and it shows the step heights. (c) AFM lateral force image of the surface of the OGA crystal. The upper right-hand corner of the image is shown after FFT filtering. The contrast covers height variations in the 0–100 nm range in (a) and (b), and it is proportional to lateral force variations in relative units in (c).

shows steps of heights 3.2 nm, 3.4 nm, and 9.7 nm, which correlate with the crystallographic b-constant of the OGA crystal (i.e., 3.243 nm) or its integral multiple [16a]. The high-resolution image of the nanocrystal (Fig. 12.6 (c)) exhibits a nearly rectangular pattern, and the repeat distances 0.52 nm and 0.47 nm correspond to the a- and c-constants of the bulk OGA crystal, respectively.

12.3.2 Layers with Crystal-Like Order

AGA adsorbates on mica, graphite and silicon form several types of layered structures with uniform thickness. They are either molecular layers with crystal-like order or aggregate layers composed of micellar structures. The molecular layers are found for thin adsorbates (thickness in the nanometer range). Thicker adsorbates possess surfaces consisting of the aggregate layers.

Three types of molecular layer (i.e., double layers, single layers and overlayers) have been identified in the AGA adsorbates [20]. Double layers are found more frequently than the others on both mica and graphite. AFM analysis reveals the molecular order in these layers, as described below. Aggregate layers are more relevant for the discussion of supramolecular assemblies, and are therefore presented later.

12.3.2.1 Thin overlayers

The AGA adsorbates show thin overlayer structures, as presented in Fig. 12.7 (a) for HGA. The cross-sectional profile across the line A–A shows that these overlayers are only 0.5–0.8 nm thick, so the overlayers consist of the molecules lying flat on the surface. An as-received image of the OGA overlayer on graphite is given in Fig. 12.7 (b), and its filtered image in the upper half of Fig. 12.7 (c). Among the peaks in the FFT power spectrum of the as-received image, one can find those corresponding to the hexagonal lattice of graphite, as revealed by the associated filtered image in the lower right-hand part of Fig. 12.7 (c). According to the local surface hardness concept (Chapter 9), the AFM image of the OGA overlayer reflects the superposition of the hardness variations of the graphite lattice onto those of the overlayer. The AFM pattern in the lower left-hand part of Fig. 12.7 (c) is obtained after removing the contribution of the graphite substrate from the power spectrum. This pattern more correctly presents the molecular structure of the OGA overlayer. Figures 12.7 (d) and 12.7 (e) show the AFM image of the NGA overlayer.

The images of the OGA and NGA overlayers resemble those found in STM studies of alkanes and alkylcyanobiphenyls on graphite (see Chapter 11). They show the individual molecules lying parallel to the substrate in lamellar structures and aligned almost perpendicular to the direction of the lamella. The repeat distance along the perpendicular direction is approximately equal to twice the length of the "extended"

254 12 Self-Assembled Structures

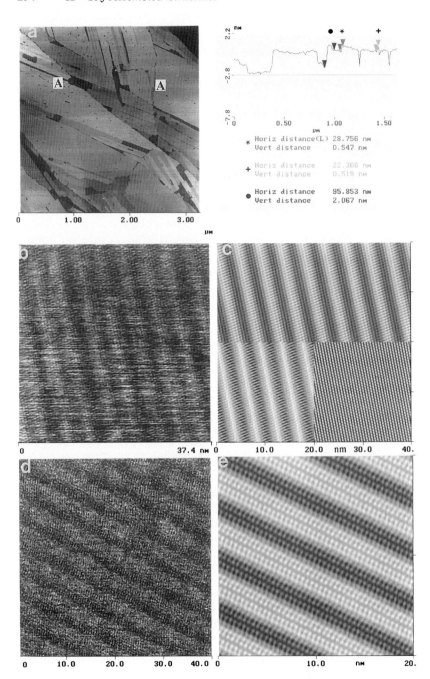

conformation of the amphiphile molecules. (Here the extended conformation of an amphiphile is defined as the one in which the molecule is stretched to become as linear as possible. For an AGA molecule, this means that the extreme end of the head is given by the terminal OH group, and the molecule avoids the bent shape found in the bulk crystal.) Each strand representing the amphiphile molecule consists of a bright and a dim part. Regardless of whether the bright parts are assigned to the heads or tails of the amphiphile molecules, it is clear that the molecules have a head-to-head order within the overlayer.

12.3.2.2 Double Layers

The layered structures found for the DGA adsorbates on graphite are shown in Fig. 12.8 (a). The cross-sectional profile indicates that the height of the layers (ca. 4.1 nm) corresponds to twice the length of the DGA molecule in its extended conformation. A similar correlation is found for the self-assembled layers of the other AGAs (e.g., ca. 3.4 nm in HGA, ca. 3.7 nm in OGA, ca. 3.9 nm in NGA). This implies that the molecules are aligned perpendicular to the layer plane and adopt the extended conformation.

A large hexagon-like defect (Fig. 12.8 (b)) on the topmost layer suggests a hexagonal packing of the molecules within the layer. As shown in Fig. 12.8 (c), the molecular-scale image of the DGA double layer indeed has a hexagonal pattern with repeat distance of ca. 0.52 nm. Similar images are detected for the double layers of OGA [20] and other homologues. The AFM measurements show that the "hexagonal" double layer of OGA has a smaller surface unit cell area than does the "rectangular" microcrystal of OGA (ca. 0.2342 nm^2 vs. 0.2427 nm^2). Thus the OGA molecules are more closely packed in the double layer than in the microcrystal. This finding is consistent with the suggestion from the layer thickness measurements that the OGA molecules in the double layer adopt the extended conformation in which the extreme end of the OGA head is given by the terminal OH group. In the bulk crystal structure of OGA, and hence in the microcrystal of OGA as well, the OGA molecules adopt the "folded" conformation in which the O–C–C end of the OGA head makes a fold.

◄ **Figure 12.7** (a) AFM height image of the overlayer structures of HGA adsorbate on graphite. The cross-sectional profile along the line A–A in (a) is shown on the right-hand side of this image. (b) Molecular-scale AFM height image of the thin OGA overlayer. (c) Top: the image pattern obtained by FFT filtering of the image in (a). Bottom: the lamellar (left) and hexagonal (right) lattices obtained by selecting different peaks from the FFT power spectrum of the image in (a). (d) Molecular-scale AFM lateral force image of the thin NGA overlayer found in its adsorbate on graphite. (e) Part of the image in (d) after FFT filtering. The contrast covers height variations in the 0–10 nm range in (a) and in the 0–7 nm range in (b), and it is proportional to lateral force variations in relative units in (d).

This folded conformation requires more space to pack the molecules than does the extended conformation. Therefore, the molecules in the double layers are likely to have a head-to-head arrangement and hence are bilayers. This suggestion agrees with the general observation that amphiphile molecules tend to have the head-to-head arrangement in self-assembled layered structures.

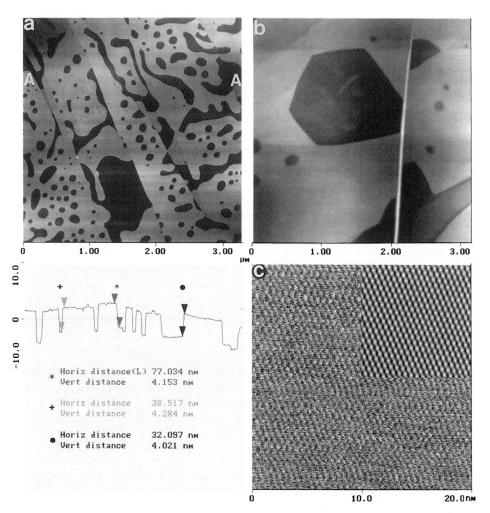

Figure 12.8 (a), (b) AFM height images of the layered structures in DGA adsorbate on mica. The contrast covers heights variation in the 0–50 nm range in (a) and in the 0–20 nm range in (b). The cross-sectional profile presented below the image was measured across the line A–A. (c) Molecular-scale AFM image recorded in the lateral force mode on the surface of the DGA double layers. The upper right-hand corner of the image is shown after FFT filtering. The contrast in these image is proportional to lateral force variations in relative units.

Double layers of the AGAs with hexagonal packing are found not only on flat surfaces of mica and graphite but also on a relatively corrugated surface of silicon wafer [20]. The latter implies that the intermolecular interactions within the AGA layers, which involve hydrogen bonding between the head-groups, are stronger than those between the AGA molecules and the substrate. For other organic adsorbates in which the interactions between the organic molecules are weak (e. g., the adsorbate layers of normal and cyclic alkanes), the molecular organization depends sensitively on the topography of the substrate (see Chapter 11).

Some self-assembled layers of AGAs have a thickness corresponding to the extended length of only one amphiphile molecule (see the cross-sectional profile in Fig. 12.7 (a)). It is easier to detect these single layers on hydrophobic graphite (where ultrathin adsorbates are found more frequently) than on hydrophilic mica. The single layers have a hexagonal packing similar to that found for the double layers [20, 21].

12.3.3 Supramolecular Assemblies

When the amount of adsorbed materials increases, the molecular layers lying directly on the substrates are covered by aggregates and aggregate layers, both made up of micellar structures. On these aggregate layers, one can find supramolecular structures such as fiber- and rod-like periodic assemblies with lengths up to several hundreds of microns.

12.3.3.1 Micellar Structures

The AFM image of the OGA adsorbate in Fig. 12.9 (a) shows a large number of circular aggregates with diameter in the 30–200 nm range and patches of layers on the right-hand side. The cross-sectional profile along the line A–A indicates that the height of these layers is ca. 4 nm, similar to that of the OGA double layers. However, traces of circular aggregates (marked by arrows) are distinguishable within these layers. The latter are different from the molecular layers with crystal-like order. They are formed from the circular aggregates by fusion and hence are aggregate layers. Small isolated aggregates are found in the upper left-hand corner. Their tendency to cluster into larger units, seen in the areas marked with arrows (Fig. 12.9 (a)), is probably responsible for the formation of the larger circular aggregates and the aggregate layers.

Circular aggregates are disk- or toroid-like, and their heights are ca. 4 nm. (Those with the brightest contrast are twice as high.) A magnified view of large aggregates in Fig. 12.9 (b) shows that the surfaces of the disks and toroids consist of "grains" (i. e., aggregates of several micelles). The grains indicated by arrowheads are ca. 15 nm in diameter. It is suggested that the disks and toroids are 2D arrangements of OGA micelles with height equal to the diameter of an individual micelle, as depicted in Fig. 12.9 (c).

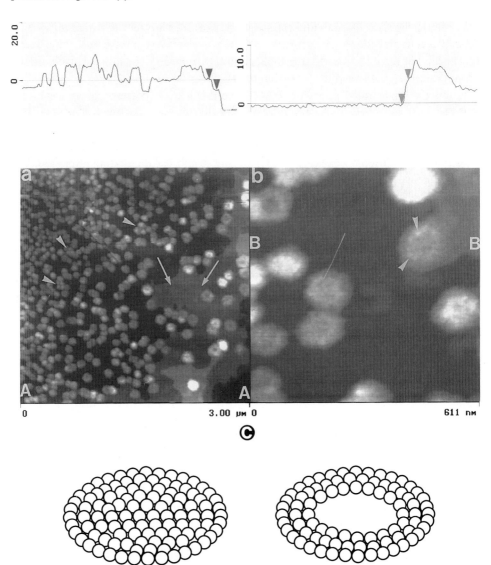

Figure 12.9 (a) AFM height image of micellar aggregates in OGA adsorbate on OGA microcrystal. (b) AFM height image of the central part of the area shown in (a). The cross-sectional profiles along the lines A–A and B–B are shown above the images. The vertical distance between the arrows is 3.7 and 4.1 nm in the left and right profiles, respectively. The contrast covers height variations in the 0–20 nm range in (a) and in the 0–10 nm range in (b). (c) Disks and toroids of OGA micelles with uniform diameter.

12.3.3.2 Fiber-Like Assemblies

AFM images of the AGA adsorbates show fiber-like assemblies (FLAs) resembling those found in TEM studies [14, 18, 19]. Figures 12.10 (a) and 12.10 (b) show parts of an amphiphile network with knots (bright spots) and branches. The branches consist of single fibers as well as bunches of fibers. Some isolated FLAs have a height of 4 nm, which corresponds to the thickness of the underlying OGA double layer, but bunched FLAs with height of 8 nm are more frequently found. Most FLAs exhibit a pitch (i.e. the repeat distance along the fiber direction) of ca. 24 nm, as shown in Fig. 12.10 (a) and in the cross-sectional profile along the line B–B of Fig. 12.10 (b). (The height corrugation along the line B–B is in the 2–3 nm range.) The cross-sectional profile along the line A–A in Fig. 12.10 (b) shows that several individual FLAs lie above the close-packed bunch of FLAs, all ca. 8 nm in height. The lateral force image (Fig. 12.10 (c)) reveals that the central part of the bunched FLAs consists of six close-packed fibers, each ca. 18 nm in width. There is a distinctive substructure along the individual fibers, and the cross-sectional profile along the line C–C exhibits a period of ca. 12 nm, half the pitch typically found for such FLAs.

Other FLAs of OGA with a pitch of ca. 24 nm are shown in Figs. 12.11 and 12.12. The large-scale images (Figs. 12.11 (a) and 12.12 (a)) show that knots composed of toroids are interconnected by FLAs. As representative examples, the three FLAs in the central part of Fig. 12.11 (a) (marked with arrows) may be examined in some detail. The FLA on the left lies on the double layer, and its height is 8 nm. A zoomed-in view of the remaining two FLAs (Fig. 12.11 (b)) shows that the pair of FLAs on the left-hand side have a pitch of ca. 24 nm and width of ca. 25 nm. The height corrugations along the central part of these structures are minimal, but the corrugations along the side are strong, so these FLAs are made by the fusion of circular grains with identical heights. The heights determined from the cross-sectional profile in the perpendicular direction are ca. 16 nm. Because these structures lie on the double-layer patch, the actual height of the FLAs is ca. 12 nm. The FLA on the right in Fig. 12.11 (b) is more complex in structure. The main part of this FLA has a pitch of 24 nm and height of 8 nm. The terminal part is represented by two smaller FLAs, each of which has a pitch of 12 nm and is composed of small grains ca. 4 nm in height and ca. 16 nm in width.

In Fig. 12.12 (b), the FLAs possess numerous perpendicular branches growing out of the topmost layer. The FLAs marked by the arrows have a pitch of ca. 24 nm, heights of 8 and 12 nm, and a width of ca. 30 nm. The perpendicular rows attached to these FLAs exhibit a smaller pitch (ca. 12 nm) and have the same height (ca. 4 nm) as the topmost layer. These rows are similar in structure to those found in the terminal parts of the FLAs on the right in Fig. 12.11 (b). Note that the topmost layer of Fig. 12.12 (b) exhibits numerous holes and a slightly corrugated surface, i.e., it is an aggregate layer. The substructure of the supramolecular aggregates incorporated in such a layer is shown in the lateral force image of Fig. 12.12 (c). The pitch

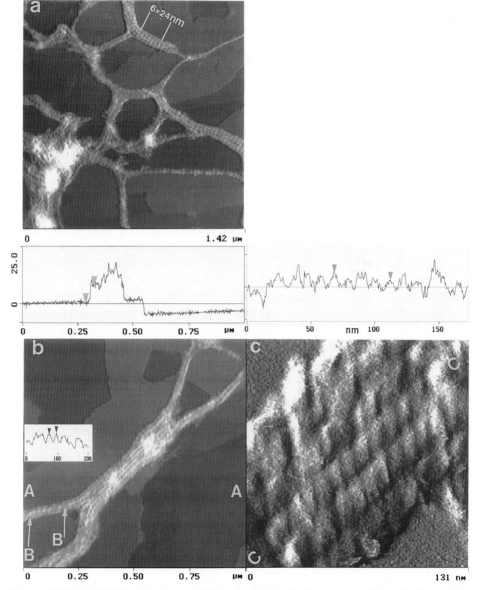

Figure 12.10 (a) AFM height image of FLAs in OGA adsorbate on graphite. (b) Zoomed-in image of part of the OGA adsorbate shown in (a). The cross-sectional profile along the line A–A is shown above the image, where the vertical distance between the arrows is 7.6 nm. The cross-sectional profile along the line B–B is shown in the insert of this image. The horizontal distance between the arrows is 24 nm. (c) AFM lateral force image of the part of the two-dimensional structure shown in (b). The cross-sectional profile along the line C–C is shown above the image. The contrast covers height variations in the 0–30 nm range in (a) and (b), and it is proportional to lateral force variations in relative units in (c).

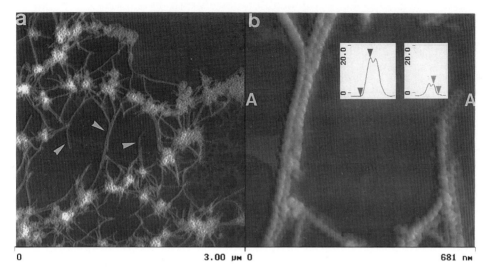

Figure 12.11 (a) AFM height image of OGA adsorbate on graphite. (b) AFM height image of the central part of the area in (a). The cross-sectional profile along the line A-A was used to determine the heights of linear structures. Parts of this profile are shown in the inserts of the image. The vertical distances between the arrows in the left and right inserts are 16.4 nm and 4.2 nm, respectively. The contrast covers height variations in the 0–100 nm range in (a) and in the 0–40 nm range in (b).

of these aggregates is ca. 12 nm, and the cross-sectional profile shown in the inset demonstrates that each grain has a periodic substructure with a repeat distance of 4 nm. Therefore, it is suggested that the small FLAs grown within the aggregate layer are also made up of micellar structures.

12.3.3.3 Rod-Like Assemblies

Aggregates of rod-like assemblies (RLAs) up to 100 μm in length are frequently observed in AFM images of OGA adsorbates. Figure 12.13 (a) shows a surface region of an OGA adsorbate with RLAs. The most characteristic feature of RLAs is their distinctive periodic corrugations with a fine pitch of ca. 9 nm. This feature has also been found from TEM micrographs of similar aggregates [14, 18]. Many RLAs have a height of ca. 8 nm and a width in the 25–30 nm range, but similar aggregates with greater height (e.g., ca. 16 nm) and width can be also found [21]. The image in Fig. 12.13 (a) shows linear structures consisting of several fused RLAs. A magnified view of the middle-right part of this image (Fig. 12.13 (b)) shows the structures of two large assemblies in detail. The cross-sectional profile taken along one of the RLAs shows a step height of ca. 8 nm and an identical pitch (i.e., ca. 8 nm) along the whole length (i.e., above and below the step). Therefore, the thick RLAs consist

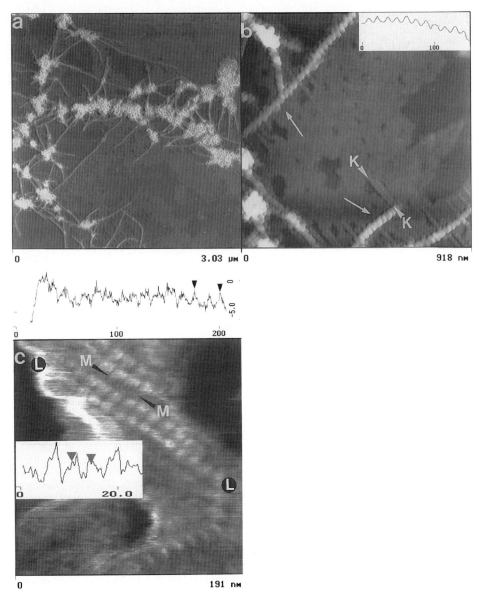

Figure 12.12 (a) AFM height image of OGA adsorbate on graphite. (b) AFM height image of part of the area in (a). The insert in the upper right-hand corner shows the cross-sectional profile along the line K-K. The period of these height corrugations is ca. 12 nm. (c) AFM lateral force image of the periodic array in the topmost layer of OGA adsorbate. The cross-sectional profile along the line L-L is shown above this image, where the horizontal distance between the arrows is 24 nm. The cross-sectional profile along the line M-M is shown in the insert on the left in the image, where the horizontal distance between the arrows is 4 nm. The contrast covers height variations in the 0-100 nm range in (a) and in the 0-50 nm range in (b), and it is proportional to lateral force variations in relative units in (c).

12.3 AFM Study of N-(n-alkyl)-D-gluconamides

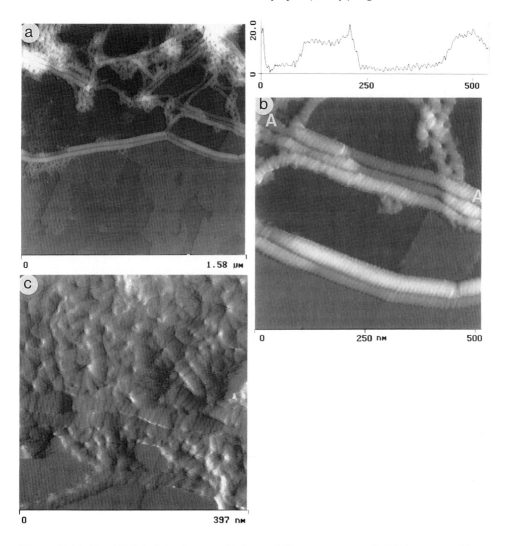

Figure 12.13 (a) AFM height image of the rod-like structures of OGA on graphite. (b) Zoomed-in part of the image in (a). The cross-sectional profile determined along the line A–A in (b) is shown above this image, where the vertical distance between the arrows is 8 nm and the small corrugations along this profile have a period of 9 nm. (c) AFM deflection image of the rod-like structures of OGA adsorbate on graphite. The contrast covers height variations in the 0–40 nm range in (a), in the 0–50 nm range in (b), and is proportional to cantilever deflections (nm) in (c).

of thin RLAs with the same pitch and height. The frequently observed RLAs 8 nm in height are the elementary assemblies of this type. The width and height of the elementary RLAs are similar to those of the toroids found in the knots. It is suggested that both assemblies consist of similar subunits, arranged into linear sequences in RLAs and circular sequences in toroids.

Additional information about the RLAs is obtained from the AFM images showing places on the surface where rod-like structures are incorporated into the arrays of micellar grains (Fig. 12.13 (c)). The latter have a diameter of ca. 25 nm and a height of ca. 8 nm. This image suggests that RLAs and their surrounding assemblies have a common substructure. Many grains present in the upper part of this image exhibit linear striations with a separation of ca. 8 nm, which corresponds to the pitch of the RLAs.

12.3.4 Structural Models

The TEM micrographs of several AGA homologues show supramolecular FLAs with a pitch of ca. 23 nm, which are referred to as "bulgy" fibers [18]. The individual bulgy fibers are characterized by a pitch of 22.4 nm and a maximum width of 18.6 nm. Analysis of the TEM micrographs led to the quadruple helix model for the bulgy fiber (Fig. 12.14 (a)) made up of four micellar rods, each ca. 4 nm in diameter [18c]. Supramolecular fibers with pitch of ca. 8 nm and width of ca. 28 nm were also found systematically in the TEM micrographs of AGAs [18]. These structures were described in terms of a hollow tubular helix (Fig. 12.14 (b)). To explain the value of the pitch, it is suggested that the tubular helix is formed from a micellar rod 7.4 nm in diameter (i.e., two double-layer thicknesses) [18c]. In such a case the height of the tubular helix should be at least 15–16 nm. According to the AFM results discussed above, the heights of the individual RLAs are only ca. 8 nm. The width-to-height ratio in the periodic self-assemblies of OGA (i.e., FLAs and RLAs) is typically larger than 2, so these "flattened" structures are not consistent with the helical strands and tubules suggested by the TEM studies.

The geometrical parameters of the periodical supramolecular structures are mostly given by integral multiples of the OGA double-layer thickness (i.e. ca. 4 nm). The fine nanostructure with a periodicity of ca. 4 nm, detected in several supramolecular assemblies, indicates that the basic unit of these assemblies has a dimension corresponding to the double-layer thickness. Consequently, the periodic supramolecular structures of OGA determined in the AFM study are explained by a simple structural model in which the supramolecular assemblies are constructed by the fusion of micelles. Micelles of diameter 4 nm may be fused to a form 1D structure, i.e., a micellar rod of diameter 4 nm (Fig. 12.15), and 2D structures such as disks and toroids (Fig. 12.9 (c)). Similarly, the growth of micellar aggregates can proceed by fusing the micelles of diameter ca. 4 nm into 3D units of dimensions 8 nm × 8 nm × 8 nm.

12.3 AFM Study of N-(n-alkyl)-D-gluconamides

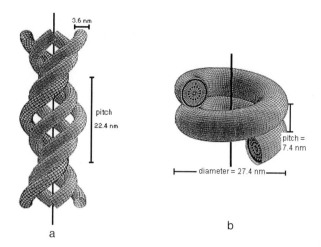

Figure 12.14 (a) Quadruple helix model describing the structure of the bulgy fiber structure of OGA. (b) Tubular helix model describing the structure of an OGA rod with a pitch of ca. 8 nm. (Reproduced from Ref. 18c).

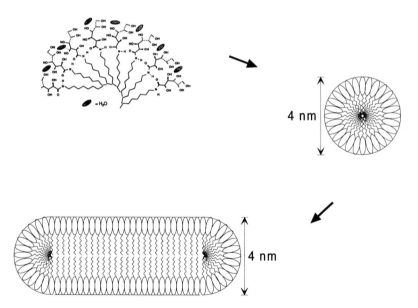

Figure 12.15 Formation of a micellar rod from an individual OGA micelle.

Such 3D units may or may not adopt a spherical shape, but their aggregation can lead to larger 1D, 2D, and 3D assemblies (Fig. 12.16). Therefore, the periodical structures of the observed FLAs and RLAs of OGA are all explained in terms of micellar units whose dimensions are given by integral multiples of ca. 4 nm. The structural model discussed in terms of OGA is also applicable to the supramolecular structures of other AGAs.

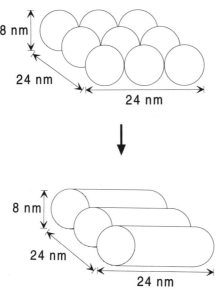

Figure 12.16 Formation of RLAs with a pitch of ca. 8 nm from 3D micellar units of dimensions 8 nm × 8 nm × 8 nm.

12.4 AFM Study of *N*-(n-alkyl)-*N'*-D-maltosylsemicarbazones

The head-group (i.e., the maltose group) of the amphiphiles 10MS and 16MS is different in structure from that of the AGAs. Thus, the self-assembled structures of 10MS and 16MS exhibit interesting differences from those found for the AGAs [22]. This is discussed below.

12.4.1 Self-Assembled Structures of 10MS

The large-scale AFM image of the 10MS adsorbate (Fig. 12.17) was recorded at the place indicated by an arrow in Fig. 12.4 (e). The adsorbate in the central part of the image consists of several layers. The lower-lying layers (**1** and **2**) are more compact than the higher-lying ones (**3–5**). Two FLAs (**6**) are present in the upper right-hand part of the image. The "dark holes" on the adsorbate probably results from the drying of water drops. Figure 12.18 (a) shows the image recorded at a place on the layer **1** where its integrity is lost and the underlying substrate is visible. When the central part of the adsorbate is scratched away with high-force scanning, a "window" is formed. The cross-sectional profile across the window indicates that the layer **1** is ca. 0.8 nm in height. The atomic-scale images of an area within the window exhibit the surface lattice of mica. Therefore, the layer **1** lies directly on the substrate surface, and the 10MS molecules lie flat on the surface. "Grains" of diameter in the 10–14 nm range and their clusters are seen in different parts of the first layer (indicated by arrowheads in Fig. 12.18 (a) [22]. The grains are most likely micellar disks of the 10MS molecules lying flat on the substrate.

Figure 12.18 (b) shows an image around the "window" made in a region containing the layers **1** and **2**. The cross-sectional profile across the window shows that the layer **2** is substantially thicker than the layer **1**. However the thickness of the layer **2** is smaller than twice the estimated length of the 10MS molecule in the extended conformation (i.e., ca. 5.8 nm). The image of Fig. 12.18 (c) shows that "grains" constitute the layer **2** (hereafter referred to as the grain-layer), and the smallest grains have a diameter smaller than 10 nm. Linear structures made up of several fused

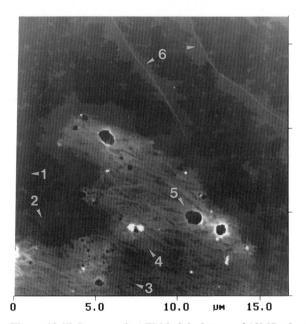

Figure 12.17 Large-scale AFM height image of 10MS adsorbate on mica. Numbers **1–6** indicate places with different self-organized structures. The contrast covers height variations in the 0–60 nm range.

268   12 Self-Assembled Structures

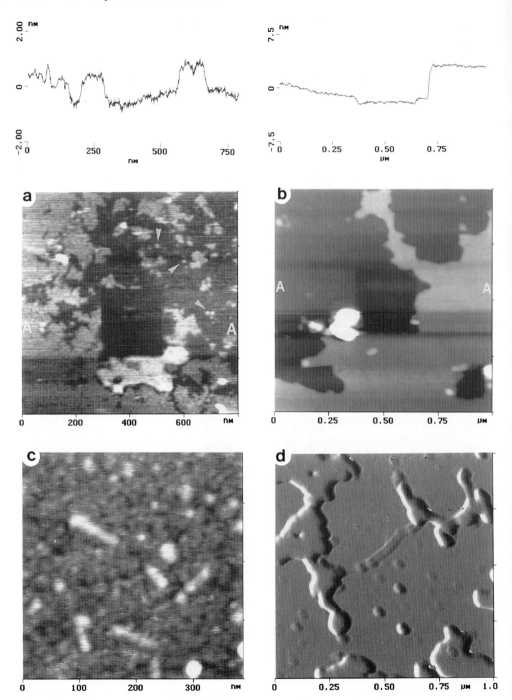

"grains", formed within the grain-layer, are also seen in this image. In several places, ribbon-like "bridges" are found to connect the neighboring islands of the grain-layer [14]. This is shown in Fig. 12.18 (d), where grains are transformed into ribbons with linear substructure. The higher-lying layers **3–5**, not as densely packed as the first two layers, consist of thin and elongated platelets (Fig. 12.19 (a)). The cross-sectional profiles (Fig. 12.19 (b) reveal that these platelets have the same thickness (i. e., 5 nm), as the grain-layer. The high-resolution image (Fig. 12.19 (c)) demonstrates that "rods" of diameter ca. 5 nm are closely packed to form the platelets. Thus, the layers of the platelets may be referred to as the rod-layers.

The AFM images of the 10MS adsorbates exhibit FLAs of different types. The image of Fig. 12.20 (a) shows that the topmost layer is composed of grains, and several FLAs come out of the layer. They exhibit a height of ca. 5 nm but can have different widths. The narrowest part of the strands (ca. 11 nm in width), indicated by an arrowhead in Fig. 12.20 (a), is embedded in the layer. These FLAs possess a grain-like structure. The "isolated" FLAs seen in Fig. 12.20 (b) are grown out of the rod-layer and represent another type of supramolecular structures. Two FLAs of 50 nm in width merged into a large one. The FLAs exhibit a periodic height corrugation with a pitch of ca. 60 nm, and their maximum height is ca. 5 nm. The vertical corrugation along the most periodic region (in the upper part of the image) is ca. 1 nm. The height and deflection images of Fig. 12.21 reveal that bundles of short "rods" (ca. 60 nm long and ca. 5 nm in width) are linked together to form the FLAs.

12.4.2 Self-Assembled Structures of 16MS

The AFM images of Figs. 12.22 (a) and 12.22 (b) show a region of the 16MS adsorbate with two periodic FLAs lying on the topmost layers. These FLAs are perpendicular to each other. The FLA running from the lower left to the upper right corners has a pitch of ca. 63 nm and consists of grains connected to each other. The other FLA (with a pitch of ca. 73 nm) consists of separated grains. The lateral force image (Fig. 12.22 (c)) reveals that the topmost layer is a rod-layer (composed of close-

◀ **Figure 12.18** (a) AFM height image of the first layer of 10MS adsorbate. The contrast covers height variations in the 0–2.5 nm range. (b) AFM height image of 10MS adsorbate. The contrast covers height variations in the 0–15 nm range. Cross-sectional profiles along the lines A–A in (a) and (b) are shown above the respective images. (c) AFM height image of the second layer of 10MS adsorbate. The contrast covers height variations in the 0.0–0.5 nm range. (d) AFM deflection image of 10MS adsorbate showing the formation of linear structures between islands of the second layer. The contrast indicates cantilever deflections in relative units.

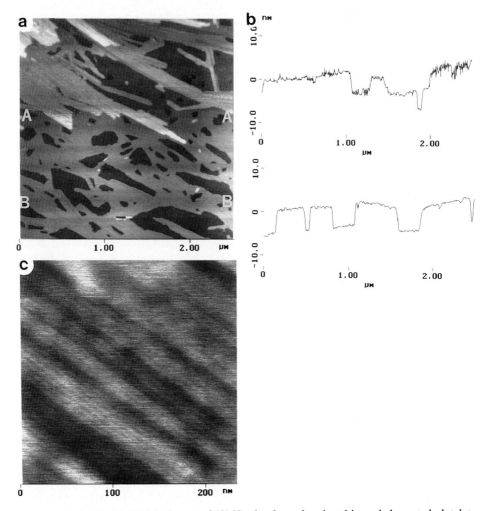

Figures 12.19 (a) AFM height image of 10MS adsorbate showing thin and elongated platelets. The contrast covers height variations in the 0–15 nm range. (b) Cross-sectional height profiles along the lines A–A and B–B. (c) AFM height images revealing the 5 nm-wide linear substructure of the platelets. The contrast covers height variations in the 0–1 nm range.

packed "rods" with a diameter in the 5.5 nm range). These rods make an angle of ca. 45° with respect to the directions of the two FLAs.

As confirmed by the images of Fig. 12.23, the topmost layers of the 16MS adsorbates consist of rod-like structures. The height image (Fig. 12.23 (a)) shows a multilayered structure of the adsorbate with steps ca. 5.5 nm in height. This indicates that the height of the 16MS rod-layers corresponds to the width of the rods. The lateral force image of Fig. 12.23 (b) shows that the rod-layers are composed of do-

12.4 AFM Study of N-(n-alkyl)-N'-D-maltosylsemicarbazones

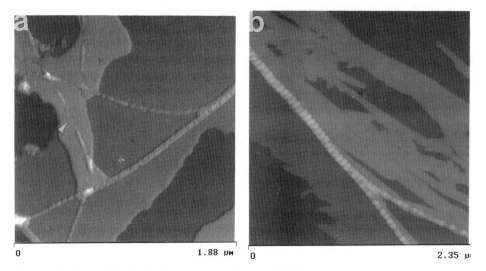

Figures 12.20 (a) AFM height image of the FLAs of 10MS coming out of the grain-layer. The contrast covers height variations in the 0–20 nm range. (b) AFM height image of the FLA of 10MS lying on the rod-layer. The contrast covers height variations in the 0–25 nm range.

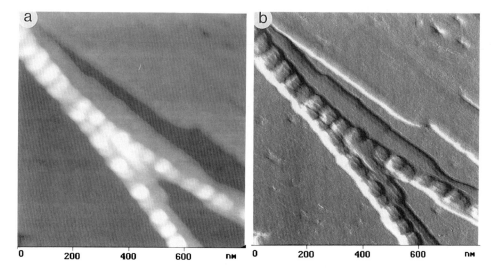

Figure 12.21 Zoomed-in image of the FLAs of 10MS shown in Fig. 12.20 (b). (a) Height and (b) amplitude AFM images. The contrast covers height variations in the 0–20 nm range in (a) and is proportional to cantilever amplitude variations in relative units in (b).

272   12 Self-Assembled Structures

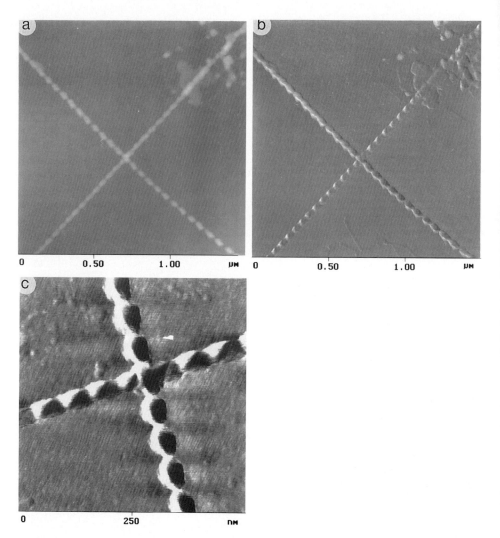

Figure 12.22 (a) Height and (b) lateral force AFM images of 16MS adsorbate which show two FLAs crossing each other at 90°. (c) Lateral force AFM image of part of the area shown in (a). This image was recorded at a different rotation angle from those in (a) and (b). The contrast covers height variations in the 0–50 nm range in (a) and is proportional to lateral force variations in (b) and (c).

mains with different rod orientations. Similar structures were found for the 10MS adsorbates. The high-resolution image of Fig. 12.23 (c) shows that each rod ca. 5.5 nm in width consists of stripes 0.5 nm-wide, which are inclined to the running direction of the rod at an angle of ca. 45–50°. The width of these stripes corresponds to the thickness of a single amphiphile molecule.

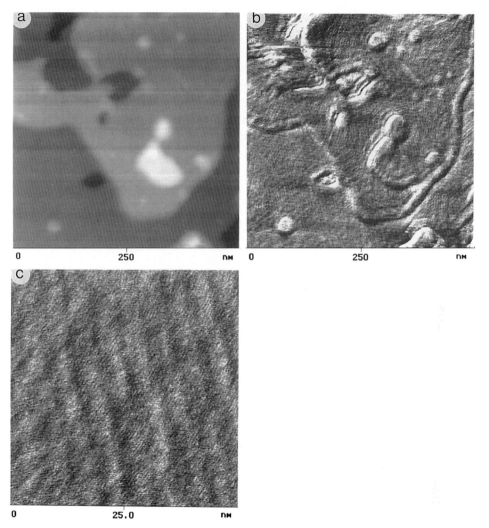

Figure 12.23 (a) Height and (b) lateral force AFM images of the layer structures of 16MS adsorbate. (c) Small-scale lateral force AFM image recorded on part of the area in (a). The contrast covers height variations in the 0–20 nm range in (a) and is proportional to lateral force variations in relative units in (b) and (c).

12.4.3 Structural Models

The supramolecular structures of the AMSs include grain-layers, rod-layers and FLAs. By analogy with the case of the AGAs, one can explain the supramolecular structures of the AMSs in terms of micelles and micellar rods. As indicated by the thickness of the grain- and/or rod-layers, however, the diameters of the micelles

274   12 Self-Assembled Structures

needed for this construction are smaller than those expected from the extended lengths of the amphiphile molecules (e. g., ca. 5.0 nm vs. ca. 5.8 nm for 10MS, and ca. 5.5 nm vs. ca. 6.5 nm for 16MS). The latter can be explained if the amphiphile molecules adopt a significantly bent conformation in the micelles, or if the micellar disks forming the rods are inclined with respect to the running direction of the rods. The latter is indeed the case, as confirmed by the molecular-scale image of the rod-layer of 16MS in Fig. 12.23 (c), where each strip of width ca. 0.5 nm represents the edge of a micellar disk of 16MS molecules.

12.5 Concluding Remarks

The self-assembled structures of the AGAs and AMSs were examined by studying the AFM images of their gel adsorbates on graphite and mica. The lattice parameters of the crystal-like layers, the lateral dimensions of the micellar aggregates, the thicknesses of the adsorbate layers and the periodical supramolecular structures were determined from these images. This structural information is vital in understanding the architecture of the self-organized assemblies, although the accuracy of the measurements is not very high due to several reasons (see Chapter 4) and the possible elastic deformation of these structures under the tip force. The AGA adsorbates lead to molecular layers with crystal-like order, which are covered with aggregates or aggregate layers made up of micellar structures. Typically, the AMS adsorbates do not form molecular layers with crystal-like order. Their aggregate layers lying directly on the substrate are thin overlayers of micellar disks, whereas the topmost layers are rod-layers. Periodic supramolecular structures of AGAs and AMSs are found on the topmost layers of their adsorbates.

In the thin overlayers and double layers of the AGA adsorbates, the amphiphile molecules adopt an extended conformation and have a head-to-head arrangement. The supramolecular structures of the AGA adsorbates have the thickness of the double layers and its integral multiples. This leads to a structural model in which these assemblies are formed in terms of micelles, micellar rods, and micellar blocks. The supramolecular structures of the AMSs are described in terms of micellar grains and micellar rods. The rod-layers of the AMSs have a thickness considerably smaller than twice the extended length of the amphiphile molecules, because the molecules are inclined with respect to their layered structures. In contrast, in the double layers of the AGAs, the amphiphile molecules are arranged perpendicular to the layers. This difference in the AGA and AMS layer structures reflects the fact that the head-groups of AGAs and AMSs are different in size and shape (i.e., small and acyclic sugar groups in AGAs versus large and cyclic sugar groups in AMSs). The inclination of the AMS molecules in the rod-layers implies that the maximum hydrogen bonding between the maltosyl groups is achieved by a slipped arrangement of these groups.

As noted, the AMS adsorbates do not form crystal-like layers but the dominant layers of the AGA adsorbates are close packed bilayers. This observation is consistent with the fact that the head-groups are larger for the AMSs than for the AGAs.

The periodic supramolecular assemblies of the AGAs and AMSs characterized by AFM possess "flattened" structures and are quite different from periodic structures such as helical strands and tubules deduced from TEM studies. To understand this difference, it is important to examine how the periodical supramolecular structures in solution are modified by dehydration. The methods of sample preparation employed in AFM and TEM studies are different, and may lead to samples with different extents of dehydration.

References

[1] (a) A. Ulman, *An Introduction to Ultrathin Organic Films: From Langmuir-Blodgett to Self-Assembly*, Academic Press, New York, **1991**. (b) G. G. Roberts, *Langmuir-Blodgett Films*, Plenum, New York, **1990**. (c) A. J. Bard, *Integrated Chemical Systems. A Chemical Approach to Nanotechnology*, J. Wiley, New York, **1994**.

[2] L. H. Dubois, R. G. Nuzzo, *Annu. Rev. Phys. Chem.* **1992**, *43*, 437.

[3] (a) C. A. Widrig, C. A. Alves, M. D. Porter, *J. Am. Chem. Soc.* **1991**, *113*, 285. (b) Y.-T. Kim, A. Bard, *Langmuir* **1992**, *8*, 1096.

[4] (a) M. Salmeron, G. Neubauer, A. Folch, M. Tomitori, D. F. Ogletree, P. Sautet, *Langmuir* **1993**, *9*, 3600. (b) U. Dürig, O. Züger, B. Michel, L. Häussling, H. Ringsdorf, *Phys. Rev. B* **1993**, *48*, 1711. (c) J. P. Bucher, L. Santesson, K. Kern, *Appl. Phys. A* **1994**, *59*, 135.

[5] (a) C. Schönenberger, J. A. M. Sondag-Huethorst, J. Jorritsma, L. G. J. Fokkink, *Langmuir* **1994**, *10*, 611. (b) J. A. M. Sondag-Huethorst, C. Schönenberger, L. G. J. Fokkink, *J. Phys. Chem.* **1994**, *98*, 6826.

[6] (a) J. A. Zasadzinski, R. Viswanathan, D. K. Schwartz, J. Garnaes, L. Madsen, S. Chiruvolu, J. T. Woodward, M. L. Longo, *Coll. Surf. A* **1994**, *93*, 35. (b) C. F. Chi, M. Anders, H. Fuchs, R. R. Johnston, H. Ringsdorf, *Science* **1993**, *259*, 213. (c) T. Nakagawa, K. Ogawa, T. Kurumizawa, *Langmuir* **1994**, *10*, 525.

[7] (a) J. Garnaes, D. K. Schwartz, J. A. Zasadinski, *Nature (London)* **1992**, *357*, 54. (b) L. Bordieu, O. Ronsin, D. Chatenay, *Science* **1993**, *259*, 798.

[8] H.-J. Butt, E. Seifert, E. Bamberg, *J. Phys. Chem.* **1993**, *97*, 7316.

[9] (a) G. S. Blackman, C. M. Mate, M. R. Philpott, *Phys. Rev. Lett.* **1990**, *65*, 2270. (b) G. S. Blackman, C. M. Mate, M. R. Philpott, *Vacuum* **1991**, *41*, 1283. (c) H. A. Mizes, K.-G. Loh, R. J. D. Miller, S. K. Ahuja, E. F. Grabowski, *Appl. Phys. Lett.* **1991**, *59*, 291.

[10] (a) R. Overney, E. Meier, J. Frommer, D. Brodbeck, L. Howald, H.-J. Güntherodt, M. Fujihira, H. Takano, Y. Gotoh, *Nature (London)* **1992**, *359*, 133. (b) C. D. Frisbie, L. F. Rozsnyai, A. Noy, M. S. Wrighton, C. M. Lieber, *Science* **1994**, *265*, 2071. (c) J. L. Wilbur, H. A. Biebuyck, J. C. MacDonald, G. M. Whitesides, *Langmuir* **1995**, *11*, 825.

[11] (a) G. U. Lee, D. A. Kidwell, R. J. Colton, *Langmuir* **1994** *10*, 354. (b) V. T. Moy, E.-L. Florin, H. E. Gaub, *Coll. Surf. A* **1994**, *93*, 343.
[12] J. N. Israelachvili, *Intermolecular and Surface Forces*, Academic Press, New York **1985**, Chapters 15–17.
[13] (a) J.M. Schnur, *Science* **1993**, *262*, 1669. (b) J.-H. Fuhrhop, J. Köning, *Membranes and Molecular Assemblies: The Synthetic Approach*, The Royal Society of Chemistry, Cambridge, **1994**.
[14] (a) B. Pfannemüller, W. Welte, *Chem. Phys. Lipids* **1985**, *37*, 227. (b) B. Pfannemüller, *Starch* **1988**, *40*, 476.
[15] (a) S. Demharter, H. Frey, M. Drechsler, R. Mülhaupt, *Coll. Polym. Sci*, **1995**, *273*, 661. (b) S. Demharter, PhD Thesis, Freiburg University, **1995**.
[16] (a) V. Zabel, A. Müller-Fahrnow, R. Hilgenfeld, W. Saenger, B. Pfannemüller, V. Enkelmann, W. Welte, *Chem. Phys. Lipids* **1986**, *39*, 313. (b) A. Müller-Fahrnow, A. Hilgenfeld, H. Hesse, W. Saenger, B. Pfannemüller, *Carbohydr. Res.* **1988**, *176*, 165.
[17] S. Svenson, J. Köning, J.-H. Fuhrhop, *J. Phys. Chem.* **1994**, *98*, 1022.
[18] (a) J.-H. Fuhrhop, P. Schnieder, J. Rosenberg, E. Boekema, *J. Am. Chem. Soc.* **1987**, *109*, 3387. (b) J.-H. Fuhrhop, P. Schnieder, E. Boekema, W. Helfrich, *J. Am. Chem. Soc.* **1988**, *110*, 2661. (c) J. Köning, C. Boettcher, H. Winkler, E. Zeitler, Y. Talmon, J.-H. Fuhrhop, *J. Am. Chem. Soc.* **1993**, *115*, 693.
[19] J.-H. Fuhrhop, W. Helferich, *Chem. Rev.* **1993**, *93*, 1562.
[20] I. Tuzov, K. Crämer, B. Pfannemüller, W. Kreutz, S. N. Magonov, *Adv. Mater.* **1995**, *7*, 656.
[21] I. Tuzov, K. Crämer, B. Pfannemüller, S. N. Magonov, M.-H. Whangbo, *New J. Chem.*, in press.
[22] S. Demharter, R. Mülhaupt, H. Frey, K. Crämer, S. N. Magonov, M.-H. Whangbo, in preparation.

13 Polymers

Invaluable information about the morphology, nanoscale structure and chain order in polymers can be obtained from STM and AFM studies. In particular, AFM has become one of the most advanced methods for the examination of polymer surfaces, because of the recent development in instrumentation. It is an important partner to microscopic and diffraction methods in the characterization of polymers. This chapter describes STM and AFM studies of polymers.

13.1 General Considerations

13.1.1 Polymer Structure

The properties of polymers are governed largely by the chemical nature of the monomer, the chain stereochemistry, and the molecular weight [1]. The most common polymers are polyethylene (PE) and polypropylene (PP) (Fig. 13.1). Due to the pres-

PE $\quad \mathrm{+CH_2-CH_2 \large)_n}$

PP $\quad \mathrm{+CH_2-CH \large)_n}$
$\qquad\qquad\;\;\;\;|$
$\qquad\qquad\;\;\;\mathrm{CH_3}$

iPP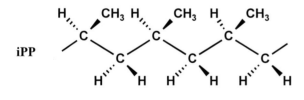

sPP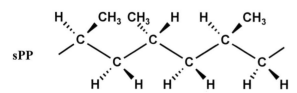

Figure 13.1 Chemical formulas of polyethylene (PE) and polypropylene (PP). Perspective views of isotactic PP (iPP) and syndiotactic PP (sPP).

ence of the methyl side groups, the PP chains can have different configurations. All repeat units have an identical configuration in isotactic polypropylene (iPP), the two configurations alternate along the chain in syndiotactic polypropylene (sPP), and there occurs a random alternation of the two configurations in atactic polypropylene (Fig. 13.1). Stereoregular polymers can be crystallized but typically lead to semicrystalline materials because a complete crystallization is hindered by the chain entanglements, structural defects (e.g., chain ends, branches), and impurities. The degree of crystallinity is an important parameter which determines the properties of polymer materials. Atactic polymers lead to amorphous materials.

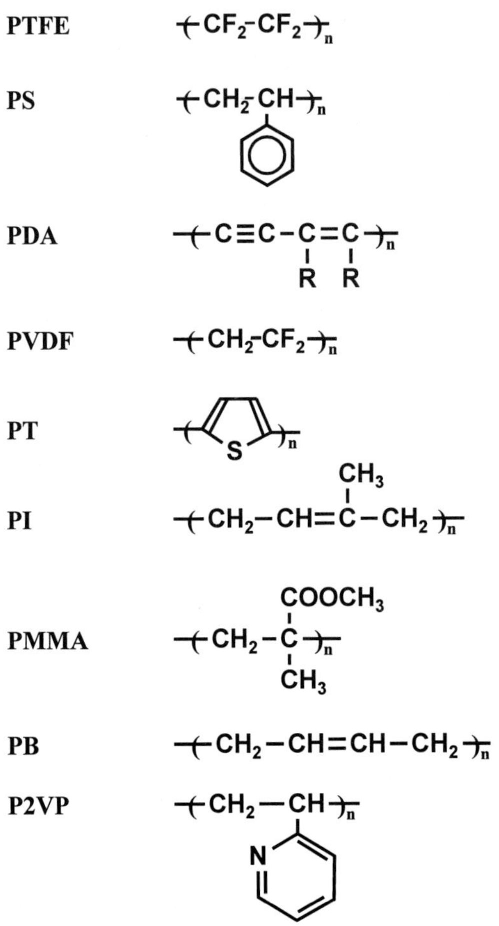

PTFE $\quad -(CF_2-CF_2)_n-$

PS $\quad -(CH_2-CH)_n-$ with phenyl side group

PDA $\quad -(C\equiv C-C=C)_n-$ with R, R side groups

PVDF $\quad -(CH_2-CF_2)_n-$

PT $\quad -(\text{thiophene})_n-$

PI $\quad -(CH_2-CH=C(CH_3)-CH_2)_n-$

PMMA $\quad -(CH_2-C(COOCH_3)(CH_3))_n-$

PB $\quad -(CH_2-CH=CH-CH_2)_n-$

P2VP $\quad -(CH_2-CH)_n-$ with 2-pyridyl side group

Figure 13.2 Chemical formulas of polytetrafluoroethylene (PTFE), polystyrene (PS), polydiacetylene (PDA), poly(vinylidene fluoride) (PVDF), polythiophene (PT), polyisoprene (PI), poly(methyl methacrylate) (PMMA), polybutadiene (PB), and poly(2-vinylpyridine) (P2VP).

Figure 13.2 presents other polymers to be discussed in this chapter. They include polytetrafluoroethylene (PTFE), polystyrene (PS), polydiacetylene (PDA), poly(vinylidene fluoride) (PVDF) and polythiophene (PT). Also studied are di-block copolymers based on PS, polyisoprene (PI), poly(methyl methacrylate) (PMMA), polybutadiene (PB), and poly(2-vinylpyridine) (P2VP). Di-block copolymers (e. g., ····-A-A-A-A-B-B-B-B-····) are formed from blocks of two chemically different monomers (e. g., A and B) and are called symmetric when the blocks have similar molecular weights.

13.1.2 Analysis of Polymer Surfaces

The important properties of polymer materials, such as adhesion, friction, wetting, swelling, penetrability, and biological compatibility, are strongly influenced by their surfaces. In general, the structure and morphology of the topmost surface layers (several nanometers thick) differ from those of the bulk polymer. Therefore, a comprehensive surface analysis is needed to correlate the surface structures with the physical/chemical properties. The depth profiles of polymer surfaces with resolution below 100 nm are analyzed by forward recoil spectroscopy, nuclear reaction analysis, secondary mass spectroscopy, X-ray and neutron reflectivity [2]. X-ray photoelectron spectroscopy is sensitive to the local environment near the surface but has a relatively low lateral resolution (ca. 1 µm). Electron microscopy has a higher resolution but requires special sample preparation, and the electron beam may damage polymer samples. Since the first application of STM and AFM to polymer films deposited on graphite substrates [3], a large number of polymer materials have been studied by STM and AFM. Important questions concerning these applications are:

(a) What properties of polymer surfaces can be examined by STM and AFM? and
(b) How comparable are the results of STM and AFM with those obtained from electron microscopy and diffraction techniques?

13.1.3 Applying STM and AFM

Polymer films prepared from melt, or cast from solutions, are most suitable for STM and AFM investigation. Industrial polymer tapes and fibers can be studied without special preparation. In some cases, flat sample surfaces can be prepared with a ultramicrotome. A difficulty in the application of scanning probe microscopy to polymers is illustrated in the AFM images of Fig. 13.3, where the rectangular "windows" are produced by the scanning tip on a soft polymer surface [4]. The grooves along the slow-scanning direction occur because the surface material is dragged by the tip in the fast-scanning direction. In studies of atactic PS, it was found that the

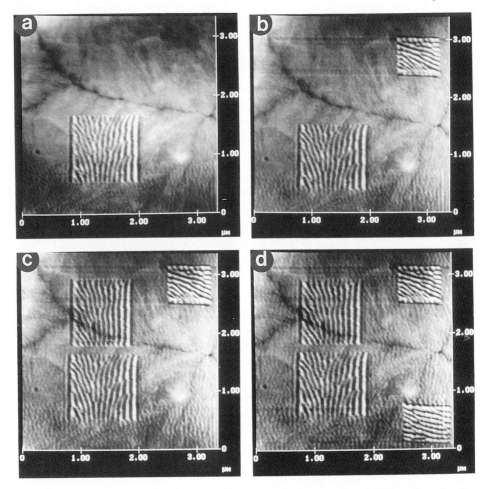

Figure 13.3 AFM height images of a poly(bisphenol-A)carbonate (Lexan 165) surface showing the surface modification produced by the tip. The "windows" with linear features were produced by scanning small areas, and each image was taken after zooming out of the scanned area. The windows at the upper and lower right corners were obtained by scanning at a rotational angle of 90°. The contrast covers height variations in the 0–50 nm range.

periodicity of the grooves correlates with the molecular weight of the polymer [4b]. To avoid such a surface modification, low-force imaging is required (e.g., STM measurements at large R_{gap}, and AFM experiments under liquid [5]). This goal is reached by carrying out sub-water low-force AFM imaging in the contact and tapping modes, which allows one to detect weakly bonded surface layers and to improve the resolution on the nanometer scale (Chapter 4) [5]. In contact-mode AFM, low-force measurements require the use of probes with small force constants (typically below 0.05 N/m). However, some polymer surfaces cannot be immersed in water because their structures are disturbed by swelling or partial dissolution. For such systems, tapping-mode AFM in air is a better choice. The applications of other force modulation techniques (Chapter 3) present new possibilities for imaging polymer surfaces. However, their potential cannot be fully exploited unless a rational interpretation is found for the observed contrast variations. One faces the same problem in the analysis of LFM images. The patterns of the lateral-force images of multicomponent surfaces are certainly related to the differences in the surface regions, but their interpretation is not straightforward [6]. For details, see Section 13.5.

13.2 STM of Polymer Samples

Most polymers are insulating, so STM applications are limited to conducting polymers, metal-coated polymers, and polymer layers adsorbed on conducting substrates. The primary use of STM lies in studies of morphology and molecular organization.

13.2.1 Conducting Polymers

Conducting polymers have poor solubility and typically possess a noncrystalline structure, so only a few methods can be used to characterize their structures. Therefore, studies of conducting polymer films with STM and AFM are of special importance [7, 8]. The STM image of a PT film grown on a conducting (indium tin oxide) glass electrode (Fig. 13.4 (a)) reveals a grain-like surface morphology (the smallest grains being 10-20 nm in diameter) and locally ordered features [9a]. The molecular-scale image in Fig. 13.4 (b) exhibits bright patterns 0.4-0.5 nm in diameter, which are assigned to the electron-rich thiophene rings of the polymer chain.

The growth of conducting polymer films and the effect of dopants on the film structure have been observed by STM [8]. In the electropolymerization of polypyrrole and PT on graphite [8c], single polymer chains adopt a helical structure and can be organized into superhelical structures. In the initial stage of the growth, islands of ordered fibrillar arrays are formed, and a further growth leads to an amorphous material. Polymerization on a solid surface is influenced by the nature of the substrate. For example, the growth of conducting polymer films from 1-naphthol and 2,6-dimethylphenol proceeds more homogeneously on gold than on graphite [9a].

A nonconducting polymer can be mixed with a conducting filler (e.g., carbon black or conducting organic crystals) to form a conducting material. When the block copolymer poly(butadiene-b-4-vinylpyridine) is doped with tetracyanoquinodimethane (TCNQ), charge-transfer complexes are formed between TCNQ and vinylpyridine. This bestows a percolative conductivity upon the material. The STM image in Fig. 13.5 shows the conducting domains containing molecular stacks characteristic of TCNQ-based charge-transfer complexes [10].

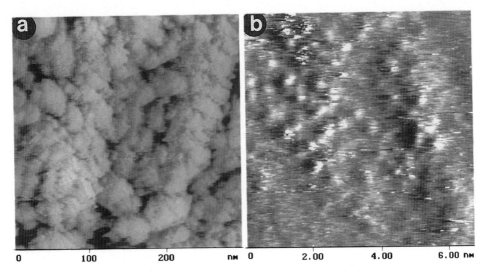

Figure 13.4 STM current images of a PT film ($I_{set} = 1$ nA, $V_{bias} = 100$ mV). The contrast is proportional to current variations in relative units.

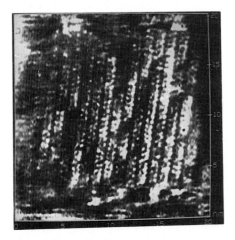

Figure 13.5 STM current image of poly(butadiene-b-4-vinylpyridine) doped with TCNQ ($I_{set} = 0.6$ nA, $V_{bias} = 117$ mV). The contrast is proportional to current variations in relative units.

13.2.2 Metal-Coated Polymer Surfaces

Before the advent of scanning probe techniques, microscopic studies of nonconducting polymer surfaces in the micrometer and sub-micrometer range were performed by scanning electron microscopy (SEM). These studies require the coating of a polymer surface with a metallic layer at least a few nanometers thick. The lateral resolution of SEM is high enough (ca. 5 nm) to observe individual PS macromolecules [11]. Metal-coated polymer surfaces can also be examined by STM, as demonstrated for PE single crystals [12a], crystalline aggregates of poly(ethylene oxide) (PEO) [12b], and various polymer fibers [12c]. The contrast of an STM image is mainly related to the surface height variation, whereas that of an SEM image is not straightforward to interpret. In addition, the vertical resolution is much higher in STM than in SEM. Therefore, fine morphological details of PE single crystals and fibers are better detected in STM images. However, in the characterization of rough surfaces (e.g., those formed by different fracture processes), SEM is more reliable because the STM/AFM imaging of steep surface features can lead to image artifacts (Chapter 4). As shown in the study of carbon fibers prepared from a polymeric precursor [13], a combined use of scanning probe techniques and SEM is useful for surface characterization.

13.2.3 Polymer Layers on Conducting Substrates

As described in Chapter 11, organic molecules deposited on conducting substrates can be imaged by STM. This has led to attempts to observe single biological and synthetic macromolecules on conducting substrates and epitaxial layers of several polymers (e.g., polyamide, poly(tetrahydrofuran), etc.) formed on graphite [14]. In the latter case, the STM images exhibit atomically resolved but complex moiré patterns. These were explained in terms of the geometry of the epitaxially polymerized molecules and the superposition of the polymer and graphite lattices [14]. To understand such images properly, it is necessary to conduct a parallel AFM study, perform imaging with different tip forces, and calculate electron density plots for appropriate models (Chapters 9 and 11).

Large- and atomic-scale STM images were obtained also for alkane crystals on different substrates [15] and thin layers of highly oriented PE, polybutene-1 and PTFE [16] deposited on graphite. Lamellar structure, nanofibrils and packing motifs of polymer chains were seen in such images. It is unclear whether the scanning tip is above the topmost layer or is "embedded" in the sample surface, thereby damaging part of the material. Another puzzling problem is the STM imaging of thick polymer layers up to 100 nm [17]. Why such an imaging occurs is not understood, although theories suggest an electrical conductivity through thick organic layers [18]. To comprehend this kind of observation, it is essential to examine systematically the dependence of the image contrast on experimental conditions. A recent finding exemplifies the need for such systematic studies: a low-current STM study of DNA deposited on mica [19] has shown that the imaging is enhanced by increasing the humidity. This is explained in terms of a lateral conductivity via ultrathin water films (i.e., the contamination layer) covering the sample and substrate.

13.3 AFM of Polymer Crystal Surfaces

13.3.1 Polydiacetylene Single Crystal

PDA crystals are prepared by a solid-state polymerization, and their structure is similar to those of monomer crystals [20]. The crystals of 2,4-hexadienylenebis(p-fluorobenzenesulfonate) (PFBS) exhibit a lozenge shape with a large surface corresponding to the crystallographic bc-plane (Fig. 13.6). In the large-scale AFM image of the bc plane of a monomer crystal (Fig. 13.7 (a)), the flat terraces have a step height of ca. 1.5 nm, which corresponds to the a-axis length. In low-force AFM measurements of a polymer crystal, fibrils oriented along the b-direction are seen (Fig. 13.7 (b)), and surface holes and linear defects are found (Fig. 13.7 (c)). The high-force image shows a more pronounced fibrillar morphology (Fig. 13.7 (d)) [21 a]. The latter is produced most likely by a strong tip force, because a PFBS crystal possesses an anisotropic mechanical property (i.e., the chains are oriented along the b-axis).

The molecular-scale AFM image of a monomer crystal (Fig. 13.8 (a)) shows [21 b] a periodic pattern corresponding to the bc-plane surface. According to the bulk crystal structure, the bc-plane surface is made up of the end groups (i.e., the most pro-

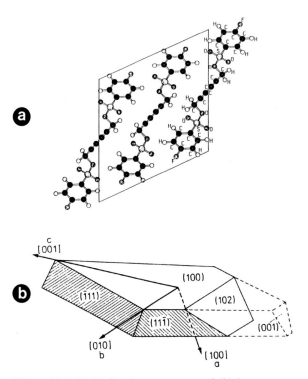

Figure 13.6 (a) Molecular structure and (b) lozenge-type monomer crystal of PFBS.

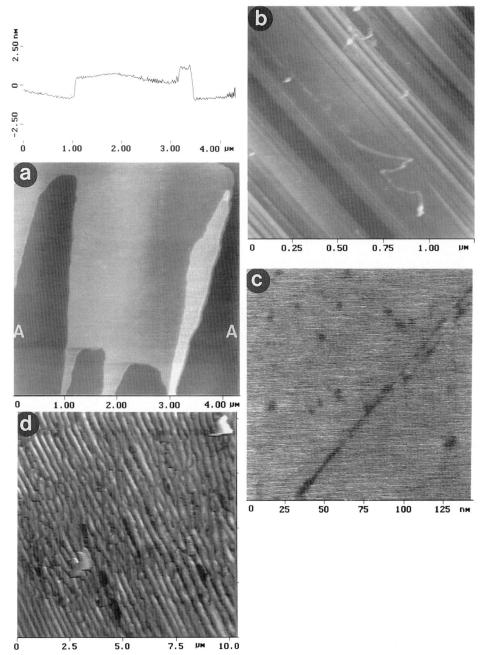

Figure 13.7 (a) AFM height image of a monomer crystal of PFBS. The cross-sectional profile along the line A-A is shown above the image. (b)-(d) AFM height images of a polymer crystal of PFBS. The images were recorded in the contact mode in air for (a) and (d), and under water for (b) and (c). The contrast covers height variations in the 0-8 nm range in (a), in the 0-12 nm range in (b), in the 0-2 nm range in (c), and in the 0-30 nm in (d).

truding H and F atoms) of the phenyl rings (Fig. 13.8 (b)). In adjacent rows of phenyl rings along the *b*-axis, the ring planes are oppositely oriented. In the AFM image, adjacent rows of bright spots exhibit a pattern generally consistent with the opposite orientations of the phenyl rings. In one row the pattern has two resolved spots for every phenyl ring (representing its two most protruding atoms, H and F). In the adjacent row the corresponding spots are not resolved. In addition, the image suggests that the relative orientations of the phenyl rings (with respect to the *b*-axis) in adjacent rows deviate somewhat from those found in the bulk crystal structure. This suggests a slight change in the positions of the phenyl rings within the surface unit cell [21 b], which may be caused either by the tip force or by the surface truncation.

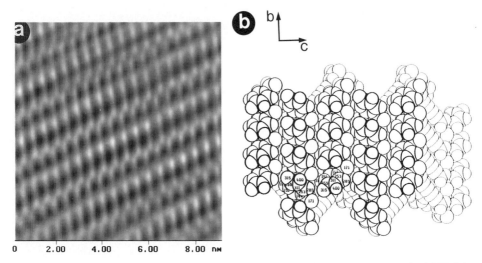

Figure 13.8 (a) AFM force image of the *bc*-plane surface of a monomer crystal of PFBS in air. The contrast covers force variations in relative units. (b) Arrangement of the PFBS molecules on the *bc*-plane surface expected from the bulk crystallographic data. The numbers within the circles representing the topmost surface atoms indicate their relative heights (pm) from a reference plane parallel to the *bc*-plane.

13.3.2 Polyethylene Single Crystal

Polymer single crystals grown from a dilute solution, where the probability of chain entanglement is low, exhibit a multiple chain backfolding and hence a lamellar structure. Numerous studies of polymer lamellar crystals have been carried out using PE single crystals [22]. Optical and electron microscopy studies show that PE crystals grow in dilute solution as hollow pyramids (Fig. 13.9 (a)). On a flat substrate, the pyramid structures are flattened and become lozenge-shaped lamellar platelets (Fig. 13.9 (b)) [23a]. In these PE crystals, the chains are oriented perpendicular to the lamellar surface, the crystal thickness (5–20 nm) represents the length of the extended part of the PE molecules, and the molecular folds are the main structural elements of the lamellar surface. Dark-field microscopy studies show that in the sectors of a relaxed crystal the molecular orientation is different (Fig. 13.9 (b)) [23 b].

Although the main morphological and structural features of PE single crystals are established, the molecular structure of the lamellar fold surface is not well understood. The AFM study of crystals of cycloalkanes $(CH_2)_{48}$ and $(CH_2)_{72}$ shows that their lamellar surfaces are characterized by a periodic arrangement of the molecular folds in perfect agreement with that expected from the bulk crystallographic data (Fig. 13.10) [24]. The lamellar surface of polymer single crystals is more complex because macromolecular chains might form tight folds (i.e., adjacent re-entry folds), loose folds (i.e., loops), and "switch board" connections of different lengths (Figs. 13.9 (c) and (d)). The occurrence of adjacent re-entry folds was deduced by

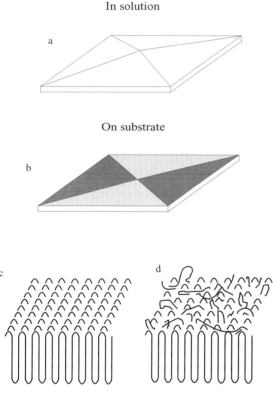

Figure 13.9 (a), (b) Schematic representations of a PE single crystal. The different sectors of the PE crystal, which correspond to the pyramid faces, are emphasized by dark and gray patterns. (c), (d) Possible molecular structures of the lamellar surface of the PE single crystal: (c) adjacent re-entry chain arrangement; (d) surface with possible defects such as loops, switchboard chains, and chain ends.

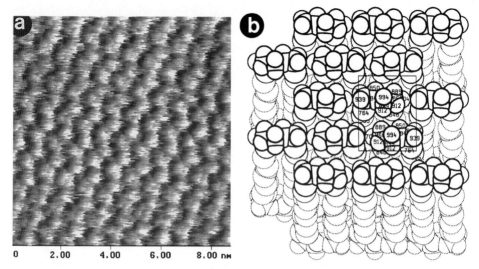

Figure 13.10 (a) AFM height image of the lamellar surface of a $(CH_2)_{48}$ crystal. The contrast covers height variations in the 0–1 nm range. (b) Structure of the lamellar surface of a $(CH_2)_{48}$ crystal expected from the bulk crystal data. The numbers within the circles representing the topmost atoms indicate their relative heights (pm) from a reference plane parallel to the lamellar surface.

spectroscopic and decoration studies of PE lamellar surfaces [23c, 25a]. However, small-angle X-ray diffraction studies suggest that on the lamellar surface there is a definite amount of amorphous phase, probably related to the surface consisting of loops, switch board chains, and chain ends [25b]. In the earlier AFM studies of PE and PEO crystals, only the morphological features were recognized [26a, b]. Recent studies on the lamellar surface of a PE single crystal show the pattern consistent with a periodical arrangement of adjacent re-entry folds [26c, 27].

The AFM images of PE single crystals show that the latter have a lozenge shape with the acute angle close to 67.5° (Fig. 13.11). The cross-sectional height profile along the line A–A shows that the central part of the lamellar crystal has a nearly uniform thickness of ca. 10 nm, and the crystal border (ca. 0.5 μm wide) is thicker. AFM allows one to observe directly the elevated diagonal striations (15–20 nm wide along the diagonals of the crystal) associated with the crystal sectorization (Fig. 13.11 (b)). These striations divide the PE crystal into four sectors and represent the misfit planes between the sectors with different molecular orientation. The sectorization lines are observed only in crystals of high-molecular-weight PE. This probably means that during the collapse of a pyramid on a flat surface, the relaxation of the pyramid edges is more hindered with an increased molecular weight.

Electron microscopy micrographs of PE single crystals usually demonstrate overlapping crystals but do not provide information about their topography. In STM images of metal-coated PE aggregates, the decorating layer screens the real topography

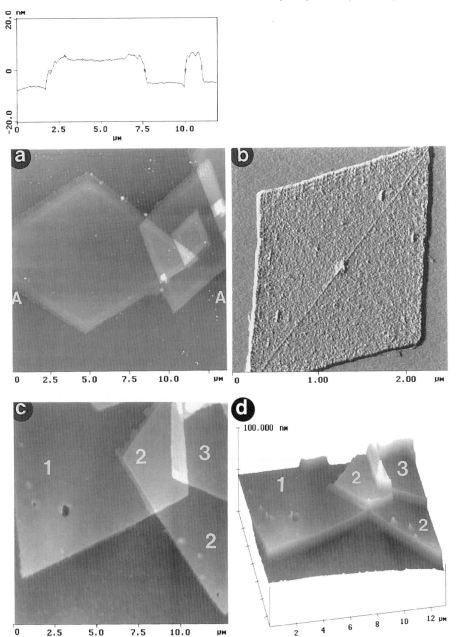

Figure 13.11 (a) AFM height image of a PE single crystal ($M_w = 4500$). The topographic profile along the cross-sectional line A–A is shown above the image. (b) AFM amplitude image of a PE single crystal ($M_w = 67000$), exhibiting the sectorization diagonals. (c) AFM height image of overlapping PE single crystals. Individual crystals are numbered **1, 2**, and **3**. (d) Surface view of the image in (c). The contrast covers height variations in the 0–50 nm range in (a) and in the 0–85 nm range in (c), and cantilever amplitude variations in the 0–10 nm range in (b).

of PE [12a]. AFM is free from these disadvantages. In the top and perspective views of the AFM images (Figs. 13.11 (c) and 13.11 (d)), one can identify three overlapping crystals with a height of 10 nm (labeled as **1**, **2** and **3**): crystal **2** is partly lying on crystal **1**, and crystal **3** on crystal **2**. The overlying crystals form sharp steps at the borders of the underlying lamella (Fig. 13.11 (d)). These features suggest the shearing of the top platelets into two parts at the edge of the underlying crystals. This is likely to occur by unfolding the chains along the direction perpendicular to the surface, since it is a low-energy process [28]. Such a chain-unfolding process is probably involved in the pyramid relaxation of PE crystals and the formation of aggregates with screw dislocation. An example of the latter is shown in Fig. 13.12, where the spiral steps are each ca. 10 nm thick.

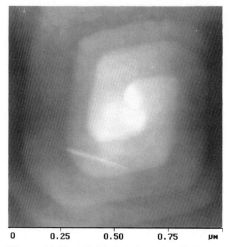

Figure 13.12 AFM height image of PE crystalline aggregate exhibiting a double screw dislocation. The contrast covers height variations in the 0–85 nm range.

The AFM images of a single crystal of low-molecular weight PE (Fig. 13.13) show the nanostructure of the lamellar surface. The image of the crystal corner (Fig. 13.13 (a)) reveals a grain-like structure in its central part and thicker borders. These features are common for crystals of PE with an M_w of 4500–50000 g/mol. The central part of the surface (Fig. 13.13 (b)) shows the ordering of small grains 10–15 nm in diameter into larger structures (70–80 nm in diameter). The small grains are found in all samples, but the large patterns are seen only in some crystals. The grain structure remains the same in measurements with different forces, so it can be assigned to the surface topography, not to the variations in surface hardness. The roughness of the lamellar surface is in the 0–1 nm range. The grains may represent locally different protrusions of the chain stems (within the adjacent re-entry model), short loops, or local switchboard connections. The grain nanostructure is found more frequently than the molecular-scale periodic pattern [26c].

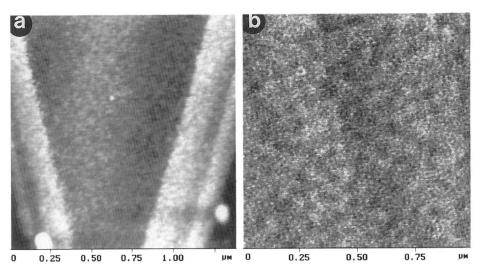

Figure 13.13 AFM height images of the nanostructure of the lamellar surface of PE single crystals (M_w = 4500). The contrast covers height variations in the 0–10 nm range in (a) and in the 0–2 nm range in (b).

13.3.3 Polymer Spherulites

Crystallization of polymers from a melt proceeds by formation of various aggregates of lamellar crystals [22]. Spherulites, observed by optical microscopy, are typical crystalline aggregates in polymer samples. The spherulite consists of a nucleus at the center and radially grown lamellar blocks (Fig. 13.14). The latter fill the space by branching, and therefore form a 3D structure. The lamellar blocks consist of crystalline and less-ordered regions, whose shapes and dimensions vary in real samples. The electron microscopy of polymer spherulites reveals thick primary and thin secondary lamellae, and the secondary lamellae appear to be epitaxially grown on the surface of the primary lamella [22].

The spherulite morphology of melt-crystallized iPP, sPP, PS, PVDF, and aliphatic polycarbonate (prepared from 2,2-dimethyltrimethylene carbonate) has been examined by AFM [29]. The surface of a melt-crystallized sample is usually covered with amorphous material, and a chemical etching of the topmost amorphous layer is required to expose the underlying crystalline aggregates for observation. Spherulites of the aliphatic polycarbonate grown from large nuclei are shown in Fig. 13.15 (a), and those of the α- and γ-spherulites of PVDF [29b] in Fig. 13.15 (b). These AFM images resemble the optical microscopy photographs. The contrast of the optical photographs is related to the birefringence of the crystallites, but that of the AFM images depends on the lateral variations of the surface topography and hardness.

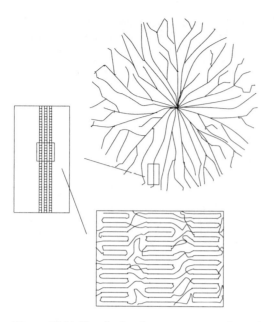

Figure 13.14 Sketch showing a structure of a polymer spherulite.

It is challenging to use AFM for the detection of the nanostructure of polymer spherulites. Figures 13.15 (c) and 13.15 (d) show images of the lamellar structures of sPP and iPP spherulites [29b]. In Fig. 13.15 (c) the numerous linear patterns of width in the 100–150 nm range present the radial lamellar blocks of the spherulites. Figure 13.15 (d) demonstrates the periodic contrast variations along the edges of the lamellar blocks, which reveal their substructure. The latter is characterized by a stack (**1**) of linear features (20–25 nm in width) related to the close-packed individual lamellae and by another stack (**2**) consisting of the lamellar blocks differently oriented from those of the stack (**1**). (In general, crystalline lamellar blocks are harder than amorphous material, so their image contrast becomes brighter.) Other image features resembling the edge-on, inclined, twisted and flat-lying lamellae were observed on spherulites of different polymers. Lamellar structures in the 1–20 nm range of the relatively rough spherulite surfaces are more difficult to characterize than those on the flat surfaces of polymer single crystals. To overcome this difficulty, methods of enhancing the image contrast and resolution should be found.

In addition to the examination of single crystals and spherulites, AFM has been applied to study the morphology and molecular structure of polymer layers grown epitaxially on different substrates [30]. Through combined transmission electron microscopy (TEM) and AFM studies of the epitaxial crystallization of iPP (α- and γ-phases) on a benzoic acid crystal, it was possible to identify the plane of the polymer crystal that was in contact with the substrate [31].

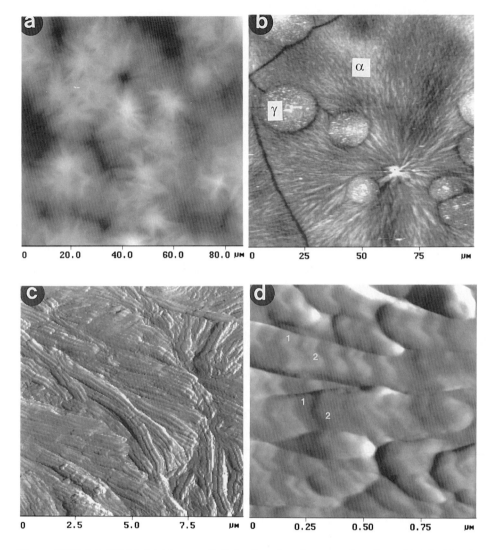

Figure 13.15 (a) AFM height image of aliphatic polycarbonate spherulites. (b) AFM height image of α- and γ-PVDF spherulites. (c) AFM amplitude image of the lamellar blocks of an sPP spherulite. (d) AFM amplitude image of the lamellar substructure of an iPP spherulite. The contrast covers height variations in the 0–3 μm range in (a) and in the 0–200 nm range in (b), and cantilever amplitude variations in the 0–60 nm range in (c) and in the 0–15 nm range in (d).

13.4 AFM of Oriented Polymers

The morphology and molecular structure of oriented polymers are of primary interest because they are used for advanced materials [32]. A uniaxial stretching of polymers leading to an extended-chain structure is applied to prepare oriented materials with mechanical properties approaching the theoretical limits. The deformation of PE and other polymers has been examined by different structure-sensitive methods (diffraction, spectroscopy, electron microscopy), and the main morphological and molecular processes accompanying the deformation are known. However, even the widely accepted Peterlin theory [33] cannot explain all the experimental findings, because of a poor knowledge of structural features on the 1–100 nm scale. Characterization of these nanoscale features can be significantly improved with AFM.

13.4.1 Imaging of Molecular Chain Order

In the first AFM studies of polymers, it was a challenge to achieve molecular resolution. Images showing individual polymer chains were obtained for the surface of a cold-extruded PE sample (Fig. 13.16) [34a]. In these images, PE chains are seen as vertical rows, which are separated by 0.5 nm and exhibit a pitch of 0.25 nm. These parameters correlate with those expected for the crystalline packing of extended all-

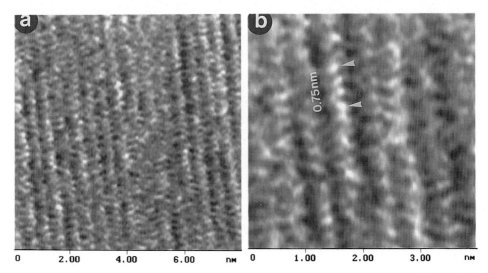

Figure 13.16 (a) AFM force image showing chain packing on the surface of a cold-extruded PE. (b) Zoomed-in part of (a). The periodicity of 0.25 nm along the chain corresponds to the all-*trans* chain conformation. The extrusion direction is indicated by arrowheads. The contrast is proportional to force variations in relative units.

trans PE molecules. Similar images were recorded for stretched films of ultrahigh-molecular-weight (UHMW) PE [34b]. The AFM image of a PTFE layer (Fig. 13.17), deposited on a hot substrate by rubbing, also shows chain packing with a separation of 0.56 nm, in agreement with the X-ray structural data [35]. The repeat distance along the rows representing PTFE chains (0.83 nm) is equal to half the period (1.69 nm) along the 13_6-helix chain (Fig. 13.17 (c)). Thus, with AFM, one can detect the conformation of polymer molecules. In studies of stretched iPP samples, it was possible to distinguish chains with left- and right-hand orientations [36].

The molecular-scale images of polymers as well as of other crystalline materials (see Chapters 6, 9, 10, 12) exhibit the surface lattices but do not show atomic- and

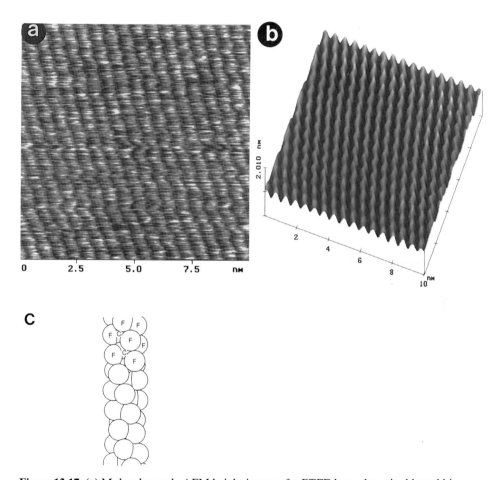

Figure 13.17 (a) Molecular-scale AFM height image of a PTFE layer deposited by rubbing on a hot glass substrate. (b) Surface view of (a) after filtering. (c) Sketch showing the 13_6-helix chain of PTFE. The rubbing direction is parallel to the linear patterns. The contrast covers height variations in the 0–0.5 nm range in (a) and in the 0–1 nm range in (b).

molecular-scale defects. Such images are likely to result from periodic modulations of the lateral motion of the tip while the tip and sample are in a multiatom contact. Because the AFM images of polymers presented above were recorded in air, this imaging was performed most probably on hard crystalline regions. The latter are present on the surface or are reached by the tip after removing weakly bonded surface layers. These aspects of molecular-scale AFM imaging raise some doubts about its importance. It is likely that the unique potential of AFM for investigating polymers lies primarily in the studies of partially-ordered surface features on the 1–100 nm scale, which are inaccessible by any other microscopic techniques.

13.4.2 Nanostructure of Polyethylene Tapes and Fibers

Oriented tapes of UHMW PE have been studied extensively by AFM. The starting material for drawing is obtained by preparing gels of UHMW PE in p-xylene and then casting to a gel film [37]. This procedure leads to a polymer sample consisting of packed lamellar crystals. The stretching of this sample to make a tape is performed at 125 °C, i.e., 5 °C below its melting point [34b]. During stretching, transformation of the morphology from lamellar to fibrillar proceeds in the necked part (Fig. 13.18). The stretched material is characterized by the draw ratio $l = 8-10$. A further stretching of this material leads to tapes with $l > 200$. The AFM study of UHMW PE tapes with l in the 10–70 range [34b] made it possible to classify their fibrillar morphology into (a) bundles of microfibrils with diameters in the 4–7 µm range, (b) microfibrils with diameters between 0.2 and 1.2 µm, and (c) nanofibrils. In tapes with $l = 10-30$, most nanofibrils have a width in the 15–25 nm range. Since these measurements were conducted in air, the nanofibril dimensions just described are over-estimated values. Indeed, in the low-force imaging performed under water, the small nanofibrils were found have widths in the 5–7 nm range [38a]. The number of such nanofibrils increases with increasing draw ratio, and their width is close to that of the nanofibrils detected in diffraction studies of a similar material [38b].

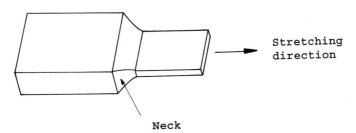

Figure 13.18 Schematic representation of the deformation of a polymer tape induced by stretching.

Recently, it has been shown that the fibrillar structure of the UHMW PE tapes under examination is coated with a weakly bound overlayer. The latter is removed by the scanning tip if the AFM imaging is not conducted at minimal force (a few nanonewtons). The required low-force conditions were achieved by optimizing subwater AFM imaging in contact and tapping mode [5]. Figure 13.19 shows the top-

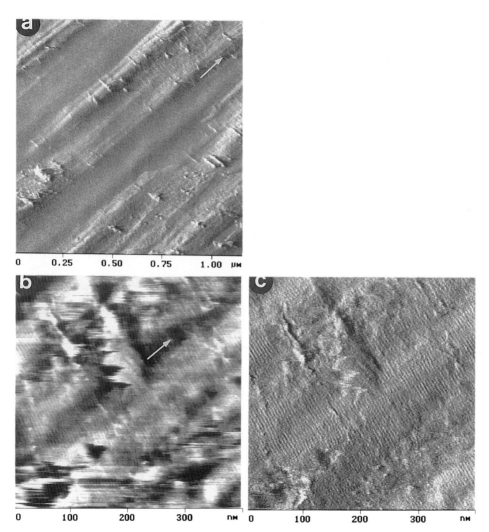

Figure 13.19 (a) AFM amplitude image of a stretched gel-cast PE film with $l = 30$. The stretching direction is indicated by an arrow. The AFM height and amplitude images simultaneously registered on part of the surface area in (a) are presented in (b) and (c), respectively. All images were recorded under water. The contrast covers cantilever amplitude variations in the 0–3 nm range in (a) and in the 0–2 nm range in (c), and height variations in the 0–3 nm range in (b).

298 13 Polymers

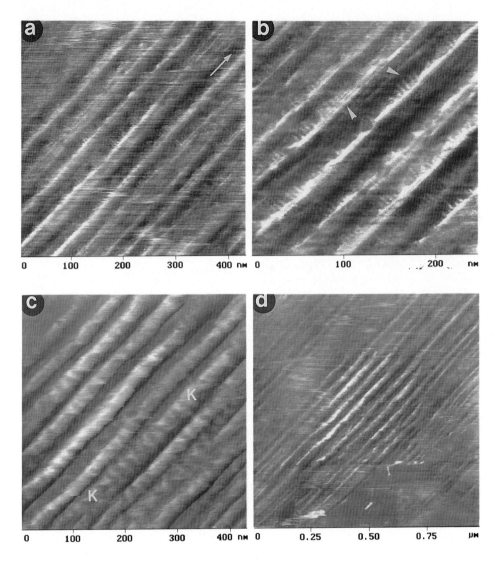

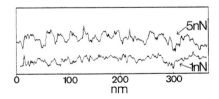

most surface layer consisting of ordered arrays of nanofibrils oriented perpendicular to the stretching direction. These "transverse" nanofibrils are 2.5–3.0 nm in width and are arranged periodically with a repeat distance of 5–6 nm. The dark triangular spots in Fig. 13.19 (b) can be related to indentations made by the tip. Therefore, the tip force should be further reduced for a nondestructive imaging of such surface structures. The transverse nanofibrils are probably formed by the stretching of part of the polymer material perpendicular to the drawing direction. Similar transverse fibrils were observed in a TEM study of uniaxially stretched PE films [39], where the contraction force acting in the neck was suggested to be responsible for the formation of the transverse nanofibrils. However, it is not clear why the transverse nanofibrils are periodically ordered.

In the contact-mode AFM image recorded with a force of ca. 1 nN under water, only the most stable transverse nanofibrils (hidden in the grooves on the surface) are found, and the "longitudinal" fibrils become more evident (Fig. 13.20 (a)) [40a]. The fine longitudinal nanofibers 5–7 nm wide and the remnant transverse nanofibrils are marked with arrowheads in Fig. 13.20 (b). Scanning of the area shown in Fig. 13.20 (a) with an elevated force of 5 nN leads to the removal of the skin layer consisting of the transverse and longitudinal nanofibrils. Thicker fibrils 20–30 nm in width are seen in Fig. 13.20 (c), and the periodic contrast variations with a repeat distance of ca. 25 nm become evident. The corrugation of the associated cross-sectional profile (shown below Fig. 13.20 (c)) diminishes when the applied force is reduced from 5 to 1 nN. This indicates that the periodic contrast variation is caused by the tip force, and that the surface deformation is elastic [40a]. By zooming out of the area scanned with the high force, it is clearly seen that the surface outside this "window" possesses fine longitudinal and transverse elements (Fig. 13.20 (d)).

The AFM observation of the force-induced periodic contrast variations along the inner fibrils (Fig. 13.20 (c)) is important. A similar observation is also presented in Fig. 13.21 [40b], which shows two images of a PE surface recorded with different forces. The periodic height corrugation along the fiber in the high-force image is in the 0.1–0.2 nm range (with a period of ca. 25 nm) and most likely represents the tip indentation depths on the hard and soft regions of the nanofibril. This is confirmed

◀ **Figure 13.20** (a) AFM lateral force image of a drawn gel-cast PE tape ($l = 30$) recorded under water in the contact mode at low force (1 nN). The stretching direction is indicated by an arrow. (b) Zoomed-in part of (a). The regions with transverse and longitudinal nanofibrils are indicated by white arrowheads. (c) AFM lateral force image recorded under water at higher force (5 nN) for the area shown in (a). The cross-sectional profiles along one of the fibers (i.e., the line K–K) at forces of 1 and 5 nN are shown below. (d) Large-scale AFM lateral force image of a PE tape recorded under water with the contact mode at low force (1 nN). The image shows a "window" of the area previously scanned in (c), where the surface material was removed during the scanning with high force. In all images the contrast covers lateral force variations in relative units.

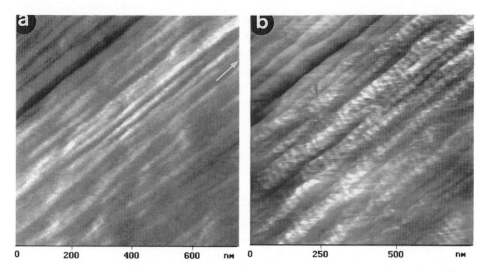

Figure 13.21 (a) AFM height image of a stretched PE tape ($l = 10$) recorded at 2 nN force. The stretching direction is indicated by an arrow. (b) AFM force image of the same area as in (a), recorded at 10 nN force. Both images are obtained in the contact mode in air. The contrast covers height variations in the 0–10 nm range in (a) and force variations in relative units in (b).

by the molecular-scale image of Fig. 13.22 (a), where the brighter spots (i.e., hard regions) contain linear striations oriented along the drawing direction. The linear striations can be assigned to oriented PE molecules, and groups of striations forming patterns ca. 15 nm in length can be assigned to crystallites with extended chain order. The darker areas between these bright domains correspond to less-ordered, and therefore softer, surface regions, which become more strongly indented by the tip. The relevant molecular organization within the longitudinal nanofibril is schematically presented in Fig. 13.22 (b). The periodic structure variations observed in the AFM images are known as the *long period* (20–25 nm), which is routinely detected in oriented PE samples by X-ray diffraction studies.

On the basis of the AFM analysis, the following model of the surface structure can be proposed for the UHMW PE tape (with a moderate draw ratio) (Fig. 13.23). The topmost layer of transverse nanofibrils (**1**) covers longitudinal nanofibrils (**2**). The latter form a surface layer of larger inner fibrils (**3**), which are the main structural elements of this material. The structure of a large inner fibril is characterized by a periodic sequence of more- and less-ordered extended chain arrays (**4**). It is likely that a further stretching splits the large fibril into smaller nanofibrils in highly oriented PE samples.

The experience gained from the AFM study of oriented UHMW PE tapes helps in analysis of the commercial gel-spun fibers (Spectra 900 and 1000; Allied Signal) [41]. The AFM images of these fibers reveal the heterogeneity of their surfaces.

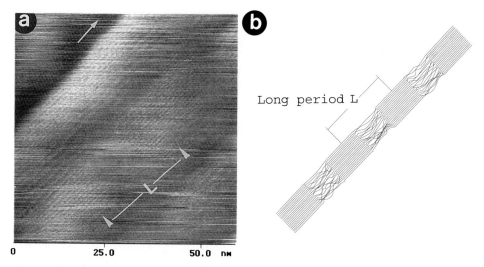

Figure 13.22 (a) Molecular-scale AFM lateral force image recorded on the surface of a stretched PE tape ($l = 30$) under water in the contact mode. The stretching direction is indicated by an arrow. The contrast covers force variations in relative units. (b) Schematic representation of the fibrillar regions with different chain order related to the long period L.

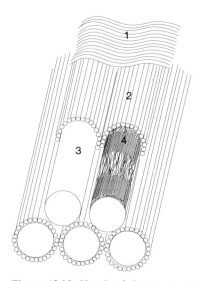

Figure 13.23 Sketch of the structural organization in oriented gel-cast PE tapes stretched to a moderate draw ratio. It shows the topmost layer of transverse nanofibrils (**1**), the underlying longitudinal nanofibrils (**2**), the main fibrils (**3**), and the internal molecular structure of the latter (**4**).

302   13 Polymers

Figure 13.24 (a) shows that part of the surface is totally covered by overgrown ribbons of width in the 30–110 nm range, which are oriented perpendicular to the fiber direction. Figure 13.24 (b) shows lamellar structures incorporated into the surface layer as well as the dark spots which appear after fluorination of the surface (fluorination is applied to improve fiber adhesion). The surface irregularities of the

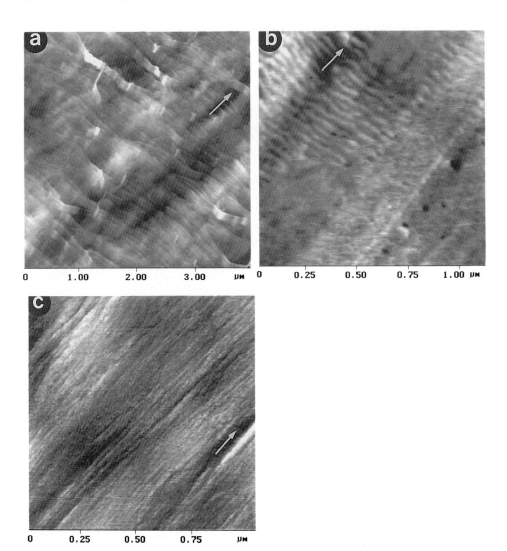

Figure 13.24 (a), (b) AFM height images of gel-spun PE fiber (Spectra 900; Allied Signal). (c) Lateral force image of gel-spun PE fiber (Spectra 1000; Allied Signal). The stretching direction is indicated by an arrow. All images were recorded under water in the contact mode. The contrast covers height variations in the 0–70 nm range in (a) and (b), and is proportional to force variations in relative units in (c).

PE fibers described above might play an important role in adhesion. The mechanical properties of PE fibers are not significantly influenced by these surface imperfections, but they are influenced by the fibrillar nanostructure (Fig. 13.24 (c)). This is similar to the nanostructure found in UHMW PE tapes and consists of large (40–50 nm wide) nanofibrils, which are composed of smaller ones with diameter in the 5–7 nm range.

13.4.3 Other Oriented Polymer Samples

To generalize the structural features observed by AFM on oriented polymers, the images of commercial microporous iPP tape as well as those of melt-drawn PE and iPP films were examined. The images of the microporous iPP film (Celgard 2400; Hoechst–Celanese) show alternating fibrillar and dense lamellar regions (Figs. 13.25 (a) and 13.25 (b)). The main morphological features of this commercial film (used as membranes) known from SEM and TEM studies [42a] are reproduced by AFM. The widths of the nanofibrils determined from the image can be overestimated due to the image artifact inherent in the scanning of sparsely packed cylinder-like structures (Section 3.4.1). However, the image reveals surface corrugations (e. g., steps of 25 nm high between the elevated lamellar regions and the fibrils [42b]) which are not accessible by electron microscopy. *Long-period* contrast variations (repeat distance of ca. 25 nm) were found along the iPP nanofibrils in high-force AFM images.

Figures 13.25 (c) and 13.25 (d) show the AFM images of melt-drawn PE and iPP films (thickness ca. 20 nm) prepared according to Peterman and Gohil [43a]. Surfaces of these films are characterized by fibrillar and lamellar structures [43b,c]. Nanofibrils 15–20 nm wide are covered by numerous bent lamellae aligned perpendicular to the stretching direction. On the PE surface, the topmost lamellae are 25–30 nm in width and extend over 200–300 nm (Fig. 13.25 (c)). The lamellae of the iPP film are ca. 30 nm wide and up to 500 nm long, and they exhibit a substructure with elevated edges (Fig. 13.25 (d)). The morphology patterns just described are frequently found on the surfaces of melt-drawn films.

13.5 AFM of Di-Block Copolymers

Block copolymers containing chemically different blocks and polymer blends with immiscible components are known to exhibit self-organized domains on the micrometer and nanometer scale [44]. The structural organization of block copolymers in bulk is examined by electron microscopy and diffraction techniques. The lamellar ordering and morphology in thin films of several block copolymers studied by AFM are discussed below.

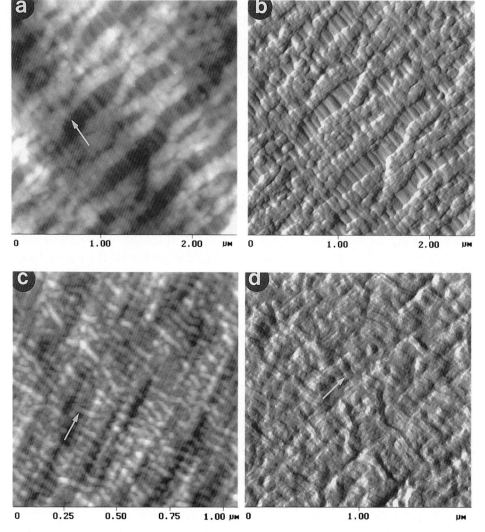

Figure 13.25 AFM height and amplitude images of a microporous iPP film (Celgard 2400; Hoechst–Celanese) recorded simultaneously are presented in (a) and (b), respectively. (c) AFM height image of ultrathin oriented PE film prepared by melt-drawing. (d) AFM amplitude image of ultrathin oriented iPP film prepared by melt-drawing. The stretching direction is indicated by an arrow. The contrast covers height variations in the 0–200 nm range in (a) and in the 0–25 nm range in (c), and is proportional to the cantilever amplitude variations in the 0–10 nm range in (b) and (d).

13.5.1 Poly(styrene-b-isoprene) Films

Thin films of symmetric di-block copolymers deposited on a glass substrate exhibit a layered lamellar morphology due to the specific interactions of the polymer blocks with the air and substrate interfaces [45a]. For example, PS has a higher affinity to the substrate, and PI to the air [45b]. As a result, the film thickness, d, of a symmetric block copolymer poly(styrene-b-isoprene), P(S-b-I), is given by the equation

$$d = (n + 0.5)L$$

where L is the lamellar period and n is an integer. Such equilibrium lamellar structures are formed in spin-coated films during annealing, and this process can be monitored by AFM [45c,d]. The image in Fig. 13.26 (a) presents the equilibrium struc-

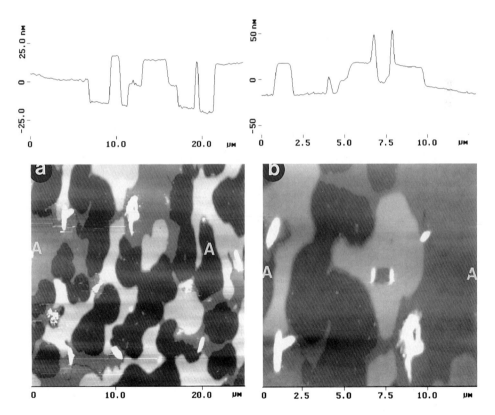

Figure 13.26 (a) AFM height image of a thin film of P(S-b-I) block copolymer deposited on glass. (b) AFM height image of the upper left part of the surface area shown in (a). The "window" in the central part was obtained by scratching the topmost layer with the tip during the scanning of this small area. The cross-sectional profiles along the lines A-A are shown above the images. The contrast covers height variations in the 0-50 nm range in both images.

ture of P(S-b-I) film on a glass substrate. The cross-sectional profile shows the surface steps associated with the first and second block copolymer layer. The height of the block copolymer layers determined from the AFM images is slightly lower than that obtained by X-ray reflectivity [45 e], probably because of the layer deformation in the AFM experiment. The image of Fig. 13.26 (b) shows a "window" produced by the tip, which scratches the material during the scanning of a small area. The cross-sectional profile across the "window" shows that only the topmost layer is destroyed. This layer is more easily damaged than the underlying one in the continuous scanning of the large area, as evidenced by the appearance of grain features in several places on the surface (Fig. 13.27 (a)).

In studying the friction of organic layers, it is necessary to consider that different organic layers may respond differently to the tip–sample force interactions. Figure 13.27 (a) shows the AFM height image of a thin film of P(S-b-I) block copolymer, and the LFM images recorded for the same area are shown in Figs. 13.27 (b) and 13.27 (c). The first and the second layer domains have different lateral force responses: they appear dark and bright in the trace image, respectively, and exhibit an opposite contrast in the retrace image. These contrast changes suggest a higher friction on the second layer (Chapter 3). However, the surfaces of both layers are identical in chemical nature, because they are composed of the PI blocks, which have the higher affinity to the air. In this case, a deeper indentation of the tip on the softer surface is responsible for a stronger lateral force on this layer.

13.5.2 Poly(styrene-b-methyl methacrylate) and Poly(styrene-b-2-vinylpyridine) Films

Block copolymers poly(styrene-b-methylmethacrylate), P(S-b-MMA), and poly(styrene-b-2-vinylpyridine), P(S-b-2VP), form different types of microdomains consisting of spheres, cylinders or lamellae, due to the microphase separation. The bulk structural organization of such systems is examined by TEM, since it can reveal the grain boundary morphology. The latter is important for rheological and diffusion properties of these materials [46a]. AFM is complementary to TEM in studying the surface morphology of these films [46b].

Two examples of the surface morphology of P(S-b-MMA) and P(S-b-2VP) are shown in Fig. 13.28. Lamellar structures ca. 60 nm wide dominate on the surface of P(S-b-MMA) film. Large superlamellar structures with widths of about 250–400 nm are detected on the P(S-b-2VP) surface. They exhibit a substructure with lamellae ca. 40 nm in width. The assignment of the AFM features to the surface topography is justified by force-dependent measurements, since they are not accompanied by contrast changes. The surface topography demonstrates phase-separation patterns similar to those found in the bulk, but it is most probable that the surface is decorated by blocks of the component with the higher affinity to air. This was shown to be

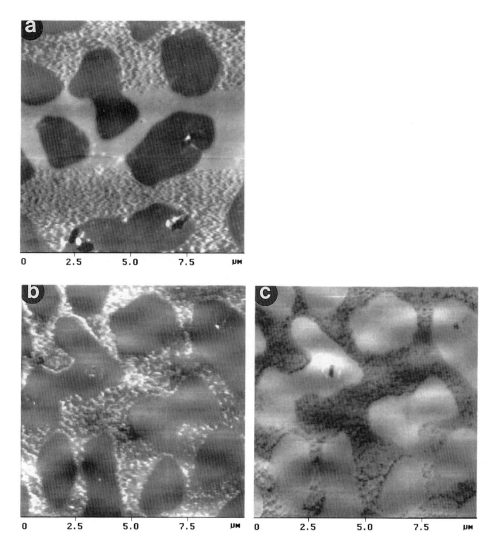

Figure 13.27 (a) AFM height image of a thin film of P(S-b-I) block copolymer. The grain structure in several places is the consequence of tip-induced damage to the topmost layer. The retrace and trace lateral-force images of part of the area in (a) are given in (b) and (c), respectively. The images in (b) and (c) were recorded with the scanning directions rotated by 90° with respect to that for the image in (a). The contrast covers height variations in the 0–30 nm range in (a) and lateral force variations in relative units in (b) and (c).

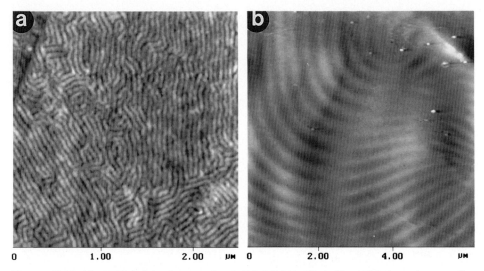

Figure 13.28 (a) AFM height image of a P(S-b-MMA) film. (b) AFM force image of a P(S-b-2VP) film. The contrast covers height variations in the 0–30 nm range in (a) and force variations in relative units for (b).

the case in a comparative AFM and TEM study of poly(styrene-b-butadiene), in which PB forms the topmost layer [46c]. In P(S-b-MMA) and P(S-b-2VP) films, the topmost layer probably consists of PS blocks [46c,d].

13.6 Concluding Remarks

The results of AFM studies on different polymer samples demonstrate the capability of this technique in characterizing the surface morphology and nanostructure. High-resolution AFM imaging allows one to visualize the nanostructural and morphological elements deduced from other indirect methods, as exemplified by the detection of the sectorization borders and nanoscale grains of PE single crystals. Low-force AFM imaging led to the detection of nanostructural elements (e.g., the transverse nanofibrils in oriented PE tapes) that are not accessible by other techniques. Furthermore, development in the dynamic AFM measurements (e.g., phase and frequency imaging modes) (Section 3.3.3) will also help to understand nanostructures of polymer surfaces. For example, amplitude and phase images of Celgard 2400 tape are compared in Figures 13.29 (a) and 13.29 (b), which demonstrate a dramatic enhancement of contrast in the phase image. By combining the tapping mode with phase imaging, one can image a soft polymer such as polydiethylsiloxane (PDES) that is difficult to image with the contact mode. The AFM images revealing the mesophase nanostructures of PDES are presented in Figs. 13.30 (a) and 13.30 (b)

[47]. The phase image distinctly shows the elongated mesophase structures embedded in amorphous material (Fig. 13.30 (b)), which are hardly distinguished in the height image (Fig. 13.30 (a)). It appears that a soft polymer material behaves solid-

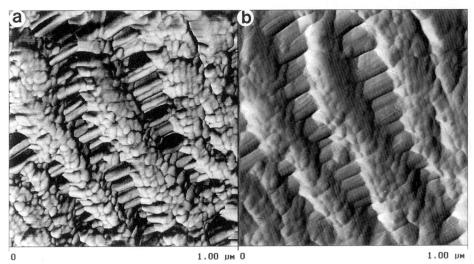

Figure 13.29 Tapping mode AFM images of Celgard 2400 tape: (a) phase image and (b) amplitude image. The contrast covers phase variations in the 0–56° range in (a), and amplitude variations in the 0–11 nm range in (b).

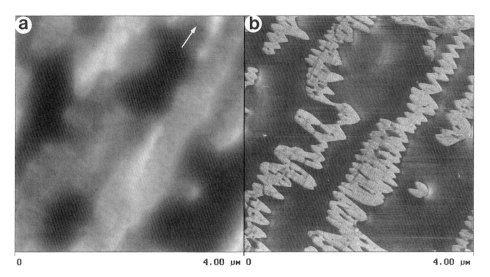

Figure 13.30 Tapping mode AFM images of PDES deposited on a silicon wafer by rubbing: (a) amplitude image and (b) phase image. The rubbing direction is indicated by the arrow. The contrast covers height variations in the 0–100 nm range in (a), and phase variations in the 0–50° range in (b).

like when probed with oscillation frequencies in the hundred kHz range used in the tapping mode. At present, it is not well understood how to relate the features of phase images to the topography and/or the mechanical properties of sample surfaces. Nevertheless, it is evident that phase imaging will broaden our knowledge about the surface nanostructures of polymers.

The discovery of the force-dependent image changes associated with the long-period structure variations in oriented PE samples indicates another important aspect of AFM applications. One may deliberately employ force-dependent imaging to probe the surface nanomechanical properties of polymers [48]. According to theoretical and experimental studies (Chapter 2), the behavior of the AFM probe in the low-force regime might be critically influenced by the surface forces, which include VDW forces in general as well as more specific tip–sample interactions (hydrophilic, hydrophobic, electrostatic, etc.). This opens the way to mapping surface regions with different specific properties if one can rationally deconvolute the contributions of weak tip–sample interactions to the AFM image.

References

[1] For general reading on polymers, see: (a) R. J. Young, P. A. Lovell, *Introduction to Polymers*, Chapman and Hall, London, **1991**. (b) J. M. G. Cowie, *Polymers: Chemistry and Physics of Modern Materials*, Chapman and Hall, London, **1991**. (c) H.-G. Elias, *Mega Molecules*, Springer, Berlin, **1983**. (d) E. L. Thomas (Ed.), *Structure and Properties of Polymers*, VCH, Weinheim, **1993**.

[2] (a) M. Stamm, *Adv. Polym. Sci.* **1992**, *100*, 379. (b) M. Tirrel, E. E. Parsonage, Ref. [1 d], p. 653.

[3] T. R. Albrecht, M. M. Dovek, C. A. Lang, P. Grütter, C. F. Quate, S. N. J. Kuan, C. W. Frank, R. F. W. Pease, *J. Appl. Phys.* **1988**, *64*, 117.

[4] (a) O. M. Leung, M. C. Goh, *Science* **1992**, *255*, 64. (b) G. F. Meyers, B. M. DeKoven, J. T. Seitz, *Langmuir* **1992**, *8*, 2330.

[5] A. Wawkuschewski, K. Crämer, H.-J. Cantow, S. N. Magonov, *Ultramicroscopy* **1994**, *58*, 185.

[6] S. N. Magonov, *Polym. Sci.* **1995**, in press.

[7] H. Siegenthaler, R. Christoph, in *Scanning Tunneling Microscopy and Related Methods* (Eds.: R. J. Behm, N. Garcia, H. Rohrer), Kluwer Academic Publishers, Dordrecht, The Netherlands, **1990**.

[8] (a) R. Yang, K. M. Dalsin, D. F. Evans, L. Christensen, W. A. Hendrickson, *J. Phys. Chem.* **1989**, *93*, 511. (b) R. Yang, D. F. Evans, L. Christensen, W. A. Hendrickson, *J. Phys. Chem.* **1990**, *94*, 6117. (c) R. Yang, X. R. Yang, D. F. Evans, W. A. Hendrickson, J. Baker, *J. Phys. Chem.* **1991**, *95*, 3765. (d) J. G. Mantovani, R. J. Warmack, B. K. Annis, A. G. MacDiarmid, E. Scherr, *J. Appl. Polym. Sci.* **1990**, *40*, 1693.

[9] (a) G. Bar, S. N. Magonov, H.-J. Cantow, T. Greczmiel, G. Kossmehl, *Solid State Sci.* (Springer Series) **1992**, *107*, 379. (b) E. Lacaze, J. Garbarz, V. Quillet, M. Schott, M. C. Pham, J. Moslih, P. C. Lacaze, *Ultramicroscopy* **1992**, *42–44*, 1037.

[10] S. Kempf, H. W. Rotter, S. N. Magonov, W. Gronski, H.-J. Cantow, *Polym. Bull.* **1990**, *24*, 325.
[11] R. A. Shelden, L. P. Meier, W. R. Caseri, U. W. Suter, R. Hermann, M. Müller, *Acta Polymer.* **1993**, *44*, 26.
[12] (a) R. Piner, R. Reifenberger, D. C. Martin, E. L. Thomas, R. P. Apkarian, *J. Polym. Sci., Polym. Lett.* **1990**, *28*, 399. (b) G. J. Leggett, K. M. Shakesheff, M. J. Wilkins, M. C. Davies, D. E. Jackson, C. J. Roberts, S. J. B. Tendler, *Polym. Prepr.* **1993**, *34*, 90. (c) L. C. Sawyer, R. T. Chen, M. G. Jamieson, I. H. Musselman, P. E. Russell, *J. Mater. Sci.* **1993**, *28*, 225.
[13] S. N. Magonov, A. Ya. Gorenberg, H.-J. Cantow, *Polym. Bull.* **1992**, *28*, 577.
[14] (a) M. Sano, D. Y. Sasaki, T. Kunitake, *Science* **1992**, *258*, 441. (b) M. Sano, D. Y. Sasaki, T. Kunitake, *Macromolecules* **1992**, *25*, 6961.
[15] M. Michel, G. Travaglini, H. Rohrer, C. Joachim, M. Amrein, *Z. Phys. B, Condens. Matter* **1989**, *76*, 99.
[16] (a) H. Fuchs, L. M. Eng, P. Sander, J. Petermann, K. D. Jandt, T. Hoffmann, *Polym. Bull.* **1991**, *26*, 95. (b) K. D. Jandt, M. Buhk, J. Petermann, L. M. Eng, H. Fuchs, *Polym. Bull.* **1991**, *27*, 11. (c) P. Bodo, Ch. Ziegler, J. R. Rasmusson, W. R. Salaneck, D. T. Clark, *Synth. Met.* **1993**, *55-57*, 329.
[17] J. R. Sheats, *Langmuir* **1994**, *10*, 2044.
[18] (a) C. Joachim, *New J. Chem.* **1991**, *15*, 223. (b) A. M. Kuznetsov, P. Sommer-Larsen, J. Ulstrup, *Surf. Sci.* **1992**, *275*, 52.
[19] R. Guckenberger, M. Heim, G. Cevc, H. F. Knapp, W. Wiegräbe, A. Hillebrand, *Science* **1994**, *266*, 1538.
[20] V. Enkelmann, *Adv. Polym. Sci.* **1984**, *63*, 91.
[21] (a) G. Bar, A. Gorenberg, H.-D. Bauer, S. N. Magonov, H.-J. Cantow, *Polym. Prep.* **1992**, *33*, 792. (b) S. N. Magonov, G. Bar, H.-J. Cantow, H.-D. Bauer, I. Müller, M. Schwoerer, *Polym. Bull.* **1991**, *26*, 233.
[22] (a) D. C. Basset, *Principles of Polymer Morphology*, Cambridge University Press, **1981**. (b) B. Lotz, J. C. Wittmann, Ref. [1 d], p. 79. (c) P. J. Barnham, Ref. [1 d], p. 153.
[23] (a) D. C. Bassett, F. C. Frank, A. Keller, *Phil. Mag.* **1963**, *8*, 1739. (b) W. D. Nieglisch, P. R. Swan, *J. Appl. Phys.* **1960**, *31*, 196. (c) J. C. Wittman, B. Lotz, *J. Polym. Sci., Polym. Phys. Ed.* **1985**, *23*, 205.
[24] W. Stocker, G. Bar, M. Kunz, M. Möller, S. N. Magonov, H.-J. Cantow, *Polym. Bull.* **1991**, *26*, 215.
[25] (a) J. H. C. Ching, S. Krimm, *J. Appl. Phys.* **1975**, *46*, 4181. (b) E. W. Fischer, G. F. Schmidt, *Angew. Chem.* **1962**, *74*, 551.
[26] (a) R. Patil, S.-J. Kim, E. Smith, D. H. Reneker, A. L. Weisenhorn, *Polym. Commun.* **1991**, *31*, 455. (b) D. Snetivy, G. J. Vansco, *Polymer* **1992**, *33*, 432. (c) R. Patil, D. H. Reneker, *Polymer* **1994**, *35*, 199.
[27] K. Crämer, A. Wawkuschewski, A. Domb, H.-J. Cantow, S. N. Magonov, *Polym. Bull.* **1995**, *35*, 457.
[28] L. G. Shadrake, F. Guiu, *Philos. Mag.* **1979**, *39A*, 785.
[29] (a) H. Schönherr, D. Snetivy, G. J. Vansco, *Polym. Bull.* **1993**, *30*, 567. (b) K. Crämer, H.-J. Cantow, S. N. Magonov, G. P. Hellmann, unpublished data. (c) K. Crämer, M. Schneider, R. Mülhaupt, H.-J. Cantow, S. N. Magonov, *Polym. Bull.* **1994**, *32*, 637. (d) C. Wang, R. Thoman, K. Crämer, R. Mülhaupt, J. Kressler, unpublished work.

[30] J. C. Wittmann, B. Lotz, *Prog. Polym. Sci.* **1990**, *15*, 99.
[31] (a) B. Lotz, J.-C. Wittmann, W. Stocker, S. N. Magonov, H.-J. Cantow, *Polym. Bull.* **1991**, *26*, 209. (b) W. Stocker, S. N. Magonov, H.-J. Cantow, J.-C. Wittmann, B. Lotz, *Macromolecules* **1993**, *26*, 5915.
[32] H. Jiang, W. W. Adams, R. K. Eby, Ref. [1 d], p. 597.
[33] (a) A. Peterlin, *J. Polym. Sci.* **1965**, *C9*, 61. (b) A. Peterlin, *J. Appl. Phys.* **1977**, *48*, 4099.
[34] (a) S. N. Magonov, K. Qvarnström, V. Elings, H.-J. Cantow, *Polym. Bull.* **1991**, *25*, 689. (b) S. N. Magonov, S. S. Sheiko, R. A. C. Deblieck, M. Möller *Macromolecules* **1993**, *26*, 1380.
[35] S. N. Magonov, S. Kempf, M. Kimmig, H.-J. Cantow, *Polym. Bull.* **1991**, *26*, 715.
[36] D. Snetivy, G. J. Vansco, G. C. Rutledge, *Macromolecules* **1992**, *25*, 7037.
[37] P. Smith, P. J. Lemstra, J. P. L. Pijpers, A. M. Kiel, *Colloid Polym. Sci.,* **1981**, *259*, 1070.
[38] (a) A. Wawkuschewski, H.-J. Cantow, S. N. Magonov, S. S. Sheiko, M. Möller, *Polym. Bull.* **1993**, *31*, 693. (b) D. T. Grubb, K. Prasad, *Macromolecules* **1992**, *25*, 4575.
[39] (a) J. M. Brady, E. L. Thomas, *J. Mater. Sci.* **1989**, *24*, 3311. (b) J. M. Brady, E. L. Thomas, *J. Mater. Sci.* **1989**, *24*, 3319.
[40] (a) A. Wawkuschewski, H.-J. Cantow, S. N. Magonov, *Adv. Mater.* **1994**, *6*, 476. (b) S. S. Sheiko, M. Möller, H.-J. Cantow, S. N. Magonov, *Polym. Bull.* **1993**, *31*, 699.
[41] (a) A. Wawkuschewski, H.-J. Cantow, S. N. Magonov, *Polym. Bull.* **1994**, *32*, 235. (b) A. Wawkuschewski, H.-J. Cantow, S. N. Magonov, J. D. Hewes, M. A. Kocur, *Acta Polym.* **1995**, *46*, 168.
[42] (a) L. C. Sawyer, D. T. Grubb, *Polymer Microscopy*, Chapman and Hall, New York, **1987**, pp. 124–125. (b) K. Crämer, H.-J. Cantow, S. N. Magonov, M. Buhk, J. Petermann, to be published.
[43] (a) J. Peterman, R. M. Gohil, *J. Mater. Sci.* **1979**, *14*, 2260. (b) K. D. Jandt, T. J. McMaster, M. J. Miles, J. Petermann, *Macromolecules* **1993**, *26*, 6552. (c) L. M. Eng, K. D. Jandt, H. Fuchs, J. Petermann, *Appl. Phys. A* **1994**, *59*, 145.
[44] B. P. M. Gallot, *Adv. Polym. Sci.* **1978**, *29*, 85.
[45] (a) G. Coulon, T. P. Russell, V. R. Deline, P. F. Green, *Macromolecules* **1989**, *22*, 2581. (b) S. Hasegawa, T. Hashimoto, *Macromolecules* **1985**, *25*, 589. (c) B. Collin, D. Chatenay, G. Coulon, D. Ausserre, Y. Gallot, *Macromolecules* **1992**, *25*, 1621. (d) P. C. M. Grim, M. Psarros, G. ten Brinke, G. Hadziioannou, *Polym. Prepr.* **1994**, *35*, 125. (e) R. Mutter, Ph.D. Thesis, Department of Physics, Freiburg University, **1994**.
[46] (a) Y. Nishikawa, H. Kawada, H. Hasegawa, T. Hashimoto, *Acta Polym.* **1993**, *44*, 192. (b) B. K. Annis, D. W. Schmark, J. R. Reffner, E. L. Thomas, B. Wunderlich, *Macromol. Chem.* **1992**, *193*, 2589. (c) D. W. Schwark, D. L. Vezie, J. R. Reffner, E. L. Thomas, B. K. Annis, *J. Mater. Sci. Lett.* **1992**, *11*, 352. (d) E. Parsonage, M. Tirrell, H. Watanabe, R. G. Nuzzo, *Macromolecules* **1991**, *24*, 1987. (e) P. F. Green, T. M. Christenson, T. P. Russell, R. Jerome, *Macromolecules* **1989**, *22*, 2189.
[47] S. N. Magonov, V. Elings, Yu. K. Godovski, V. S. Papkov, to be published.
[48] (a) E. W. Stroup, A. Pungor, A. S. Lea, V. Hlady, J. D. Andrade, *Polym. Prepr.* **1992**, *33*, 74. (b) M. Radmacher, R. W. Tillmann, H. E. Gaub, *Biophys. J.* **1993**, *64*, 735. (c) J. P. Aime, C. Elkaakour, C. Odin, T. Bouhacina, D. Michel, J. Curely, J. Dautant, *J. Appl. Phys.* **1994**, *76*, 754.

14 Future Outlook

Various experimental and theoretical issues concerning how to use STM and AFM, and what information to extract from the observed images, have been discussed in this book. In the application of STM and AFM it is necessary to distinguish genuine features from experimental artifacts in the observed images, to improve the image resolution, and to interpret the observed images. According to the STM and AFM results on various materials described in the previous chapters, it is essential to record and analyze images as a function of the tunneling parameters and applied forces in order to learn about the surface morphology, nanostructure, and atomic-scale features. It is also necessary to consider the topographic, electronic, and mechanical properties of the surfaces in order to interpret the observed images properly.

The present book points out the importance of the tip–sample force interactions and the associated surface relaxation in STM and AFM. For many crystal surfaces, the surface relaxation is weak, and the associated changes of the lattice parameters are not distinguishable in the STM and AFM images. Thus, the images of these systems are explained in terms of the partial and total electron density plots calculated on the basis of the bulk crystallographic data. However, the images of a number of surfaces cannot be accounted for unless the tip force induced surface deformation is taken into consideration. The importance of tip–sample force interactions can hardly be overemphasized. STM and AFM applications and instrumentation in the future should deal with minimization of the tip–sample interactions to extract topographic information on least-perturbed sample surfaces, or the deliberate use of the weak to strong tip–sample interactions to learn about the nanomechanical properties of sample surfaces.

The experimental findings presented in this book indicate that in STM the tip–sample mechanical interaction occurs prior to their electronic contact. Consequently, the tip and sample are subject to a geometric deformation, in particular for ambient-condition measurements with I_{set} in the nanoampere range. The same may occur in UHV measurements, depending on the nature of the samples. These observations indicate the need for STM measurements with large R_{gap} (e.g., imaging with currents in the picoampere and subpicoampere range). However, imaging at high R_{gap} diminishes the resolution in atomic-scale STM images. Thus, the choice of R_{gap} depend upon the nature of the problem to be solved. By performing STM measurements at low R_{gap}, it is possible to examine the mechanical properties of sample surfaces and conduct lithography. However, for such experiments, one needs to know the forces applied to the samples, and this requires the use of a combined STM/AFM set-up.

The minimization of the applied force in AFM is necessary for a high resolution and nondestructive imaging. For this purpose, AFM modulation techniques such as the tapping mode are especially important. Recent achievements in the modulation techniques include the detection of surface atomic lattices and point defects by performing the modulation experiments in UHV, in which the sensitivity of the vibrating cantilever (i.e., the Q-factor) can be extremely high. The high Q-factor of the cantilever allows one to probe surfaces in the attractive force regime, in which only a few atoms of the tip come into contact with the sample surface while the other apex atoms engage in attractive VDW interactions. In these AFM measurements under UHV, it seems that tapping of the surface is performed with an extremely small contact area, thereby leading to a true atomic resolution. Such experiments can broaden the scope of AFM applications.

Another promising direction of low (or moderate) force AFM imaging lies in detecting the surface regions with different functional groups in terms of surface forces. In such a study, the difference in the surface forces is seen as the contrast variation in the images recorded in the contact and tapping modes, and the contrast can be enhanced by using chemically functionalized tips. However, it is a difficult problem to interpret such images for unknown sample surfaces, because several factors may contribute to the observed image contrast. One should search for a rational approach to correlate the image contrast with the intrinsic properties of the sample surface (e.g., chemical constitution, molecular structure and packing).

With AFM one can record and vary the forces applied to a sample surface. Thus, AFM provides unique possibilities for studying surface nanomechanics. So far, the macroscopic deformation of a sample surface caused by the tip force has been described by the continuum theory in which the interacting solids are treated as materials with uniform density distributions. For a surface with several different chemical environments, the extent of the surface atom depression is not uniform. The surface local hardness variation introduces an atomic-scale corrugation into the macroscopically depressed surface; such atomic-scale corrugations are detected by STM and AFM, as shown for a number of layered compounds. The variation of the surface local hardness originates from the atomic arrangements within each layer and/or from those between adjacent layers. Knowledge of the atomic-scale surface relaxation obtained from AFM and STM experiments is essential in developing the theoretical description of a mechanical deformation beyond the continuum theory and hence in understanding the mechanical properties of materials on a nanometer scale.

For polymer materials, application of the modulation techniques can provide a dynamic mechanical analysis on the nanometer scale and hence help to identify the scale of molecular motions in polymers. This information, not accessible so far, is vital for understanding the molecular origin of the viscoelasticity of polymer materials. The use of strong forces damages the sample at a particular point or leads to the peeling of the surface materials during the scanning (wear process). These phenomena are also of practical interest. Nanoindentation experiments (i.e., production of indentation marks and successive evaluation of their dimensions) can be performed with a modified atomic force microscope to characterize the surface hardness of different materials.

STM and AFM have become important tools for surface characterization. They are attractive for those engaged in applications in a variety of different fields, for those involved in developing the scanning probe instrument, and for those interested in the simulation of observed images.

The invention of the scanning tunneling microscope had had a revolutionary influence on the development of materials characterization on the nanometer scale. The capability of recording atomic-scale features with a scanning tunneling microscope has boosted research efforts in other types of scanning probe microscopes, which provide information about the topography and the mechanical, magnetic, and electrical properties of surfaces. At present, the development of new types of scanning probe microscopes is proceeding rapidly, and their industrial applications are broadening. In semiconductor industry, AFM-based profilometers are used to check the roughness of silicon wafers and the quality of integrated circuits. Atomic force microscopes are adapted for use as nanoindenters, which will expand our capability for measuring the hardness of technologically important surfaces and coatings. Recently developed magnetic force and scanning capacitance microscopes are also employed in industrial applications. With the progress in technological use of new types of scanning probe microscopes, there is no doubt that the need for a comprehensive image interpretation will arise, and the knowledge gained from the analysis of STM and AFM images presented in this book will be useful.

Acknowledgements

The following figures have been previously published or have been adapted from previously published material.

Fig. 2.5 adapted from *Langmuir* **1993**, *9*, 3618.
Fig. 3.4 adapted from *J. Appl. Phys.* **1987**, *61*, R1.
Fig. 4.3 reproduced from *Ultramicroscopy* **1995**, *58*, 185.
Fig. 4.4 reproduced from *Polym. Bull.* **1993**, *31*, 693.
Fig. 4.5 kindly supplied by A. Wawkuschewski.
Fig. 4.6 reproduced from *Surf. Sci. Lett.* **1994**, *318*, L1175.
Fig. 4.7 reproduced from *Surf. Sci. Lett.* **1994**, *321*, L170.
Fig. 4.8 reproduced from *Adv. Mater.* **1994**, *6*, 649.
Fig. 4.9a reproduced from *Polym. Bull.* **1992**, *28*, 577.
Fig. 6.5 reproduced from *Adv. Mater.* **1994**, *6*, 355.
Fig. 6.6 reproduced from *New J. Chem.*, in press.
Figs. 6.7-6.14 reproduced from *New J. Chem.*, in press.
Figs. 6.15-6.19 reproduced from *J. Am. Chem. Soc.* **1993**, *115*, 2495.
Figs. 6.20-6.25 reproduced from *J. Am. Chem. Soc.* **1993**, *115*, 3760.
Figs. 6.26, 6.28, Table 6.1 reproduced from *J. Phys. Chem.* **1994**, *98*, 762.
Fig. 6.27 reproduced from *Critical Rev. Surf. Sci.* **1992**, *1*, 217.
Fig. 7.1b reproduced from *J. Am. Chem. Soc.* **1993**, *115*, 3760.
Fig. 7.2 reproduced from *J. Am. Chem. Soc.* **1993**, *115*, 2495.
Fig. 7.3 reproduced from *J. Am. Chem. Soc.* **1993**, *115*, 3760.
Figs. 7.5-7.10, 7.15-7.19 reproduced from *Surf. Sci.* **1995**, *326*, 311.
Table 7.1 adapted from *Surf. Sci.* **1995**, *326*, 311.
Fig. 8.1 reproduced from *Phys. Rev. Lett.* **1990**, *65*, 448.
Fig. 8.2 reproduced from *J. Vac. Sci. Technol. A* **1988**, *6*, 313.
Fig. 8.3 adapted from *Phys. Rev. B* **1986**, *34*, 9015.
Figs. 8.4, 8.8 reproduced from *J. Vac. Sci. Technol. B* **1991**, *9*, 1347.
Fig. 8.6 adapted from *Phys. Rev. Lett.* **1990**, *65*, 349.
Fig. 8.7 adapted from *Phys. Rev. B* **1993**, *48*, 1711.
Fig. 8.9 reproduced from *Phys. Rev. B* **1992**, *46*, 10411.
Figs. 9.1-9.3, Table 1 reproduced from *J. Phys. Chem.* **1994**, *98*, 762.
Fig. 9.5 reproduced from *Appl. Phys. Lett.* **1991**, *58*, 472.
Figs. 9.7, 9.8 reproduced from *J. Phys. Chem.* **1993**, *97*, 4764.
Fig. 9.9a reproduced from *Angew. Chem. Int. Ed. Engl.* **1990**, *29*, 537.
Figs. 9.6, 9.9b, 9.9c, 9.10-9.14 reproduced from *J. Phys. Chem.* **1993**, *97*, 4764.
Figs. 9.16, 9.17 reproduced from *Chem. Mater.* **1994**, *6*, 678.
Figs. 9.18-9.21 reproduced from *Surf. Sci.* **1994**, *321*, L170.
Fig. 9.22 reproduced from *Adv. Mater.* **1994**, *6*, 649.
Figs. 9.23-9.25 reproduced from *Adv. Mater.* **1995**, *6*, 483.
Figs. 9.27, 9.28 reproduced from *Surf. Sci. Lett.* **1994**, *318*, L1175.
Figs. 10.6-10.8 reproduced from *Adv. Mater.* **1993**, *5*, 453.

Figs. 10.9, 10.19, 10.21 reproduced from *J. Phys. Chem.* **1993**, *97*, 9170.
Figs. 10.12, 10.14-10.16 reproduced from *Synth. Met.* **1994**, *92*, 159.
Fig. 10.17 reproduced from *Ber. Bunsenges. Phys. Chem.* **1993**, *97*, 1478.
Figs. 10.18, 10.20 reproduced from *Synth. Met.* **1994**, *62*, 83.
Fig. 10.22 reproduced from *Synth. Met.* **1995**, *72*, 189.
Figs. 10.24a, 10.24b reproduced from *Ultramicroscopy* **1992**, *42-44*, 109.
Figs. 10.25-10.27 reproduced from *New J. Chem.* **1993**, *17*, 439.
Figs. 11.3, 11.4 reproduced from *Langmuir* **1993**, *9*, 2778.
Figs. 11.5c, 11.5d, 11.8, 11.11, 11.13, 11.14 reproduced from *Appl. Phys. A* **1994**, *59*, 119.
Figs. 11.6, 11.7, 11.9, 11.10 reproduced from *Adv. Mater.* **1993**, *5,* 821.
Figs. 11.15-11.19 reproduced from *Thin Solid Films* **1994**, *243*, 419.
Figs. 12.4, 12.8-12.13 reproduced from *New J. Chem.* **1995**, in press.
Figs. 12.6, 12.7 reproduced from *Adv. Mater.* **1995**, *7*, 656.
Fig. 12.14 reproduced from *J. Am. Chem. Soc.* **1993**, *115*, 693.
Figs. 12.17-12.23 reproduced from *New J. Chem.* **1995**, in press.
Fig. 13.4 reproduced from *Solid State Sci.* (Springer Series) **1992**, *107*, 379.
Fig. 13.5 reproduced from *Polym. Bull.* **1990**, *24*, 325.
Fig. 13.8 reproduced from *Polym. Bull.* **1991**, *26*, 233.
Fig. 13.10 reproduced from *Polym. Bull.* **1991**, *26*, 215.
Figs. 13.11, 13.13 reproduced from *Polym. Bull.* **1995**, *35*, 457.
Figs. 13.15c, 13.15d reproduced from *Polym. Bull.* **1994**, *32*, 637.
Fig. 13.16 reproduced from *Polym. Bull.* **1991**, *25*, 689.
Fig. 13.19 reproduced from *Ultramicroscopy* **1994**, *58*, 185.
Fig. 13.20 reproduced from *Adv. Mater.* **1994**, *6*, 476.
Fig. 13.12 kindly supplied by A. Wawkuschewski.
Figs. 13.21, 13.22a reproduced from *Polym. Bull.* **1993**, *31*, 693.
Fig. 13.24 reproduced from *Acta Polym.* **1995**, *46*, 168.

Index

acceptor level 125
- trapped-hole state 126
acceptor substitution
- at the ligand site 125
- at the metal site 127
adhesive force 13, 18
adjacent re-entry folds 287
adsorbate
- double layers 255
- molecular layers 253
- on graphite 252
- thin overlayers 253
adsorbate structure
- influence of substrate 233
alkanethiols
- on Au(111) 243
4-alkyl-4'-cyanobiphenyls
- on HOPG 235
- on β-Nb_3I_8 237, 240
N-(n-alkyl)-D-gluconamides 246
- crystal structures 250
N-(n-alkyl)-N'-D-maltosylsemicarbazones 246
- self-assembled structures 266, 269
amphiphiles 243
- self-organization 245
angle-resolved photoelectron spectroscopy 106
atomic-scale corrugations 314
atomic-size defects 4

ballistic regime 11
band bending 118
- electron accumulation 71
- electron depletion 71
- possible effects 71
BEDT-TTF 189
BEDT-TTF salts
- anion-layer images 212
- cation-layer images 206, 212
- images 206
β-(BEDT-TTF)$_2$AuI$_2$ 192, 194, 195
κ-(BEDT-TTF)$_2$Cu(SCN)$_2$ 193, 213, 214, 216
κ-(BEDT-TTF)$_2$Cu[N(CN)$_2$]Br 213, 215

κ-(BEDT-TTF)$_2$Cu[N(CN)$_2$]Cl 192, 215
α-(BEDT-TTF)$_2$I$_3$ 60, 207, 209
- dynamic surface processes 196
β-(BEDT-TTF)$_2$I$_3$ 211
- HOMO density 209
κ-(BEDT-TTF)$_2$I$_3$ 194
α-(BEDT-TTF)$_2$TlHg(SCN)$_4$ 208, 210
- variations of image patterns 196
(BEDT-TTF)$_2$X 189
bias voltage
- polarity 29
bilayers 246
Bloch orbital 65, 66
block copolymer 279, 281, 303
bulgy fibers 264

(CH$_2$)$_{48}$ 221
(CH$_2$)$_{72}$ 221
C$_{36}$H$_{74}$ 221
cantilever
- buckling 33, 34
- constant k 36
- resonance frequency 36
- torsion 33, 34
cantilever deflection
- set-point 33, 50, 51
capillary force 6, 13, 18
- hysteresis 52
charge density waves 85
- MC$_8$ (M = K, Rb, Cs) 105
coarse mechanism 22
cold-extruded PE 294
compounds
- with metal-clusters 113
conductance
- normalized 31
contact-mode AFM 1
- optimization 50
contamination layer 4, 73, 138
cycloalkane adsorbates
- on HOPG 228
cycloalkanes 219

DNA 41, 42
– low-current STM study 283
donor level
– trapped-electron state 123
donor substitution
– at the ligand site 123
– at the metal site 120
draw ratio 296
dynamic AFM
– force-modulation techniques 37
– tapping-mode AFM 37

effective Hamiltonian 66
electron transfer
– sample-to-tip 29
– tip-to-sample 29
electron transfer process 9 ff.
– conventional tunneling regime 9, 10
– electronic contact regime 9, 10
– mechanical contact regime 9, 10
electron-trapped area 123, 124
electronic band structure 3
– tight-binding method 65
electronic contact 73
electronic density of states 67
electronic structures
– intractable by band structure calculations 123
– tractable by band structure calculations 118
4EP(TCNQ)$_2$ 189, 191, 202
etching
– surface layers 49
extended Hückel method 68

fast Fourier transform 26
feedback control 22, 33
first primitive zone 66
force interaction
– in ambient conditions 138
– in STM 137
– in UHV 140
– on the atomic scale 145
force-modulation mode 38
force-modulation technique
– nanomechanical properties 39
force-vs.-distance curve 2, 12, 50, 51
– jump-in-contact point 12
– jump-out point 12
– nontouching line 13
– touching regime 12

force-vs.-distance relationship 13
frictional force microscopy 34

gain parameter 24
gap resistance 5
geometric deformation
– induced by tip force 201
graphite intercalation compounds
– stage-1 105
– stage-2 105

hole-trapped area 125, 126
HOPG 61
– A-site carbon 77
– B-site carbon 77
– hexagonal moiré patterns in STM images 154
– hexagonal superstructure 151
– macroscopic bending 152
– microscopic relaxation 152
– simultaneous STM/AFM measurements 138, 151
– STM images 80
– three-for-hexagon pattern of the AFM image 77, 151
– tip force induced deformation 151
hydration force 17
hydrophilic force 6
hydrophilic repulsion 16, 17
hydrophobic attraction 16, 17
hydrophobic force 6

I–V curve
– linear 139
– nonlinear 140
image artifacts 58
image imperfection
– atomic-scale images 129
– d^2 2H–MX$_2$ systems 128
– nanometer-scale images 131
image interpretation 2
image modification
– erasing scan lines 26
– flattening 26
– highpass filtering 26
– lowpass filtering 26
– plane fit adjustment 26
– zooming 26
image resolution 40
image simulation
– AFM 76
– density plot calculation 74

- energy window 74, 75
- STM 74
image variation
- tip–sample force interaction 57
images
- amplitude shift 37
- frequency shift 37
- phase shift 37
intermediate-state tunneling 11
jump-out point
- set-point level 52
junction
- metal–insulator–metal 68
- semiconductor–insulator–metal 69

KC_8 106
- nonuniform charge distribution 107
- surface CDW 107
KC_{24} 106

lamellar surface
- of polymer single crystals 287
Langmuir–Blodgett films 243
lateral force microscopy 34
lattice sum 67
layered inorganic materials 83
layered tellurides 5
layered transition-metal compounds 83
layered transition-metal tellurides
- atomic-scale deformation 169
- commensurate tellurides 169
- incommensurate tellurides 176
- MA_xTe_2 169
- superstructure modulations 169
lift mode 40
liquid/solid interfaces
- STM imaging 221
local density of states 72
local hardness 5
- atomic-scale corrugation 184
- microscopic deformation 147
long period 300, 301
long-range force 16
- Derjaguin–Landau–Verwey–Overbeck theory 17
- Lifshitz theory 16
- VDW force 16
long-tube scanner 23
loose folds 287

macroscopic deformation 5, 314
magnetic force microscopy 17, 39
measurements
- force dependent 56
- optimization 48
- partial aspects 47
mechanical contact 73
- between the tip and sample 142
mechanical profilometer 21
metal clustering 84
metrological applications 43
micelles 246
microscopic deformation 5
α-$MoCl_3$ 159
- AFM images 167
$MoCl_2$ 88
molecular aggregates
- bilayers 245
- micelles 245
morphological information
- LFM images 35
MoS_2 86, 129, 131
MX_2 layer 84
MX_3 layer 85
M_3X_8 layer 85

nanofibrils
- longitudinal 299, 301
- transverse 299, 301
nanoindenter 147
nanomechanical properties
- thin films 244
nanoscale ring structures
- electron-trapped area 180
- MoS_2 180
- WSe_2 180
$NbGe_{3/7}Te_2$ 57
$NbTe_2$ 92
- tip force induced changes 177
β-Nb_3I_8 61, 94, 113, 220
nonconducting polymer
- conducting filler 281
noncontact-mode AFM 37, 38
- long-range attractive force 38
normal alkanes
- all-*trans* conformation 219
- on HOPG 223
- on MoS_2 233
- on β-Nb_3I_8 233

4-*n*-octyl-4′-cyanobiphenyls 220
ohmic contact regime 11
on-site repulsion 127
organic adsorbates
– at liquid/solid interfaces 219
organic conducting salts 189
oriented polymers 294, 303
oscillating cantilever
– Q-factor 38

partial electron density plot 2
PE crystals
– lamellar surface 290
– screw dislocation 290
PE single crystal
– lamellar surface 288
– sectorization 288
piezoceramic actuator 23
piezoceramic scanner 22
point defects
– acceptor atom 118
– donor atom 117
– semiconductor 2H-MoS_2 116
– STM images 113
– substitutional 117
Poisson's ratio 147
polybutadiene 278
polydiacetylene 278
– single crystal 284
polydiethylsiloxane 308
polyethylene 277
– fibers 296
– single crystal 286
– tapes 296
poly(ethylene oxide) 283
polyethylene tape 5, 53
polyisoprene 278
polymer 277
– conducting 281
– crystal surface 284
– low-force imaging 280
– metal-coated 282
– on conducting substrates 283
– partial dissolution 281
– spherulites 291
– stretching 296
– surface 279
– swelling 281
– uniaxial stretching 294
poly(methyl methacrylate) 278

polypropylene 277
polystyrene 278
poly(styrene-b-isoprene) 305
poly(styrene-b-methylmethacrylate) 306
poly(styrene-b-vinylpyridine) 306
polytetrafluoroethylene 278
polythiophene 278
poly(vinylidene fluoride) 278
poly(2-vinylpyridine) 278
PTFE chains 295
pull-out force 13, 50

Qn(TCNQ)$_2$ 189, 191, 201
quantum-dot contact regime 144

ReSe$_2$ 94
retrace image 25
retrace profile 34
retrace scan 24
rigid cantilevers
– dynamic-mode AFM 37
– nanomechanical properties 37
α-RuCl$_3$ 159
– images at high applied force 163
– images at low applied force 160
– tip force induced surface deformation 163

scanning
– optimum direction 35
scanning directions
– fast 24
– slow 24
scanning electron microscopy 282
scanning frequency 25
scanning modes
– constant-height mode 25
– constant-interaction mode 25
scanning tunneling spectroscopy 3
– I–V curve 30
Schottky emission 10, 11
self-assembled monolayers 143
self-assembled structures 243
– drying speed 248
– substrate surface 248
semiconductor
– magnetic 29, 67
– normal 30, 67
Sharvin resistance 11, 143
short-range force 13
short-tube scanner 23
Slater-type orbitals 68

SrTiO$_3$ 37
- (305) surface 43, 44
stepper motor 24
supramolecular assemblies 245
- fiber-like assemblies 259
- micellar structures 257
- rod-like assemblies 261
- structural models 264, 273
surface corrugation
- tip-force induced 15
surface deformation
- Hertz theory 13, 14
- local hardness 14
- macroscopic deformation 14, 15
- microscopic deformation 15
- surface hardness 146
- surface stiffness 146
- tip-sample contact area 13
surface force 6
surface force apparatus 21
surface local hardness 314
surface processes
- during imaging 194
surface relaxation
- in STM and AFM images 151
switch board connections 287

TaGe$_{0.355}$Te$_2$ 57, 59
tapping-mode AFM 4, 38, 39
- lateral force 53
- optimization 53
- under water 54
- WSe$_2$ 54
TaSe$_2$ 98
TCNQ 189
TCNQ salts
- with substituted phenylpyridines 205
TEA(TCNQ)$_2$ 189, 191, 204
Tersoff-Hamman theory
- assumption 72
thin organic films 243
tip
- effective sharpening 42, 58
tip-sample force interaction 1, 2, 4
- surface relaxation 313

tip-sample interaction
- conventional tunneling regime 136
- electronic-contact regime 136
- mechanical-contact regime 136
- tip electronic states 135
- tip-induced local states 136
topografiner 21
total electron density plot 3
trace image 25
trace profile 34
trace scan 24
transmission electron microscopy 243, 292
trapped electron state
- donor dopant 55
- ring 56
1T-TaS$_2$ 113
- simultaneous STM/AFM study 75
TTF-TCNQ 189-191, 193, 199
tubular helix 264
tunneling
- sample-to-tip 75
- tip-to-sample 75
- transmission probability 68
tunneling current filament 143
tunneling density of states 31
tunneling parameters
- bias voltage 27
- gap resistance 27
- set-point current 27

vacancy
- ligand-atom 119
- metal-atom 120
vibrational noise
- isolation 27
- rubber cord isolation 27
- suspension-string system 27

wagon-wheel patterns
- MoSe$_2$ epilayers on MoS$_2$ 157
WSe$_2$ 129
WTe$_2$ 61, 89

Young's moduli 147

VCH Reference Sets

Materials Science

Encyclopedia of Applied Physics

Edited by George L. Trigg

"Exclusively devoted to the technical and industrial applications of physics"

The 20-volume *Encyclopedia of Applied Physics*, edited by G.L. Trigg, contains 500 alphabetically arranged, in-depth articles on the most relevant current and future applications of physics.

More technical than a general purpose encyclopedia, this reference work will prove invaluable not only to physicists and engineers, but also to scientists in associated areas.

A major scientific undertaking, the *Encyclopedia of Applied Physics* is sponsored by the physical societies of various countries, e.g., the American Institute of Physics.

Coverage ranges from general devices and laboratory methods through condensed matter physics to aeronautics and space physics.

Cumulative subject indexes are published after every three volumes. The series can only be purchased as a complete set.

Sensors

A Comprehensive Survey

Edited by W. Göpel, J. Hesse, J.N. Zemel

"Meeting the varying needs of the sensor community"

Sensors are a key element in the rapidly envolving field of measurement and instrumentation.

Each of the 8 volumes of the series *Sensors - A Comprehensive Survey* consists of three parts: specific physical and technological fundamentals and relevant measuring parameters; types of sensors and their technologies; and the most important applications with a discussion of emerging trends.

The volumes concentrate on Fundamentals (Vol. 1), Chemical & Biochemical Sensors (Vols. 2/3), and Thermal (Vol. 4), Magnetic (Vol. 5), Optical (Vol. 6), Mechanical (Vol. 7) Sensors, and Market and Technological Trends (Vol. 8).

Series editors are W. Göpel, J. Hesse, and J.N. Zemel, and the volume editors and authors are internationally renowned experts.

The series can be purchased as a complete set or by volume.

Materials Science and Technology

A Comprehensive Treatment

Edited by R.W. Cahn, P. Haasen, E.J. Kramer

"Sure to establish itself as a seminal work."

The phenomenal turnout of new products in all industries, from conventional to high-tech, owes its success largely to intensified efforts in materials research and development. The 18-volume series *Materials Science and Technology - A Comprehensive Treatment* is an in-depth, topic-oriented reference work created specifically to further these efforts.

The series covers the most important classes of materials: metals, ceramics, glasses, polymers, semiconductors, and composites.

Each volume deals with properties, processing, applications, or general phenomena associated with the above-mentioned materials.

Edited by R.W. Cahn, P. Haasen, and E.J. Kramer, the series began publication in 1990 and is proceeding at a rate of four to five volumes per year, each volume containing 10 to 20 contributions and averaging 600 pages. The series can be purchased as a complete set or by volumes.

To order please contact your bookseller or:
VCH, P.O. Box 10 11 61, D-69451 Weinheim, Fax: 06201-60 61 84
VCH, Hardstrasse 10, P.O. Box, CH-4020 Basel
VCH, 8 Wellington Court, Cambridge CB1 1HZ, UK
VCH, 303 N.W. 12th Avenue, Deerfield Beach, FL 334421788, USA (toll-free: 1-800-367-8249)
VCH, Eikow Building, 10-9 Hongo 1-chome, Bunkyo-ku, Tokyo 113